Studies in Global Justice

Volume 21

In today's world, national borders seem irrelevant when it comes to international crime and terrorism. Likewise, human rights, migration, climate change, poverty, inequality, democracy, development, trade, bioethics, hunger, war and peace are all issues of global rather than national justice. The fact that mass demonstrations are organized whenever the world's governments and politicians gather to discuss such major international issues is testimony to a widespread appeal for justice around the world.

Discussions of global justice are not limited to the fields of political philosophy and political theory. In fact, research concerning global justice quite often requires an interdisciplinary approach. It involves aspects of ethics, law, human rights, international relations, sociology, economics, public health, and ecology.

Studies in Global Justice takes up that interdisciplinary perspective. The series brings together outstanding monographs and anthologies that deal with both basic normative theorizing and its institutional applications. The volumes in the series discuss such aspects of global justice as the scope of social justice, the moral significance of borders, global inequality and poverty, the justification and content of human rights, the aims and methods of development, global environmental justice, global bioethics, the global institutional order and the justice of intervention and war.

Volumes in this series will prove of great relevance to researchers, educators and students, as well as politicians, policymakers and government officials.

More information about this series at http://www.springer.com/series/6958

Gottfried Schweiger
Editor

Migration, Recognition and Critical Theory

Editor
Gottfried Schweiger
Centre for Ethics and Poverty Research
University of Salzburg
Salzburg, Austria

ISSN 1871-0409 ISSN 1871-1456 (electronic)
Studies in Global Justice
ISBN 978-3-030-72734-5 ISBN 978-3-030-72732-1 (eBook)
https://doi.org/10.1007/978-3-030-72732-1

This Springer imprint is published by the registered company Springer Nature Switzerland AG
The registered company address is: Gewerbestrasse 11, 6330 Cham, Switzerland

Contents

Part II Recognition, Migration Policies and the State

Part III Recognition and Refugees

About the Editor

Gottfried Schweiger has been working at the Centre for Ethics and Poverty Research of the University of Salzburg as a Senior Scientist since 2011. He was a guest researcher in St. Gallen and Bochum. Schweiger (co-)authored several peer-reviewed articles and chapters, (co-)edited volumes, and wrote two monographs on the philosophy of childhood (both published by Palgrave Macmillan) together with Gunter Graf. His latest publications are "Absolute Poverty in Europe: Interdisciplinary Perspectives on a Hidden Phenomenon" (Policy Press 2019, co-edited with Helmut P. Gaisbauer and Clemens Sedmak), the Special Issue "Global Justice for Children" (*Journal of Global Ethics* 2019, co-edited with Johannes Drerup), and "Poverty, Inequality and the Critical Theory of Recognition" (Springer 2020).

About the Contributors

Heiko Berner, PhD, is an educational scientist and social worker. He is a researcher and lecturer in the BA Social Work and MA Social Innovation study programmes at Salzburg University of Applied Sciences. His main areas of interest are racism and discrimination, empowerment, and migration society.

Rizza Kaye C. Cases is an assistant professor at the Department of Sociology, College of Social Sciences and Philosophy, University of the Philippines Diliman. Her research interests include migration studies, mixed-methods social network analysis, and relational and comparative sociology.

Hilkje C. Hänel works as an assistant professor of political theory at Potsdam University. From 2018 to 2020, she worked as an assistant professor in practical philosophy at the Freie University in Berlin. In January 2018, Hänel successfully defended her PhD in philosophy at the Humboldt-University of Berlin, Germany. Her research is on the intersection of recognition theory, epistemic injustice, and ideology theory from a feminist point of view. Further interests are philosophy of disability, migration, and nonideal theory. In 2018, Hänel published her first book *What Is Rape?* with Transcript Publishers. She has published articles in *Ergo, Hypatia,* and *Journal of Social Philosophy*.

Benno Herzog is an associate professor for Sociological Theory, Research Methods, and Sociology of Language at the Department of Sociology and Social Anthropology at the University of Valencia. He is director of the research group on Critical Theory and Frankfurt School. His latest book was *Invisibilization of Suffering: The Moral Grammar of Disrespect* (Palgrave Macmillan 2020).

Sabine Hirschauer is an assistant professor with the Department of Government, New Mexico State University, Las Cruces, NM, USA. She has published about critical security studies, human security, migration, gender, and identity politics. She is the book author of *The Securitization of Rape: Women, War and Sexual Violence* by Palgrave Macmillan in 2014.

Onni Hirvonen is a senior lecturer in philosophy at the Department of Social Sciences and Philosophy in the University of Jyväskylä. His main interests include Hegelian philosophy of recognition, critical social philosophy, and contemporary social ontology. Hirvonen's recent publications include *Grounding Social Criticism: from Understanding to Suffering and Back*, *Populism as a Pathological Form of Politics of Recognition* (with Joonas Pennanen), and *Recognitive Arguments for Workplace Democracy* (with Keith Breen).

Martin Huth has been graduated from the University of Vienna with a dissertation thesis on biomedical ethics from a phenomenological perspective. Since 2008, he is a lecturer at the Department of Philosophy at the University of Vienna; since 2011, he holds a post doc position at the Messerli Research Institute in Vienna. His research priorities are ethics, applied ethics (biomedical ethics, ethics of human-animal relations), political theory, and phenomenology.

David Ingram is a professor of philosophy at Loyola University. His MA and PhD degrees in philosophy are from the University of California at San Diego. He has taught at Loyola since 1987, before which time he taught at the University of Northern Iowa. Ingram's areas of specialization are social and political philosophy, philosophy of law, philosophy of social science, critical race theory, and contemporary German and French philosophy.

Simon Laumann Joergensen is an associate professor in political theory at the Department of Politics and Society, Aalborg University. He has published in journals and books on policies of democratic integration and political theorists with a special focus on critical and democratic theory.

Alyssa Marie Kvalvaag is a PhD research fellow in Sociology at Nord University, Bodø, Norway where she is researching the integration of immigrants in the North of Norway with a regional focus on policy and immigrant perspectives. She is a member of the Social Work Research Group and Network for Ethnicity, Gender Equality, and Social Responsibility (ELSA-Nettverket) at Nord University and a Book Review Editor at the Nordic Journal of Migration Research. She received her MSc in Human Rights and Multiculturalism from the University of South-Eastern Norway, Drammen, Norway, and BA in Social Work from Cornerstone University, Grand Rapids, USA.

Odin Lysaker is a professor of ethics at the University of Agder. He holds a PhD in philosophy from the University of Oslo. He has specialized in moral philosophy, social philosophy, political philosophy, and philosophy of nature. His publications include "Democratic Disagreement and Embodied Dignity: The Moral Grammar of Political Conflicts" (in Odin Lysaker and Jonas Jakobsen (eds.) *Recognition and Freedom: Axel Honneth's Political Thought*, Brill 2015), "Miracles in Dark Times:

Hannah Arendt and the Refugee as 'Vanguard'" (with Cindy Horst, in *Journal of Refugee Studies*, online first 2019), and "Ecological Sensibility: Recovering Axel Honneth's Philosophy of Nature in the Age of Climate Crisis" (in *Critical Horizons*, 2020).

Gonçalo Marcelo holds a PhD in Moral and Political Philosophy from NOVA University in Lisbon. Currently, he is a researcher hired by the Centre for Classical and Humanistic Studies (CECH, University of Coimbra) and an invited Lecturer at Católica Porto Business School. He is also a group leader in the COST Action 16211 (RECAST) on civic rights and democracy in Europe and a translator of philosophy. Formerly, he was a FCT posdoctoral fellow (SFRH/BPD/102949/2014) and a visiting scholar at the Université Catholique de Louvain (2009–2010), Fonds Ricœur (2011), and Columbia University (2011 and 2016). His main research interests are in the fields of social and political philosophy, ethics, hermeneutics, and critical theory.

Gabriela Mezzanotti is an associate professor at the Master Program in Human Rights and Multiculturalism at the University of South-Eastern Norway. She is a member of the research group in Human Rights and Diversities (HRDUSN), a member of the research group Human Rights and Vulnerabilities (Universidade Católica de Santos, Brazil), and a member of the Human Rights and Reconciliation in a Post-conflict, Multicultural Society project (NORPART). She is a lawyer. She coordinated Unisinos University UNHCR Sergio Vieira de Mello Chair from 2011 until 2019.

Zuzana Uhde has specialized in critical social theory, feminist theory, and research on global interactions. Her research has especially focused on the political economy of transnational migration and EU-Africa relations, transnational migrants' claims for global justice, the commodification of care, migrant care workers, and global gender justice. She works as a researcher at the Institute of Sociology of the Czech Academy of Sciences and is editor-in-chief of the academic journal *Gender a výzkum/Gender and Research* (www.genderonline.cz). She held a Fulbright fellowship at UC Berkeley, USA, and a visiting research fellowship at Makerere University in Kampala, Uganda, and has had other fellowships in Brazil, Hungary, Slovakia, Lithuania, and France.

Chapter 1
Recognition and Migration: A Short Introduction

Gottfried Schweiger

Abstract In this short introduction, I highlight the relation between recognition and migration and explore the potentials of a critical theory of recognition. Migration in all its forms and its causes is by no means a new phenomenon, but it has become more intense in some parts of the world, and, especially in Europe, its perception by politics and the population has changed. Migration is also widely discussed in ethics and political philosophy. So what contribution can a critical theory of recognition make here? Recognition is an analytical concept, which can also be employed in empirical research to better understand the claims and living conditions of migrants. Recognition is also a normative and ethical concept, which helps to formulate what a just society owes migrants and to conceptualize their moral rights. Finally, I will also say a few words about the relation of recognition and migration before the background of the COVID-19 pandemic.

Keywords Recognition · Migration · Political philosophy · Justice

1.1 Introduction

All books are also documents of their time. From 2015, after the so-called refugee crisis, during which millions of people set out to find a new home in Europe and which shook European politics, migration and flight became the most discussed topics—in public, politics, and science, and also in political philosophy (Duarte et al. 2018), migration, the rights and duties of migrants, their vulnerability and suffering, and what responsibility the states and the world community has toward them were debated. In real politics, European leaders reached an agreement relatively quickly with neighboring states in Asia (Turkey) and North Africa to close the borders and stop hundreds of thousands of people on their way to a better life. The

G. Schweiger (✉)
Centre for Ethics and Poverty Research, University of Salzburg, Salzburg, Austria
e-mail: Gottfried.schweiger@sbg.ac.at

G. Schweiger (ed.), *Migration, Recognition and Critical Theory*, Studies in Global Justice 21, https://doi.org/10.1007/978-3-030-72732-1_1

suffering in the refugee camps and the silent deaths of thousands in the Mediterranean, which initially shocked and moved people, were noticed more and more hesitantly and less by the European public and the media. The spirit of populism had taken root in the politics of many European states (such as Germany, Austria, or Hungary) and with it racism and xenophobia (Borneman and Ghassem-Fachandi 2017).

Certainly, migration is a global phenomenon and not one that only occurs in the migration movement from North Africa and the Middle East to Europe. People from all over the world are trying to get to the rich countries of the Global North. Geographic proximity is one factor, but migration movements are global. Women from South America and Southeast Asia work in the care sector in North America and Europe, and men go to the Gulf States as construction workers. The refugee crisis of 2015, which is also the implicit starting point for this book, is only one example of global migration, but one that has attracted a lot of attention—but attention economies are not fair either. For decades, people have been fleeing war and hunger and their fates have been largely ignored, but when a few hundred thousand people managed to enter Germany because the borders in the Balkans were opened, the public and politicians were awakened. Even in a globalized world, which urgently needs global ethics and global solutions for such challenges as war, poverty, or flight, those events which directly affect the population of the Global North are receiving much more attention. One can think here of Peter Singer and the drowning child. When the refugees from Syria arrived at the German train stations, they were cared for and welcomed by volunteers. As long as they were in the refugee camps at the border of Europe, in their homeland, or in Lebanon or Jordan, almost nobody was interested in them. In the meantime, the culture of welcome has changed into a partly open rejection of migrants or at least has been pushed out of public awareness again.

In political philosophy, debates on the ethics of migration were carried on in the aftermath and under the impression of the events of the year 2015, and these debates were actually neither particularly new nor were their substance significantly changed by the refugee crisis. The refugee crisis was just a hook to talk about open borders, nationalism, and human rights, and everyone knew that these issues were present for a long time (Fine and Ypi 2016; Sager 2016). The radical inequality of life chances in this world and the many difficulties that migrants face on their journey and in their new homes are not new. Global injustices are not new. The war in Syria, which forced millions to flee, was just one conflict among many in recent decades, but one that brought to light what was actually known to everyone who wanted to know, namely, that poverty and misery and fear cause people to flee and that this migration movement is directed toward Europe, should not come as a surprise either.

Since the academic world, especially philosophy, always needs some time to produce texts, to prepare publications and events, and to overcome long peer review processes, it is understandable why the trend toward questions of migration ethics, which was triggered by the refugee crisis in 2015, is still noticeable today, 5 or 6 years later. However, it is becoming apparent that a new major subject, the COVID-19 pandemic, will take hold of philosophy, especially ethics and political philosophy, and will be the new big talking point in the coming years (Schwartz

2020). This pandemic is also a truly global challenge. The number of publications on questions of migration and flight will decrease and move back toward the quantity before 2015. This does not, of course, mean that the many philosophical questions of migration have been answered. They are not. The problem of migration has certainly not been solved politically or socially, so there would be no longer any need to reflect on it philosophically or politically. The problems are almost all unsolved, and the suffering of many migrants in the camps and during their journey, their misery before they even become refugees, remains. Exclusion and racism, poor pay, and exploitation as well. The Global North still profits from cheap labor, which is provided because of misery and poverty. One does not have to be a prophet to see that the global economic crisis and social crisis that will follow the COVID-19 pandemic will have an impact on migration movements and public perception of migration. If millions of people in the Global North become unemployed and impoverished, it is to be feared that this will lead to even less attention being paid to the problems and injustices that people in the Global South suffer from and because of which they migrate. Xenophobia is a more likely consequence than solidarity and cosmopolitanism in times of economic tension and the existential stress that many people have to live through (Jay et al. 2019). This book is thus standing at the dawn of a new crisis—the consequences of the COVID-19 pandemic—and thus of a new major theme in political philosophy and global ethics as well, while it addresses another one of its major themes: migration as an ethical and political-philosophical problem.

This book discusses migration from the perspective of a particular normative theory: recognition theory. In recent years, the concept of recognition has found an astonishing resonance in social and political philosophy and ethics, but also in the social sciences (O'Neill and Smith 2012). The claim is made that social relations and processes can be better understood through the reference to recognition and misrecognition, which opens up potentials for criticism and overcoming injustices and distortions in modern, capitalist societies. Critics, on the other hand, often argue that the focus on recognition is misguided and obscures the view on the actual social problems and their causes and is therefore not suited to pointing the way out. Instead, political philosophy should focus on issues of distribution and redistribution of material resources. In particular, global justice and ethics should deal with the vast inequalities in the world and not focus on issues of recognition and cultural identity. Proponents of recognition theory will answer to these objections that recognition is not only concerned with issues of culture and identity but that recognition comes in material, social, and symbolic forms. Global poverty, for example, is then also an issue of misrecognition (Schweiger 2014). Central to many discussions about the value of recognition theory is its application to real-world issues and the extent to which it is able to understand and analyze emerging social phenomena and developments. Migration in all its forms and its causes is by no means a new phenomenon, but it has become more intense in some parts of the world, and, especially in Europe, its perception by politics and the population has changed. So what contribution can a critical theory of recognition make here? Is the concept of recognition appropriate to answer the political, social, ethical, and socio-theoretical questions posed by migration, flight, and integration? To what extent can global migration movements and their causation through displacement, war, poverty, hunger, or climate change be analyzed in terms of recognition theory, or is there a need for other conceptual approaches and theories? And finally, the question what distinguishes the perspective of recognition from the many other theories and normative concepts in social and political philosophy that deal with migration, and what additional insights or critique it has to offer? The chapters of this book try to provide answers to these and related questions and show the potential of recognition theory.

Against this background, I would like to develop two thoughts that relate to the theme of this book, namely, the relationship between recognition and migration, both on its own and against the background of the COVID-19 pandemic.

1.2 Exploring Recognition and Migration

In this book, recognition is understood primarily as a normative concept of political philosophy and ethics, but also as a concept of social theory and empirical social research that helps to better understand political and social phenomena—such as migration. The fact that recognition has both a descriptive and explanatory perspective and a normative-ethical one makes it such an attractive concept for a critical theory of society and justice alike. Yet the concept of recognition is not precise. It has been interpreted in different ways by various theorists, beginning with Fichte and Hegel and, more recently, by Nancy Fraser and Axel Honneth. Most of the contributions in this volume critically refer to the variant of recognition theory popularized by Axel Honneth (Honneth 1996; Schmitz 2018). Here too, many aspects can be distinguished. Honneth uses the concept of recognition to describe psychological and social phenomena as well as to criticize injustices, which he characterizes as forms of misrecognition. Recognition is demanded by subjects and collectives, it is experienced in interpersonal relationships, and it is institutionalized through social norms and practices. The relationship between description, analysis, and critique and normative claims of ethics and justice is not always clearly resolved. This is also due to the demand of the theory of recognition that the standard of its critique should derive from the analyzed society itself. One can call this immanent critique (Stahl 2017).

All people strive for recognition, and they need it to lead a good life (Pilapil 2011). But in all societies, there are also forms of misrecognition and nonrecognition that are experienced as painful. A good life is therefore a life with sufficient recognition, and a just society should not only pay attention to a fair distribution of material goods but also make recognition possible for all and protect against experiences of misrecognition and nonrecognition. Many unresolved questions are hidden in the details, and the chapters of this book shed light on some of them. If one now wants to make the insights of the theory of recognition fruitful for the problems of migration and flight, then different perspectives offer themselves.

First, one can try to describe migration as a relationship of recognition. Migrants make demands for recognition, often already the most basic one to be recognized as people with rights. Very quickly one will find out that many migrants have experienced misrecognition and nonrecognition. They are not seen as people with rights, but as a threat to national security that must be averted. They are not treated with respect and esteem, but their demands, their suffering, and their—quite legitimate—claims are ignored. These are not new insights. They can be summarized in different conceptual languages such as liberal theories of distributive justice, non-domination, or human rights. The perspective of recognition theory becomes fruitful when it is able to show how processes of recognition—the social interplay of claims and responses—unfold dynamics that help to understand essential characteristics of the situation of migrants. Not everything that migrants do is a claim for recognition, but they do constitute an important aspect. This perspective of analysis and description, which must also be strongly based on empirical research, can be applied at various levels. The micro-sociological level of individual migrants' attitudes, actions, and experiences is always embedded in the macro-sociological level, where social structures, norms, and practices are defined. The two levels are not fully overlapping—there is resistant action that is directed against the dominant norms and practices, reinterprets them, and locates them in the subject. Struggles for recognition can be initiated by individuals, but they can also become collectivized and thus ultimately change the social and political framework. In this respect, everyday experiences of racism and exclusion that many migrants experience can be both individual and collective.

However, the perspective of analysis and description is not enough for a critical theory of recognition—it wants, secondly, to criticize the conditions, to show what is wrong and unjust about them, and to suggest ways out. Most of the contributions in this volume are normatively ethical in this sense, in that they seek to better understand and criticize the injustices inherent in reality on the basis of an analysis of them and to develop ethical-moral and ethical-political claims on the basis of this criticism. This is the approach of the theory of recognition. It does not design a just society on the drawing board and does not make an ideal theory of what the perfect world should look like, but it wants to develop its ethical and political claims from within the real world and the prevailing conditions. In this respect, the theory of recognition is always unfinished and on the way, not at the end. Of course, a whole series of problems are connected with this method and this claim. It is neither clear how much recognition is needed to lead a good life, nor can the concrete forms of recognition that are important be so easily determined. Honneth early on distinguished three forms of recognition (Honneth 1996): love, respect, and social appreciation, but it is unclear whether these three are exhaustive and in what concrete material, social or symbolic goods, and practices they should be expressed. It is well-known that there is a great individual and social diversity of how social esteem can be experienced—material goods, money, praise, and status can all express social esteem, but they can also stand for love and respect. On the one hand, this flexibility is an advantage for the theory of recognition, as it allows it to describe and analyze many different social phenomena as relations of recognition—from a normative point of view, on the other hand, this makes it difficult to define exactly what people have a moral claim to. This is also the case with migration.

The ethical theory of recognition will criticize migration in three ways and examine it in terms of moral demands. Firstly, it will ask what suffering is found in the lives of migrants and to what extent this can be seen as misrecognition and nonrecognition. Since this is a normative question, the description alone is not sufficient. Not all subjectively experienced suffering is unjust. Not all experiences of misrecognition are morally reprehensible. It may well be the case that cultural practices of migrants are justifiably rejected—for example, the circumcision of girls—and that these are banned, even if some migrants feel that their cultural self-determination has been violated by this. Explaining to someone that they are doing something wrong and immoral can be experienced as degrading and shameful and yet it can still be justified. For this very purpose, criteria need to be developed to show which forms of misrecognition and nonrecognition are morally problematic and unjust. The perspective of the victims is important for this; they also have a right to be heard and to put forward their demands for recognition and articulate their suffering.

Secondly, the ethical theory of recognition will not be able to stop at the level of individual suffering and individual experience. It will have to include the level of collectives, society, and politics. On the one hand, much suffering of subjects is generated by structures. The injustices experienced by migrants, especially refugees, are not random events between individuals, but are generated and supported by policies and social norms and practices and the structures in which migration takes place. The experience of racism and exclusion, discrimination in the labor market, and the harassment and degradation by state authorities that affect the lives of migrants can only be adequately understood and criticized in structural terms. Not every experience of misrecognition is indicative of a structural problem, but if it turns out that these experiences are made by many and again and again, then it is obvious to look at the social and political level here. Migration is driven by social and global injustices, and it leads migrants into societies where injustice prevails. Even if life improves for many through migration, they are still disadvantaged in their new (temporary) home and are misrecognitioned in their status as migrants. They are denied or at least made more difficult to experience sufficient recognition. Of course, this is not true for all migrants. The wealthy manager who moves temporarily from Europe to China or the USA finds many advantages there through his economic and social status and certainly does not have the same experiences as a Filipino cleaning woman who moves to these countries.

Thirdly and finally, criticism alone is not enough (Schweiger 2020). The theory of recognition will try to develop solutions. It is not clear whether the theory of recognition will arrive at similar conclusions here to those demanded by other normative theories. For example, it is not clear whether, from the perspective of recognition, open borders or forms of migration control by individual states would be morally legitimate. Perhaps this question cannot be answered by a theory of recognition alone. A fair treatment of migrants will require that they can experience recognition and demand that they can lead a life without disregard. However, this is extremely vague and needs to be further elaborated and substantiated. The theory of recognition is primarily related to the context of individual, Western societies. It does not have its origins in cosmopolitanism (Heins 2008; Schweiger 2014). Nor can the ethics of migration escape questions of social justice. This concerns integration into the labor market as well as access to the health system or the experience of love and care during childhood. Here we are dealing with social arrangements and institutions that are often national, sometimes even deeply local. At the political level, migrants' rights play a major role: social rights as well as rights of political participation and codetermination. But the ethics of migration is also transnational and global; it cannot be limited to individual societies and states. This poses a methodological and substantial challenge for the theory of recognition. Methodologically, it is difficult to transform the form of immanent critique, which refers to normative potentials in a given social norms and practices, to the global level. In terms of substance, it is a challenge to say which transnational or global rights to recognition migrants should have. Can the mode of social esteem, which is oriented toward performance-oriented market societies of modern capitalism, become a global value—should it be? Above all, it is a previous weakness of the theory of recognition that it has not considered transnational and global institutions as actors of justice, in the sense of institutionalizing the conditions of good recognition. But global migration regimes need such a perspective.

1.3 Recognition, Migration, and the Social Impact of the COVID-19 Pandemic

This brings me back to the triangle of recognition, migration, and COVID-19 pandemic, which I mentioned at the beginning of this introduction. Today (summer 2020), it is not foreseeable how the COVID-19 pandemic will develop. In some countries, the first wave of the pandemic and the lockdowns seem to be over, but in others, the momentum is unbroken. It is already clear that the COVID-19 pandemic is not only a medical and health problem, but a deeply social one. Not all people can protect themselves equally well against infection and not everyone has access to adequate medical care. The lockdown is particularly difficult for poor and disadvantaged groups to survive and deprives many of the financial basis for their livelihoods. Above all, however, the economic downturn will lead to mass unemployment and poverty and will perpetuate them over the years unless radical countermeasures are taken. The pandemic has many aspects to do with recognition and migration. Firstly, global migration regimes and the living conditions of migrants worldwide will be affected by the pandemic and the measures to combat it. Reducing migration and travel is one measure to contain it. The claims and rights of migrants threaten to fade into the background and to lose further political support. If the scheme of the theory of recognition I have outlined above is correct, it now has the task of analyzing and criticizing these social distortions and injustices. The pandemic will create new injustices and inequalities, but above all it will deepen and reinforce existing ones. Refugees will find it even more difficult to get recognition for their suffering, and their demands for recognition will fizzle out. The pandemic will probably not trigger global solidarity, but nationalism and closed borders. In the first months of the pandemic, the rich societies and states have only looked to themselves, and, unfortunately, there are no indications at present that this will change. Perhaps the world order will be a better one when this book is published, but that is unlikely.

Secondly, now would be the time to reflect on the globalization of recognition and a shift in its meaning. The economic crisis will cut off not only migrants, but millions of people in rich countries from resources of recognition and make them vulnerable to experiences of misrecognition and nonrecognition. For many people, especially if it lasts longer, unemployment is associated with a loss of material and social and symbolic recognition. It could be that the mass unemployment caused by the economic crisis that will accompany or be partly caused by the COVID-19 pandemic will make it possible to partially reverse the common individualization of unemployment and thus make it easier for the people affected to process it. However, this is not yet clear. Unemployment, poverty, and their consequences for material, physical, and psychological well-being and social status are all produced in modern, capitalist working societies and are also part of the prevailing recognition regimes and recognition hierarchies. The poor social protection against unemployment and poverty increases the importance of gainful employment and is therefore also an instrument of power (Wacquant 2010). This diagnosis is by no means new either; it is only brought to the fore by the mass unemployment that a pandemic is generating. Against this background, the old myths about unemployment and poverty being the result of one's own fault and the need to cut social benefits in order to keep up morale seem to have been finally refuted and have become obsolete. If recognition is important from a normative point of view and also such recognition generated by gainful employment, then a new approach is needed in the post-COVID-19 world to guarantee this for all people. Otherwise, millions of people who are now losing their jobs will fall by the wayside through no fault of their own and unfairly. The migration pressure on the rich countries of the Global North will not diminish, but migrants will encounter increasingly precarious and unequal social classes competing with each other. The struggle for social esteem in the form of jobs and income will become tougher and with it the vulnerability of migrants.

1.4 About the Chapters in this Book

This book thus pursues two interlocking goals: On the one hand, the philosophical and socio-theoretical debate on migration and integration is to be promoted using the instruments of recognition as a normative and social-scientific category. On the other hand, the theoretical and practical implications of recognition theory will be reflected through the case of migration. After all, a critical theory of recognition always wants to refer back to social reality and derives its normative concepts and theories from it. In order to realize these two goals, this book brings together philosophical, social-theoretical, and empirically oriented contributions that are divided into four parts.

The first part deals with questions of recognition theory as normative theory and how it positions itself on general questions of migration and the political-philosophical debate on migration. The aim here is not only to elaborate the normative basic concepts of recognition, autonomy, and vulnerability in their application to migration but also to shed light on the relationship of recognition theory to other theories of ethics and political philosophy. What is the added value of a theory of recognition over theories of political liberalism that dominate the philosophical debate on migration?

Using examples drawn from gender-based asylum cases, David Ingram examines in his chapter how far recognition theory (RT) and discourse theory (DT) can guide social criticism of the judicial processing of women's applications for protection under the Geneva Convention Relating to the Status of Refugees (1951) and subsequent protocols and guidelines put forward by the United Nations High Commissioner for Refugees (UNHCR). He argues that these theories can guide social criticism only when combined with other ethical approaches. In addition to humanitarian and human rights law, these theories must rely upon ideas drawn from distributive, compensatory, and epistemic justice. Drawing from recent literature on epistemic injustice, this chapter shows how DT and RT illuminate the failure of asylum courts to respect the credibility of women's testimony and understand their trauma. Ingram argues that the institutional privileges accorded to asylum boards and the interpretative frameworks available within immigration law impose a burden of proof on women asylum applicants that they cannot meet. He concludes that this injustice need not reflect an irremediable tension between competing epistemic and hermeneutical standpoints.

In the second chapter, Martin Huth argues that Axel Honneth's theory of recognition provides us with some crucial prerequisites for an ethical analysis of the dealing with migration in the Western world. His consideration of the inevitable dependency of individuals on recognition, particularly the respect for rights and the esteem of achievements, is crucial for an evaluation of current practices and the detection of social pathologies. However, there are some pitfalls in this approach that can be illustrated through the application of his theory to migration. First, Honneth assumes a teleology of recognition. He, thus, cannot answer the question why differential treatments of various groups emerge (despite of interpreting it as pathological deviation from a normal state); he does not take into account specific power relations; and he cannot target limits of social inclusion. Second, he assumes reciprocity in recognition and is unable to tackle the question of asymmetries between myself and the Other, but also between members of a society and individuals who want to immigrate in this society. Third, Huth challenges the idea that it is the full-blown person or their achievements, which is to be recognized. Instead, he argues that vulnerability should be at the center of recognition, particularly of solidarity. Drawing from Judith Butler, he argues that the structural recognizability of vulnerability helps us to rectify these pitfalls. However, in turn, Butler tends to leave idle the basis for a critical evaluation of current practices. This becomes visible through an analysis of her conceptualization of solidarity. Thus, Huth proposes to complement her approach with the strategy of immanent critique, which has been developed within the critical theory.

The chapter of Odin Lysaker explores the condition of statelessness. In the last 15 years, scholars have increasingly applied Axel Honneth's recognition theory to global issues such as justice, poverty, solidarity, peace, cosmopolitanism, and climate change. UNHCR estimates that there are currently 12 million stateless people worldwide. Their citizenship rights and human rights are misrecognized. As humans, their human rights and human dignity should be protected. However, this requires being a citizen of a state, which, by definition, excludes stateless people. Although representing a fairly large group among today's irregular migrants, within the above Honneth scholarship, statelessness is seldom investigated. And if being explored, the primary focus is rights-based recognition within states. In contrast, therefore, Lysaker reconstructs what he perceives as a Honnethian idea of a "transnational struggle for recognition" of stateless people. First, he reframes the concrete universalism of the early Honneth's philosophical anthropology, which includes stateless persons. It is concrete by being grounded in subjects' bodily experiences, and it is universal by understanding recognition as a human condition. Second, building on his philosophical anthropology, Lysaker argues that embodied, vulnerable humans are motivated by misrecognition. Stateless peoples' transnational struggle for recognition is driven, then, by the lack of love, respect, and esteem. In some instances, however, statelessness even generates "refugee patients," where stateless persons may be subjected to bodily experiencing, e.g., extreme traumatization due to the limbo situation often caused by statelessness. By being nonrecognized, and not merely misrecognized, stateless persons may be hindered from struggling for recognition. Finally, Lysaker emphasizes the relevance of the entire Honnethian framework in the case of statelessness based on the suffering caused by the experience of misrecognition as well as nonrecognition. This approach involves applying all three forms of recognition: respect, love, and esteem. Lysaker concludes in his chapter that this framing invoke what I conceptualize as a "transnational recognitive demand" that stateless people be recognized—as humans—in order to live a full life with dignity.

Based on a larger study on mobility projects and networks of 134 Filipino nurses, domestics, and care workers in New York and London, the chapter of Rizza Kayes Cases examines how Filipino migrant workers "claim" their "rightful" place in the place of destination within the context of underappreciation for the kind of (care) work that they do. Research participants tend to position themselves as valuable by differentiating themselves from the stereotypical image of migrants who are just after the benefits they can get from the "host" country. Their feelings of belongingness and "deservingness to be there" are also validated by the recognition that participants get from their work and perceived greater purpose of their role as care providers. At the same time, the financial security that they were able to attain allow them not only to become a part of the "host" society but, more importantly, of their "home" country. Being able to afford the "good life" means that they are not (or no longer) located at the periphery in their homeland. However, while care work allows participants to see themselves as deserving to be in the place of destination given their contribution, being employed in what is deemed as low-status job can also make migrants feel that they are living on the margins of the "host" society and that they do not really belong. Thus, these ambivalences concerning migrants' claims-making and struggles for recognition through care work could be conceptualized as Irene Bloemraad's notion of "structured agency." On the one hand, claiming recognition conforms to certain normative ideals, such as being a "good migrant" or a "good citizen." It is also constrained by immigration regimes and practices of "controlled inclusion." On the other hand, claims-making is also agentic as migrants try to fulfill their migration projects and assert their right to have better futures for themselves and their families.

The second part of this book then focuses on questions of law and politics, which are two important points of reference for recognition and migration. Law and politics want to regulate migration and integration and thus also provide a framework for how recognition becomes achievable for migrants. A special dimension of this is the political mobilization of migrants themselves and how it drives struggles for recognition, which ultimately lead to legal changes. Furthermore, chapters in this part of the book deal with the inclusion and exclusion of migrants in the welfare state and the role migrants play for the state in providing care work. On the one hand, migrants are often portrayed in the media and politically as a threat to the welfare state; on the other hand, the withdrawal of the welfare state in the first place provides niches in which migrants are all too willing to be accepted in order to perform care work. Other aspects of welfare and state functions, such as the education system, are of great importance for the experience of recognition and disregard of migrants, and this can lead to intertwining of disadvantages and discrimination.

The first chapter of this part is the one of Simon L. Joergensen, who explores the relevance of Hegel's theories of recognition, freedom, and social integration for contemporary immigration debates. In democratic welfare states, a growing number of immigrants form a permanent part of the citizenry but are not recognized as full members of the demos as they fail to meet the demands set for full political membership. For instance, states may demand work and economic self-support as proof of integration. Joergensen's chapter aims to show that the theory of recognition developed by Hegel could further the contemporary demos-debates concerning legitimate justifications of in- and exclusion substantially. Hegel articulated a theory of recognition, freedom, and social recognition of great relevance for evaluating contemporary policies of integration such as demands of work and self-support as a prerequisite for naturalization. Normative political theoretical debates about policies of in- and exclusion of immigrants seldom confront whether work could be understood as a relevant driver and marker of integration. Much can be learned from Hegel's attempt to describe what he took to be a realistic utopia of integrated citizens reproducing freedom through recognition. This analytical model asks whether policies of integration help linking individual and communal identities and types of freedom in ways that promote rather than undermine the freedom of the individuals and the community. The central motor of integration is here taken to be recognition once this takes institutional forms that promote both individual and communal freedom while transforming both in manners that allow them to link.

The chapter of Sabine Hirschauer explores securitization and recognition theories as lenses into conflict, emerging from specific migration practices. In current critical scholarship, these practices are, for the main part, interpreted as disrupting and breaking down Axel Honneth's institutionalized recognition order. They are fundamentally undermining migrants' self-trust, legal rights, and abilities for self-determination and autonomy. By approaching the intersection of recognition and migration—and global mobility generally—from an interdisciplinary, critical, post-structural security studies perspective, this chapter makes two contributions to existing scholarship: Firstly, it highlights the linkages between misrecognition and the logic of subjective intentionality of security, also understood as the securitization of subjectivity through negative securitization logic; and secondly, it draws attention to a normative opening, an inclusive, positive securitization, which moves misrecognition injuries as emancipation toward a more advanced, more complete social and moral progress. By utilizing two similarly constructed "crises" environments—the migration discourse in Germany and the USA—this chapter takes its epistemological point of departure from both countries' containment and deterrence migration policies and practices. Specifically, it interprets the US Remain-in-Mexico and Germany's ANKER Center policies as liberal dispositifs of security. The struggle for recognition is a struggle of becoming "a self." If the processes and lived experiences of becoming a self are disrupted, we speak of misrecognition. Instead of facilitating trust, respect, and esteem through undifferentiated rights, agency, and positive security for migrants, German and US migration practices violate the two countries' own liberal and moral aspirations of universal rights, justice, and social and moral progress. A way out of the negative security and misrecognition logic can be achieved through emancipation. The proposed interdisciplinary recognition and securitization lens, including the concept of thick recognition, provides an opening toward alternative approaches. It is an approach toward positive securitization as the maintenance of just, core values lived and realized through diversity, self-determination, and multi- and transcultural agency. This chapter evaluates these claims of a new normative opening increasingly asserted by a twenty-first century, progressive era of transcultural, hybrid identities.

Onni Hirvonen's chapter is concerned with the issue of civic selection. Large-scale immigration and the refugee crisis have caused many states to adapt ever-stricter civic selection processes. Hirvonen discusses the challenges arising from civic selection from the perspective of recognition theories. The argument is that recognition theories provide good conceptual tools to critically analyze civic selection and immigration. However, his chapter also aims to highlight that many current institutional practices are problematic from the perspective of recognition. In the context of civic selection, it is helpful to understand recognition as something that comes in two analytically distinct modes: horizontal (or interpersonal) and vertical (or institutional). Many rights depend on the institutionally given statuses (skilled worker, refugee, permanent resident, etc.). For a person to have a relevant social standing, she needs to be recognized by a relevant governmental institution.

However, in vertical relationships, immigrants are faced with a lack of reciprocity. They need to one-sidedly recognize the institutions, which, in turn, have full power to withhold recognition. Migrants face also challenges on the interpersonal horizontal spheres of recognition. Even if the institutional status is granted, it does not guarantee interpersonal solidarity or care. As recognition is tied to a particular institutional setting and a particular lifeworld, large-scale immigration sets two challenges. The first is the challenge of multiculturalism and recognition of diverging cultural practices of esteem. The second is the challenge of integration and getting recognition from the preexisting cultural context. Hirvonen argues here that from the perspective of esteem-recognition, this is very much a question of working rights and providing opportunities for contributing in the new context. From the perspective of care-recognition, in turn, rights to healthcare and family unifications are central. Thus, achieving meaningful personal relationships is not guaranteed by giving rights, but it is nevertheless dependent on the institutional recognition.

The chapter of Benno Herzog then explores processes of invisibilization. By aiming at the recognition of normative claims contained in affective reactions to misrecognition or social suffering, recognition theory points—at least implicitly—also toward the visibilization of groups and individuals who suffer from misrecognition. Social invisibilization is therefore understood as hindering recognition and impeding even the perception of legitimate normative claims. However, since Foucault, we also know that visibilization could be a mechanism of control and domination, thus pointing to a form of misrecognition as the opposite of recognition. In this chapter, the diverse forms of migrants' struggle for recognition will be analyzed with regard to the question of how these struggles negotiate processes of public visibility. If visibilization can be a mechanism of control that in some cases can even lead to detention and deportation, then self-invisibilization, e.g., of undocumented migrants, could help to escape this kind of disrespect. In the first part, and with the help of Axel Honneth's theory of recognition, Herzog explores the significant key elements of a theory of invisibilization (I). He then describes direct struggles for the recognition of migrant's normative claims (II). Using the approach by Dimitris Papadopoulos et al. in a second part, subversive struggles of escaping mechanisms of visibilization and control will be shown, and proposals of derecognition, as an emancipatory strategy, will be discussed (III). Not granting certain social institution the right to recognize also means not granting them the power to promote misrecognition.

The last chapter in this part is by Alyssa Marie Kvalvaag and Gabriela Mezzanotti and explores migrant interactions with Child Welfare Services in Norway. The Norwegian Child Welfare Services (NCWS) has faced intense criticism regarding their interactions with migrant families, with international human rights monitoring mechanisms expressing concern regarding ethnic discrimination over the past decade. The aim of this chapter is to contribute to the academic discussion around migrant interactions with NCWS through exploring the suitability and relevance of Nancy Fraser's theory of social justice, with a particular focus on recognition. The authors utilize the narratives of two migrant parents and two child welfare practitioners supplemented by critiques from international human rights monitoring mechanisms to bridge the gap between the theoretical level, institutions, and daily practices. Three areas regarding the suitability of recognition in the case of NCWS are discussed: misrecognition as institutionalized subordination; equality, sameness, and difference in the Nordic welfare state; and the dynamic nature of culture. While Kvalvaag and Mezzanotti find recognition to be an essential element to be considered in the case of NCWS, they emphasize recognition must also be considered within Fraser's larger understanding of social justice, alongside redistribution and representation.

In the last part of this volume, the focus will finally be on a certain group of migrants who have been in the special focus of the social, media, and political public in recent years, namely, refugees. Refugees are not only particularly vulnerable, but also experience multiple instances of misrecognition by the state and the host society, which can be analyzed and criticized through a theory of recognition. However, refugees also have a special legal status, which is intended to protect them from persecution and which can be based on their moral claims to recognition as a mode of global and social justice.

The chapter of Hilkje Hänel starts from the premise that Western states are connected to some of the harms refugees suffer from. It specifically focuses on the harm of acts of misrecognition and its relation to epistemic injustice that refugees suffer from in refugee camps, detention centers, and during their desperate attempts to find refuge. She discusses the relation between hermeneutical injustice and acts of misrecognition showing that these two phenomena are interconnected and that acts of misrecognition are particularly damaging when (a) they stretch over different contexts, leaving us without or with very few safe spaces, and (b) they dislocate us, leaving us without a community to turn to. Hänel then considers the ways in which refugees experience acts of misrecognition and suffer from hermeneutical injustice, using the case of unaccompanied children at the well-known and overcrowded camp Moira in Greece, the case of unsafe detention centers in Libya, and the case of the denial to assistance on the Mediterranean and the resulting pushbacks from international waters to Libya as well as the preventable drowning of refugees in the Mediterranean to illustrate the arguments. Finally, Hänel argues for specific duties toward refugees that result from the prior arguments on misrecognition and hermeneutical injustice.

Heiko Berner's chapter describes Axel Honneth's concept of reification, which differs significantly from the original version by Georg Lukács. The theoretical approach is adapted and used for sociological consideration of the reifying conditions under which people with a history of flight suffer in Austria. The empirical considerations are divided into two parts. In the first part, the reifying conditions faced by asylum seekers and recognized refugees are presented. These are dominated by the influence of authorities on almost all areas of life. Next, the effects of these reifying conditions on people with experiences of flight are discussed. Finally, an outlook shows how, with the support of social work institutions, they can respond in order to mitigate or even prevent reification.

The structural misrecognition of migrants is the topic of Zuzana Uhde's chapter, which is the final chapter of this book. Her chapter builds on the theoretical and empirical arguments of the critical cosmopolitan perspective and proposals for methodological cosmopolitanism, which shifts the angle from which the social sciences look at social reality. Who is regarded as a relevant social actor to put forth cosmopolitan claims is crucial. Nevertheless, the author suggests that equally important is what struggles are taken into consideration. Uhde suggests that cosmopolitan critical social theory can be usefully oriented by the concept of recognition toward the experiences of harms and wrongs as pre-political motivations for social struggles and the related articulation of claims. Migrants' lived critique is an expression of their struggles against structural misrecognition that is mediated by the geopolitics of borders and the structures of global capitalism, and the claims they voice that arise from these struggles need to be taken into consideration in the process of articulating cosmopolitan norms. In the first part of the chapter, Uhde offers a critical explanation of the geopolitics of borders within capitalist globalization in order to outline the social relations and practices that bring about the structural misrecognition of forced transnational migrants. In the second part, she examines the lived critique of forced transnational migrants through the concept of recognition. She argues that while forced transnational migrants do not necessarily share a cosmopolitan consciousness, they can be defined as cosmopolitan actors if conceptualized as a structural group. In the concluding part, she compares the viewpoint of migrants' lived critique with that of organized migrant protests that have obtained political visibility but may provide only partial foundations for cosmopolitan critical social theory. Uhde suggests that the claims arising from migrants' lived critique expand the normative horizons of cosmopolitan imaginaries to include a more radical critique of global capitalism. In this sense, it engages in struggles also for the benefit of those who do not migrate.

References

Borneman, John, and Parvis Ghassem-Fachandi. 2017. The Concept of Stimmung: From Indifference to Xenophobia in Germany's Refugee Crisis. *HAU: Journal of Ethnographic Theory* 7 (3): 105–135. https://doi.org/10.14318/hau7.3.006.

Duarte, Melina, Kasper Lippert-Rasmussen, Serena Parekh, and Annamari Vitikainen, eds. 2018. *Refugee Crisis: The Borders of Human Mobility*, 1st ed. London: Routledge.

Fine, Sarah, and Lea Ypi, eds. 2016. *Migration in Political Theory: The Ethics of Movement and Membership*, 1st ed. Oxford: Oxford University Press. https://doi.org/10.1093/acprof:oso/9780199676606.001.0001

Heins, Volker. 2008. Realizing Honneth: Redistribution, Recognition, and Global Justice. *Journal of Global Ethics* 4 (2): 141–153. https://doi.org/10.1080/17449620802194025.

Honneth, Axel. 1996. *The Struggle for Recognition: The Moral Grammar of Social Conflicts*, 1st ed. Cambridge, MA: MIT Press.

Jay, Sarah, Anatolia Batruch, Jolanda Jetten, Craig McGarty, and Orla T. Muldoon. 2019. Economic Inequality and the Rise of Far-Right Populism: A Social Psychological Analysis. *Journal of Community & Applied Social Psychology* 29 (5): 418–428. https://doi.org/10.1002/casp.2409.

O'Neill, Shane, and Nicholas H. Smith, eds. 2012. Recognition Theory as So*cial Research: Investigating the Dynamics of Social Conflict*, 1st ed. Basingstoke/New York: Palgrave Macmillan.

Pilapil, Renante. 2011. Psychologization of Injustice? On Axel Honneth's Theory of Recognitive Justice. *Ethical Perspectives* 18 (1): 79–106.
Sager, Alex, ed. 2016. *The Ethics and Politics of Immigration: Core Issues and Emerging Trends*, 1st ed. London/New York: Rowman & Littlefield International.
Schmitz, Volker. 2018. *Axel Honneth and the Critical Theory of Recognition*, 1st ed. Basingstoke: Palgrave Macmillan.
Schwartz, Meredith. 2020. *The Ethics of Pandemics*, 1st ed. Peterborough: Broadview Press.
Schweiger, Gottfried. 2014. Recognition Theory and Global Poverty. *Journal of Global Ethics* 10 (3): 267–273. https://doi.org/10.1080/17449626.2014.969439.
———. 2020. Recognition and Poverty: An Introduction. In *Poverty, Inequality and the Critical Theory of Recognition*, Philosophy and Poverty, ed. Gottfried Schweiger, vol. 3, 1st ed., 1–34. Cham: Springer. https://doi.org/10.1007/978-3-030-45795-2_1.
Stahl, Titus. 2017. Immanent Critique and Particular Moral Experience. *Critical Horizons*, October, pp. 1–21. https://doi.org/10.1080/14409917.2017.1376939.
Wacquant, Loïc. 2010. Crafting the Neoliberal State: Workfare, Prisonfare, and Social Insecurity. *Sociological Forum* 25 (2): 197–220. https://doi.org/10.1111/j.1573-7861.2010.01173.x.

Part I
Recognition, Normative Theory and Migration

Chapter 2
What an Ethics of Discourse and Recognition Can Contribute to a Critical Theory of Refugee Claim Adjudication: Reclaiming Epistemic Justice for Gender-Based Asylum Seekers

David Ingram

Abstract Using examples drawn from gender-based asylum cases, this chapter examines how far recognition theory (RT) and discourse theory (DT) can guide social criticism of the judicial processing of women's applications for protection under the Geneva Convention Relating to the Status of Refugees (1951) and subsequent protocols and guidelines put forward by the United Nations High Commissioner for Refugees (UNHCR). I argue that these theories can guide social criticism only when combined with other ethical approaches. In addition to humanitarian and human rights law, these theories must rely upon ideas drawn from distributive, compensatory, and epistemic justice. Drawing from recent literature on epistemic injustice, this chapter shows how DT and RT illuminate the failure of asylum courts to respect the credibility of women's testimony and understand their trauma. I argue that the institutional privileges accorded to asylum boards and the interpretative frameworks available within immigration law impose a burden of proof on women asylum applicants that they cannot meet. I maintain that this burden of proof is unjust because it violates the implicit discursive procedures of argumentative fairness and, in addition, disrespects women as privileged witnesses to their own criminal victimization. I conclude that this injustice need not reflect an irremediable tension between competing epistemic and hermeneutical standpoints.

Keywords Asylum · Gender · Discourse · Recognition · Epistemology

---

D. Ingram (✉)
Department of Philosophy, Loyola University Chicago, Chicago, IL, USA
e-mail: dingram@luc.edu

G. Schweiger (ed.), *Migration, Recognition and Critical Theory*, Studies in Global Justice 21, https://doi.org/10.1007/978-3-030-72732-1_2

2.1 Introduction

In the past few years, recognition theory (Honneth 1996, 2007, 2012, 2014) has joined discourse theory (Habermas 1996, 1998) in the arsenal of normative perspectives that critical theorists have brought to bear on the problem of migration (Ingram 2009, 2018b; Thompson 2019; Schweiger 2019). Critical theories provide important supplements to mainstream ethical approaches to migration studies. Human rights and value-maximizing approaches, when applied as a form of ideal, philosophical theorizing, disregard or downplay the practical constraints imposed by legal, economic, and political institutions. Like familiar cosmopolitan applications of Rawlsian social contract theory (Carens 1995, 2013), these mainstream approaches generally endorse an "open borders" policy of migration.[1] Such cosmopolitan approaches must confront the communitarian challenge posed by institutional life, most basically, that nation-states remain the primary addressees for legally securing human rights and welfare (Walzer 1983; Miller 2016; Buchanan 2013).

Critical theories of discourse and recognition share this communitarian insight to the degree that they derive their normative standards from the values and norms commonly understood to underwrite institutional practices. However, unlike most communitarian theories, the values and norms they find to be regulative of liberal democracies straddle cosmopolitan and communitarian rationales. Critical theories of discourse and recognition thus postulate a link between universal human rights and local democratic lawmaking and adjudication that is simultaneously necessary and conflict-ridden (Benhabib 2004; Habermas 1998; Honneth 2015: 278). While endorsing less restrictive border policies for the most vulnerable migrants, these theories remind us of the costs that migration imposes on migrants who are forced to uproot themselves from institutionalized relationships in which they achieve recognition as givers and recipients of love and care, claimants to and co-legislators of rights of citizenship, and esteem-worthy contributors to society (Ingram 2018b, 2020b).

This preliminary discussion about the institutional ambivalence traversing theories of discourse and recognition needs further elaboration. Before undertaking this task, a good way to situate my argument is by asking how these theories might guide our critical thinking about one specific institutional practice that bears on the ethics of recognizing migrants: the judicial hearings in which women plead for refugee protection in fleeing various forms of gender oppression.

[1] Interestingly, Rawls himself eschews any cosmopolitan application of his theory in his *The Law of Peoples* and instead asserts that under ideal conditions of a "realistic utopia," individuals would have no political or economic incentive to migrate. Appealing to Walzer's communitarian defense of territorially bounded states as essentially for fostering conditions of civic solidarity, he denies that migrants have a right to enter a people's territory without their prior consent (Rawls 1999: 39), but also notes that states have a qualified duty to assist refugees.

2.2 Gender Asylum Cases in the United States: Reconceptualizing Persecution and Its Social Recognition

Should women who seek asylum from gender-related domestic violence be recognized as refugees? The question is important, because being granted refugee status triggers the non-refoulement conditions of the 1951 Geneva Convention Relating to the Status of Refugees and provides more protection than lesser forms of temporary or qualified asylum that are extended to persons whose lives are endangered by natural disasters, civil wars, and other circumstances not related to the targeted persecution of specific groups.[2] The question thus hinges on whether women and indeed all persons who suffer gender-related oppression can be recognized as distinctive, targeted groups within their nation and how this recognition is to be conceptualized in the absence of official government persecution.

To begin answering this question, I propose that we consider the following cases in American refugee law, as discussed by Karen Musalo, an attorney who represented some of the applicants:

Case 1. In the *Matter of Kasinga* (21 I. & N. Dec. 357 [BIA 1996]), the US Board of Immigration Appeals (BIA) overruled an immigration judge (IJ) who had previously determined that a women, Fauziya Kassindja (FK), who had entered the country seeking asylum from female genital cutting (FGC) and a forced marriage in Togo, should be deported (removed). The IJ had ruled that FK's testimony lacked "rationality," "internal consistency," and "inherent persuasiveness": FK had managed to avoid FGC until her father's death, despite her claim that FGC was pervasive. In addition to this alleged contradiction, the IJ noted that FGC could not be considered persecutory, because it was a cultural practice endemic to FK's tribal group. Although Immigration and Naturalization Services (INS) intervened on FK's behalf against the IJ's ruling, they agreed with the IJ that the cultural practice of FGC did not fit the definition of persecution as set forth in a ruling precedent, *INS v. Zacarias* (1992), which required proof of "malignant and punitive intent." The INS sought to waive this requirement in the case of FK, arguing that in most cases, practitioners of FGC "believe they are simply performing an important cultural rite that binds the individual to society." In its stead, the INS proposed a new framework that would allow anyone to seek asylum if they suffered a harm that "is so extreme as to shock the conscience of

[2] As a part of the Immigration Act of 1990, the United States began admitting asylum seekers entering from designated countries suffering from natural disasters or civil violence under its Temporary Protected Status (TPS) Program. This program does not provide permanent refuge (beneficiaries must apply for extensions), and in 2017, the DHS decided to terminate TPS for six of the ten listed countries (El Salvador, Nicaragua, Sudan, Haiti, Honduras, and Nepal). This decision has been challenged in the courts. Once TPS is terminated, beneficiaries from delisted countries (such as Liberia) can apply for Deferred Enforced Departure (DED). Sweden also provides subsidiary protection to those who do not qualify for refugee status, but beneficiaries of this program, unlike refugees, do not have their costs for family reunification and travel documentation covered.

the society from which asylum is sought," viz., a harm that is extreme, was resisted, and was inflicted not as punishment for simply refusing, but in the course of actually suffering, the conscience-violating action. The BIA, by contrast, rejected the INS proposal to waive the intent (or nexus) requirement, partly because the INS construction would not provide a rationale for granting asylum to children, who often do not resist FGC.[3] The BIA instead reasoned that social context, such as a cultural pattern of patriarchal domination, apart from the conscious motivation of the individual inflicting the harm, could suffice to establish persecutorial intent (Musalo 2010: 54).

Case 2: In the *Matter of R-A-* (1996), an IJ ruled that a Guatemalan woman fleeing from spousal beatings who had been denied protection by Guatemalan police authorities merited asylum. The IJ found that Rody Alvarado belonged to a persecuted social group possessing "immutable and fundamental characteristics" as required by ruling precedent (*Matter of Acosta* [1985]), namely, "Guatemalan women who have been involved with Guatemalan male companions, who believe that women are to live under male domination." It further noted that Alvarado had resisted her husband's acts of domination, thereby expressing a political opinion for which he persecuted her (Musalo 2010: 56). In 1999, the BIA reversed the IJ's grant of asylum. In doing so, it appeared to reject its own reasoning in *Matter of Kasinga*. First, it denied that social context, combined with government toleration of violence against women, was sufficient to constitute an act of intentional government persecution. Second, in addition to the *Acosta* requirements for defining a social group possessing immutable and fundamental characteristics (in this instance, one defined by gender and married status), the BIA now required that a persecuted group also be "recognized and understood to be a societal faction" possessing "social visibility" and "particularity." Because in 1999 Guatemalan women who resisted spousal violence did so privately, in the confines of their homes, and were diffusely spread throughout all regions, classes, and subgroups of Guatemalan society, the BIA reasoned that they could not count as a persecuted group under this new requirement (Musalo 2010: 57). Fortunately for Alvarado, her case was eventually taken up by the newly created Department of Homeland Security (DHS) and resolved favorably under the Obama administration. The DHS affirmed that the *Acosta* requirement for membership in a social group possessing immutable and fundamental characteristics

[3] The INS requirement that the harm in question not be inflicted as punishment is also problematic from the standpoint of the UNHCR 2002 guidelines (see below). The guidelines recognize two types of gender asylum claims: one in which the *form of persecution* is uniquely or disproportionately inflicted on women (such as FGC, forced marriage, and rape), apart from being related to one of the 1951 Geneva Convention grounds, and one in which the *reason for persecution* is based on gender (or sexual orientation), apart from whether the persecution in question takes a gendered form. Thus, a woman can be denied equal rights to work outside the home, testify in courts of law, or attend school because of her gender. In many cases, both types of gender claims are raised (as when women are raped because of their political activism on behalf of women's equality). The INS requirement does not appear to recognize as persecution cases in which women are raped or killed as punishment for their activism on behalf of women's equality.

> was sufficient. Besides rejecting social visibility and particularity as requirements, it also affirmed that the applicant's burden of proof regarding demonstration of persecutorial intent could be established by circumstantial evidence showing that such patterns of violence are (1) supported by the legal system or social norms in the country in question and (2) reflect "a prevalent belief within society, or within relevant segments of society, that cannot be deduced simply by evidence of random acts within that society." Thus, the abuser's intention to persecute—and the government's collusion in furthering that aim—can be inferred from either his statements or actions, in combination with circumstantial evidence regarding social context and government inaction (Musalo 2010: 59).

These cases bring to the fore several distinct questions that are pertinent to the application of recognition theory, broadly understood. The first question addresses the philosophical (or conceptual) problems that attend any legal classification: Should women who suffer domestic violence and cultural oppression be legally recognized as refugees under US law? The second, quite different, question addresses the ethical problems that attend any effort to conceptualize harms and injustices as instances of specific types of misrecognition: Should women, men, children, and transgender persons who suffer gender-based restrictions and identity-shaping socialization be recognized as wrongfully oppressed, or disrespected? If so, should this type of misrecognition be understood as a form of persecution, or violation, of a basic human right; or should it be understood as a lesser harm, such as a denial of equal social esteem or equal social inclusion (as distinct from unequal social esteem and unequal social inclusion)? If it should be understood as a form of persecution, how is persecution as a distinctive type of human rights violation be understood? Must persecution always be undertaken or at least officially supported by the state, or can it be undertaken by non-state members of society in a way that is permitted, or at least not prevented, by the state? Must persecution always target a religious, racial, national, political, or otherwise publicly visible subgroup? Or can it target a more amorphous and less publicly visible subgroup, such as the subgroup of all who are deemed to have suffered a severe diminution of their human rights on account of being compelled to conform to socially imposed gender norms?

This latter question—which arises within an ethics of recognition as it grapples with how restrictive identity-based socialization can constitute an ethically regrettable form of misrecognition that can rise to the level of a significant human rights injury—clearly has a bearing on how the US law should classify asylum seekers fleeing persecution. However, the US law has been remarkably inconsistent in how it has been applied, since the jurisprudential, recognitive ethics according to which persecution should be interpreted has not been consistently worked out. By 2009, the BIA had determined that victims of domestic and gender violence could qualify for asylum, but in the *Matter of C-A-* (2006), the BIA reaffirmed its requirement that persons qualifying for asylum must be members of a social group that is both socially visible and particular. Particularity is needed to recognize the discriminatory targeting of subgroups within a population, the entire membership of which may be seen to suffer from untargeted cultural, economic, or political oppression.[4] Social visibility, by contrast, seems unnecessary and overly exclusive. Uncomplaining and otherwise publicly silent victims of FGC as well as of other forms of domestic violence, not to mention "closeted" homosexuals and nonbinary persons, are not socially visible, even though they are socially (mis)recognized within the culture as undeserving of respect and protection. Furthermore, as Musalo remarks, the BIA appears to be mistaken in thinking that the UNHCR Social Group Guidelines support its insistence on visibility as a necessary feature of an oppressed group.

In its *amicus* brief in the *Valdiviezo-Galdamez* (2009) case, the UNHCR proposes two *disjunctive* tests for identifying an oppressed social group: the "protected characteristics" tests (based on immutable and fundamental characteristics) and the "social perception" test (based on societal recognition). The social perception test is not required by the UNHCR, and the UNHCR does not understand it to mean that "the common attribute [of the social group] be visible to the naked eye in the literal sense of the term [or] that it be one that is easily recognizable by the public (*Amicus Brief*, p. 11). What counts is that the social group in question be recognized by the society in question as a group that is not only subject to discriminatory societal treatment but is also deserving of this treatment, and also that the victim and his or her abuser can be assumed to know this. Thus, in the absence of direct forms of state persecution or in the absence of abuser's stated persecutorial intent, social recognition of social group discrimination suffices to establish the nexus between victim harm and persecution.

2.3 Gender Asylum Cases in Sweden: Disrespecting Testimony and Misunderstanding Social Context

Swedish gender asylum cases raise similar questions about the role of social recognition in establishing the category of "persecuted group." However, they also raise additional questions of social recognition pertaining to the processing of asylum

---

[4] North Koreans fleeing into China are economically and politically oppressed by their government. To regard them as if they were also persecuted—and thus as meriting refugee status—would eviscerate the meaning of persecution and overburden the mechanisms of humanitarian relief available to them (China would be compelled to process thousands of "refugees" while abrogating its treaty responsibilities to the North Korean government). That said, China's current refusal to recognize North Koreans fleeing economic destitution as refugees (the South Korean government also recognizes them as its own citizens) has led to their repatriation and subsequent imprisonment, torture, and execution as "traitors." One remedy to this tragic situation would be a change in China's policy—highly unlikely given China's reliance on North Korea as a buffer against US geopolitical power—allowing for North Korean escapees to enjoy Temporary Protected Status while transiting to South Korea. I thank Thomas Arms, S.J., for bringing to my awareness the desperate plight of North Koreans escaping to China and the inadequacy of current refugee law to remedy their situation. See his MA Thesis: *Escape from China: A Legal and Philosophical Understanding of the North Korean Escapee's Journey for Freedom* (unpublished ms).

cases. In particular, they raise questions about the fairness of argumentation and deliberation within court proceedings, questions that DT is especially well equipped to answer, as well as questions about the epistemic sources that immigration boards and appellate courts use in determining the harshness of discriminatory treatment, questions that properly fall within the province of RT. In particular, they raise questions about how much applicants' firsthand testimony of their own persecution should be weighed in comparison to countrywide reports prepared by government "experts" in constituting knowledge of what counts as socially recognized persecution within a specific context. How should an applicant's testimony regarding her privileged *experience* of the unique circumstances surrounding her trauma be properly recognized by immigration judges in order to fairly determine the extent to which her social group should be recognized as suffering societal persecution?

In order to assess the recognitive epistemic dilemmas encountered in these cases, it is necessary to understand the UNHCR's 2002 gender-related and social group guidelines for interpreting persecution within the framework of the 1951 Geneva Convention Relating to the Status of Refugees and its 1967 Protocol. The new guidelines suggest that the UNHCR has undertaken the ethical labor of reinterpreting gender identity-based harms as a form of society-wide misrecognition rising to the level of a human rights injury. Article 1 A(2) of the Convention Relating to the Status of Refugees defines a refugee as anyone who "owing to a well-founded fear of being persecuted for reasons of race religion, nationality, membership in a particular social group or political opinion, is outside the country of his nationality and is unable or, owing to such fear, unwilling to avail himself of the protection of that country ... is unable or, owing to such fear, unwilling to return to it." The barriers against defining gender-related harms as harms worthy of asylum protection under this article are apparent. This article makes no mention of gender in its listing of five grounds for persecution. Furthermore, gender-related harms, as noted in the case of FK, typically reflect cultural norms and so are not always inflicted with the kind of malicious intent commonly associated with persecution. Finally, the abusers are typically domestic partners and family members, not the state, and the harm they inflict typically occurs in private, not public, settings.

Beginning in 1985, the UNHCR permitted governments to regard "women asylum-seekers who face harsh or inhumane treatment due to their having transgressed the social mores of the society in which they live" as a "particular social group" within Article 1 A(2). The 2002 guidelines draw upon a decade of case law, state practice, and academic writing in unequivocally affirming that the definition of refugee includes gender-related claims. They expressly appeal to human rights norms as guidelines for defining persecution, so that rape, sexual violence, domestic violence, FGC, dowry-related violence, or trafficking can count as persecution, when either legally mandated or legally permitted or legally outlawed but officially unenforced. As noted above, the guidelines' "bifurcated nexus" analysis establishes persecutorial intent regardless of the non-state actor's reasons for inflicting harm, so long as the harm is related to one of the five grounds listed by the convention or the government's "inability or unwillingness" to offer protection is so related.

The cases discussed above involving US gender asylum processing illustrate the difficulty in applying the updated UNHCR guidelines, specifically as it touches upon recognizing women and non-cisgender-conforming persons as legally and ethically harmed groups. How does one relate the relatively clear legal (and ethical) concept of publicly recognized targeted discrimination of visible groups to commonly accepted practices of gendered socialization and policing that potentially restrict the freedom of all members of society and whose effects might not be personally experienced as a form of disrespect by many of those affected or might not be felt directly by those who are "closeted?" As we shall shortly see, even countries, like Sweden, that expressly incorporate these guidelines into their asylum law still have difficulty answering this question consistently. What kind of gender-based discrimination amounts to persecution, what kind of government inaction counts as a lack of protection, what kind of harms arise specifically because of gender, and what it means to belong to a gender- or sexual orientation-based group that merits protection according to their "protected characteristic" or "social perception" are all questions that resurface in Swedish gender asylum processing. And they resurface in a way that illustrates how the epistemic sources available for interpreting them are divided and ranked, with those enjoying the imprimatur of legal authority being imposed with little respect for those proffered by asylum applicants.

Hanna Wikström's study exposes the arbitrariness in the way Sweden's migration boards and appellate courts apply these guidelines in processing gender-based asylum cases. Using Miranda Fricker's model of testimonial and hermeneutical injustice, Wikström documents how applicants' credibility is routinely questioned by authorities in light of what is presumed to be more reliable country-specific and culture-specific databases and how failure to incorporate their testimony into the legal description of their cases results in misunderstanding their predicament. Citing Fricker, Wikström observes that the skepticism of Sweden's migration boards and courts directed toward applicants is fueled by cultural and gender-based stereotypes: "epistemic injustice becomes evident as the reasoning of authorities shows how arguments of culture are primarily used to negate individual claims," so that "resistance to and deviations from dominant cultural norms are not seen as credible" (Fricker 2007; Wikström 2014: 14).

The epistemic injustices Wikström documents might be described as ancillary to the injustices that theories of discourse and recognition highlight. Discursive procedures that exclude, marginalize, or otherwise diminish the experiential narratives of some participants in comparison to the statistical data advanced by authorities violate egalitarian norms of discursive deliberation and argumentation, partly by defining what counts as argumentation in a presumptive way that discounts personal experience as a reason that possesses the same persuasive force as quantifiable date. This discursive injustice is accompanied by a recognitive injustice. Discounting testimony disrespects the witness by impugning her credibility as a reliable knower. This misrecognition of persons as less than rationally accountable agents often reflects deeper prejudices against the witness owing to her gender, race, culture, age, educational attainment, or economic status. These forms of discursive and recognitive injustice are illustrated in Wikström's study.

Case 3: Wikström's study highlights the ease with which Sweden's migration boards and courts routinely misinterpret the 2002 UNHCR guidelines along three axes of analyses: culture, gender, and protection in the country of origin. One case involves a Kurdish woman (Sara) who fled Iraq after both her brother and the police from whom she sought protection threatened violence against her for having sexual relations with a colleague at work, in violation of an impending marriage her family had arranged for her against her will.

In denying her asylum in favor of temporary subsidiary protection, the migration board argued:

> It is odd and not very likely that she would initiate a sexual relation with another man when she knows she is going to marry her cousin ….That she would be so blinded by love and disregard the consequences is not a reasonable explanation, with the culture that is prevalent in Northern Iraq and with her family traditions in mind … Furthermore, it must be considered striking that, at her age, an arranged marriage has not occurred earlier (Verdict 15: 3—cited in Wikström: 214).

In commenting on this case, Ezgi Sertler remarks that "when the board listens to Sara, they do so through a frame in which 'non-white' women seem intelligible insofar as they are without any agency, and their cultures seem intelligible insofar as they are all oppressive" (Sertler 2018: 10). The parallel with the IJ's ruling in the American case involving FK is striking. Because "persecution" is understood as an extraordinary deviation from cultural norms, the board, in characterizing all of Kurdish culture as monolithically oppressive for both men and women who face criminal punishment for violations of family honor, can find no basis for Sara's claim to have suffered discriminatory persecution. Culture, rather than membership in a persecuted group, is the alleged source of harm. By blaming culture, as if it were the "agent" responsible for inflicting harm, the MB absolves everyone else of responsibility, thereby denying agency to both perpetrators/facilitators of culturally mandated violence and victims who resist it.

Case 4: Another case discussed by Wikström highlights the ease with which the migration board misapplies the UNHCR's guidelines regarding gender-related persecution. A Kurdish woman (Nesrin) flees Iraq after her family kills her lover, a man who proposed marriage to her, because she was promised to a cousin of hers. Because her brother, uncle, and cousin are members of the police force, she seeks help from a women's organization to arrange her escape. The written verdict of the migration board (MB) notes that:

> Nesrin alleges a well-founded fear of persecution due to her gender and that she should therefore receive protection as a refugee. However, it has not emerged that the threat she refers to has occurred due to her gender. She has herself stated that even the man she has had a relationship with has been subjected to violence and was eventually killed because of their relationship (Verdict 12:9, cited by Wikström: 213).

Here, the MB argues that because a man is the principal victim of honor violence (HV), Nesrin cannot claim to be a refugee based on membership in a specific gender-identified group which was targeted with gender-specific discrimination.

This parallels the BIA's ruling in the American case involving Alvarado insofar as domestic violence, or gender-related violence, like HV, is not exclusively directed to women, but relates to all forms of nonbiologically understood gender oppression, including (for instance) discriminatory treatment meted out to non-binary-identified persons and homosexuals. However, the Swedish Aliens Act (2006), which was intended to incorporate a broad spectrum of gender-related harms under the Swedish provision for refugee protection as stipulated in the UNHCR's 2002 guidelines, contradicts the MB's interpretation of gender-based refugee claims. The Aliens Act states that:

> [i]n refugee law, the term gender [kön] should be used in its broadest sense and as such include not only the biological difference between men and women, but also the socially and culturally determined and stereotyping notions concerning how men and women should behave. (Prop. 2005:21)

The Aliens Act adds that "[w]hen the forms of persecution take different shapes depending on whether the persecuted individual is a man or a woman, refugee law usually refers to gender specific persecution (Pop. 2005:22). But gender-*specific* persecution, which the MB had in mind when making its determination that Nesrin was not the one targeted with HV execution, is not required for determining gender-*related* persecution based on "socially and culturally determined and stereotyping notions concerning how men and women should behave."

Another aspect of the MB's ruling in Nesrin's case touches on the MB's assessment of the degree to which she herself (and not her boyfriend) could have found protection within her native community. According to the MB, "for abused women there is generally a possibility of good enough protection" and "Nesrin did not make contact with the police, other authorities or alternative mediation institutions in Iraq [for her protection] before seeking international protection" (Verdict 12: 4–5—cited by Wikström: 213). These statements flatly contradict Nesrin's own testimony, in which she claimed that the very police authorities that the MB says could have provided her with protection were members of her own family who had perpetrated the killing of her lover.

Case 5: Two other cases also reinforce the impression that the MB routinely disregards the credibility of women's firsthand experience of the risks they face within the unique social context of their native community in deferring to its own data-based reports about the risks women living in their country generally face in not being afforded adequate official protection. These cases illustrate the discretion taken by the MB in drawing from conflicting assessments of countrywide risk. Two Syrian women, Almas and Afya, apply for asylum at about the same time in their effort to escape from repeated acts of HV. In the case of Almas, who is awarded refugee status, the migration court relies on six national and international reports to argue that "HV and murders take place all over Syria" and that Syrian law recommends "impunity or punishment mitigation in honour-related violations" (Verdict 21: 7). In Afya's case, by contrast, the migration court relies on only the board's internal report, which asserts that "the number of honour killings has gone down in the country in recent years" because "it is said that people have become more enlightened" and men know that "they will be sentenced to harsh prison terms" (Verdict 24:4). Once again, the discretion of Swedish migration boards in issuing inconsistent rulings based on different sources of knowledge parallels the discretion of their American counterparts in choosing different cases and different constructions of common legal texts.

Not only does the same migration court use cultural and national stereotypes in rendering its verdicts, but it uses vastly different reports containing diametrically opposed assessments of the Syrian "culture of honor violence." In this instance, migration courts presume (in Wikström's word) a positivistic conception of knowledge in which numerical data provide the only reliable touchstone for determining facts on the ground. Here the presumption seems to be that the data determine a uniquely coherent description of the extent of HV and official protection countrywide, which is then presumed to describe the situation of all women and men, regardless of the uniqueness of their circumstances. There appears to be little awareness of the hermeneutical complications involved in synthesizing conflicting reports and deploying finer-grained methods in understanding the unique circumstances surrounding each applicant's risks and opportunities for protection. Indeed, to undertake this kind of finer-grained analysis would require that the migration courts concede that the applicant's own firsthand experience of her situation plays a necessary role in constituting the relevant "facts on the ground." The willful hermeneutical ignorance of the migration board thus inflicts not only testimonial and hermeneutic injustices but also what Kristie Dotson calls "contributory injustice," whereby marginalized persons are prevented from contributing to the production and revision of what is always socially constructed knowledge (Dotson 2011: 32).

As we have seen, studies by Musalo and others show that the women pleading gender-based asylum in the United States and elsewhere experience similar kinds of testimonial, hermeneutic, and contributory injustice as those applying in Sweden. Whether it be reliance on expert data or legal terminology for interpreting persecution, socially recognized group-based oppression, or political resistance, courts routinely underappreciate women's complex gendered and (multi-)cultured identity/agency. No doubt, some of this ignorance reflects straightforward kinds of misrecognition, based on cultural and gender stereotypes, that disrespect the individuality of the asylum seeker. But some of this ignorance reflects a generalized misrecognition of asylum seekers as less competent epistemic agents—due to the trauma they have suffered and the less verifiable nature of their testimony—in comparison to expert data gatherers. This leads Sertler to ponder whether the very institution of gender-based asylum contains *structural* obstacles preventing recognition of applicants as competent interpreters of their situations.[5] Her own examination of

[5] By presuming the innocence of the accused until proven guilty, criminal trials in Anglo-American jurisdictions already impose a greater burden of proof on the state. As I noted earlier, this burden of proof still inclines courts to give exceptional weight to the eyewitness testimony of living victims. However, the scale of credibility can shift in favor of the accused in some circumstances. For instance, the standard of reasonable fear invoked in determining whether a woman with a long history of suffering spousal abuse can justifiably claim to have injured or killed her assailant in

the entrenched nature of social stratification and epistemic hierarchy suggests an affirmative answer. Challenging the dominant epistemic resources poses a dilemma, for the only way this can happen would be if those in authority were to find intelligible the counter-knowledge that their own hermeneutical standpoint renders unintelligible (Pohlhaus Jr. 2012: 731).

There is an additional structural explanation for the migration board's willful ignorance that does not directly refer to structural misrecognition of asylum seekers as less credible sources of knowledge but rather reflects an antagonism directed against asylum seekers as opportunists or security threats. Asylum courts presume the legitimacy of the existing nation-state system which rests upon privileging national security. The default in the court's reasoning lies with the state's interest in protecting the freedom, prosperity, and cultural identity of its citizens (Rusin and Franke 2010). As Musalo notes, this default in the court's reasoning finds support in a kind of textual fundamentalism: refugee law was only originally intended to apply to public forms of discrimination, not private domestic abuse, so that extending its application allegedly threatens to unleash a tidal wave of asylum seekers, which can only lead to overwhelming the scarce economic, political, cultural, social, and legal resources of the state (Musalo 2010: 47–8).[6]

2.4 Grounding the Critical Theory of Gender Asylum Adjudication: A Preliminary Overview of RT and DT

Discourse theory (DT) and recognition theory (RT) overlap and complement each other when applied to a critical assessment of asylum claim processing. When properly supplemented by supporting evidence, both theories, I argue, require that judges reverse the institutional privilege accorded to officers of the state. In requiring that the state assume at least an equal (and perhaps higher) burden of proof in denying relief to asylum seekers, RT, in particular, requires judges to exercise not only *epistemic justice* by conceding credibility and authority to applicants' stories; it requires a humble effort at understanding the totality of coercive circumstances that have compelled the individual applicant to seek refuge in the first place. By

self-defense or out of "irresistible impulse" (as in the case of so-called battered women syndrome) has at times been understood to require courts to adopt the standpoint of the accused, given her unique circumstances as a victim. I discuss the advantages and disadvantages of adopting such a psychopathological conception of reasonableness in Ingram (2006). As I noted earlier, the impact of trauma can distort victim's memory of their circumstances which, in turn, can cast doubt on the credibility of their testimony (Schweiger 2019).

[6]As I remarked in discussing the case of North Korean economic "refugees" (see note 4), this concern about overextending the category of refugees cannot be dismissed, even if it is exaggerated when discussing women as a discrete class of persons who have suffered a distinctive type of persecution.

contrast, DT focuses on institutional procedures for justifying and applying laws and subsidiary principles.

So construed, DT is better equipped to illuminate the court's reasoning than its understanding. As Habermas and Günther understand it (Habermas 1996: 172, 220–32; Günther 1993; Ingram 2010: 206–211), judicial decision-making reflects a process of discursive reasoning that enters deliberation retrospectively. Choosing the most appropriate statutes, principles, and precedents requisite for doing complete justice to the unique circumstances of the case alters the meaning of this body of law.[7] Accurately describing the case depends on properly understanding, weighing, and combining different sources of evidence in a coherent narrative, with the first-person, eyewitness testimony of the asylum seeker normally weighing very heavily. In this way, proper recognition of the asylum seeker as a credible witness to her circumstances can become the catalyst for a potentially far-reaching reinterpretation of law, as can be evidenced by changes in refugee, asylum, and deportation policy over the last 20 years (Ingram 2009).

DT and RT are social—not merely philosophical—theories that take their bearings from actual practices that are far from what any ideal theory might condone. Among other things, they articulate norms and values that underwrite the popular acceptance of what are essentially imperfect legal institutions.

Beginning with this premise, DT defines a legitimate law as a law that all whom it significantly affects could freely (viz., rationally) consent to, after they, as equally positioned interlocutors, have collectively discussed and, if necessary, transformed their different interests in light of the probable impact that accepting the norm will have on everyone's welfare.[8] Human rights, associational duties, and the common good are here seen as having their ultimate justification in ongoing consensus-oriented dialogue that aspires to be fully inclusive, egalitarian, and unconstrained by the effects of social power and internal bias.

---

[7] For further discussion of this hermeneutical circle as it applies to judicial decision-making (application), see Hans-Georg Gadamer's *Truth and Method* (1960), Ronald Dworkin's *Law's Empire* (1986), and Ingram (2009, 2010).

[8] DT reconstructs the general normative expectations of persons in performing social roles embedded in social practices that are essential to social reproduction, social cooperation, and socialization. It takes its bearings from social practices that aim to coordinate action based on voluntary mutual consent, what Habermas calls "communicative action." Communicative action involves the reciprocal raising of claims that arise whenever actors offer and accept invitations to cooperate with each other. Should any claim be challenged, cooperation must be restored by redeeming it with convincing justification or by re-establishing it based on an alternative claim that all parties find rationally compelling. The special kind of action associated with the argumentative disputation and redemption of claims—what Habermas calls *discourse*—sets in motion mutual normative expectations regarding the freedom, equality, and impartiality of the participants to reach an unconstrained, viz., rationally motivated, agreement that would also include the perspective of all third parties who are also significantly impacted by the terms of the agreement. In sum, what distinguishes discourse theory from mainstream social contract theory is its emphasis on discourse as a collective medium for critical self-reflection and autonomous action (or rational self-determination).

RT takes its bearings from a broader set of institutionalized practices than those associated with the argumentative justification of action norms. It draws its normative orientation from the rights and duties implicit in all forms of human relationship---nurturing, working, and deliberating---insofar as these are necessary for realizing two complementary aspects of human fulfillment: free agency, on one side, and psychological integrity (which depends on persons being accorded respect, care, and esteem for their humanity and individuality), on the other.[9]

Seen in this light, both DT and RT are frequently put forward by their respective proponents as though they each exclusively provided the metaethical foundation for all practical and theoretical ethics. However, it is immediately apparent that this presumption is mistaken. As my earlier discussion of Swedish gender asylum cases shows, DT and RT essentially complement each other. One cannot give an adequate account of rational deliberation and argumentation that focuses exclusively on the procedural justice of making and responding to assertions. One must also attend to the epistemic justice of lending credibility to assertions that are made and seeking, whenever possible, a common conceptual and experiential framework for understanding their significance. Conversely, one cannot give an adequate account of proper recognition that focuses exclusively on persons' experiencing or not experiencing feelings of disrespect. Given that disrespect may have occurred even when it is not experienced by the one disrespected, and that disrespect may not have occurred even when it is experienced, one must have recourse to a public exchange of arguments in establishing whether someone has in fact suffered justifiable misrecognition or lack of proper recognition.

In sum, neither DT nor RT can guide critical assessments of institutional practices apart from each other, not to mention other ethical theories, spanning, as we have seen, human rights, but also theories of distributive, compensatory, value-maximizing, and epistemic justice.[10] Nevertheless, DT and RT are particularly well suited to framing harms associated with the processing of asylum claims. These injustices involve not only placing a higher burden of proof on asylum seekers and denying them equal time to prepare and make arguments—a failure for which DT is preeminently designed to diagnose. They also involve a failure to understand these claims in the first place. By failing to properly respect the individuality of the migrants that appear before them, viz., by failing to properly attend to the stories they tell about themselves and the unique circumstances that mark their desperation,

[9] RT takes its bearings from a broader range of social practices that go beyond those associated with communicative action to include noncommunicative forms of social interaction, such as relationships of care and concern among intimates and relationships requiring mutual understanding *simpliciter*. It bears a closer resemblance to virtue ethics, emphasizing the intrinsic psychological and agential good that is realized upon performing institutionalized social roles, insofar as the practices in which these roles are embedded live up to their implicit normative expectations (Ingram 2020a).

[10] Resentment toward immigrants and minorities is closely linked to feeling disrespected by government policies that allegedly favor these groups. Showing that such feelings are misplaced requires a critical theory of structural distributive injustices encompassing a critique of capitalist (and racist) globalization (Fraser and Honneth 2003).

judges and administrators misrecognize and misunderstand those who plead before them. *This injustice may occur even when discourse ethical norms regarding fair procedures of argument are superficially respected; indeed, they occur prior to formal argument, insofar as they reflect deep-seated prejudices.*[11]

2.5 DT: Reversing the Burden of Proof

DT and RT provide ample resources for framing social pathologies and social injustices, including those associated with the institutional imposition of an unfair burden of proof. Speaking more generally, their diagnosis of social injustices is relatively straightforward. For DT, forms of "voluntary" cooperation, whose presumed consensual basis in shareable reasons is contradicted by the irrational compulsions and hidden, background constraints of social domination, belie their own implicit embodiment of a norm of *egalitarian reciprocity* and so are, on the surface, unjust.[12] RT is less insistent that all recognitive relationships should involve linguistic reciprocity (a loving relationship between a parent and infant child involves expressive forms of communication that cannot be fully reciprocal, simply because the child has not developed full-fledged communicative competence and, along with it, an implicit expectation of linguistic reciprocity). Instead, RT identifies as unjust forms of institutional recognition that misrecognize or fail to recognize persons' proper sense of self to the point of denying their agency and causing harm to their personal identity as a credible witness.[13]

---

[11] Earlier in his career, Habermas defended therapeutic discourse as a privileged means for addressing the pathologies of "systematically distorted communication" including ideologically biased forms of understanding that prejudice (constrain) argument subconsciously, despite outwardly conforming to procedures of rational dialogue (Habermas 1971: 218-45). This view, which Habermas has never abandoned, supports my contention that neither discourse theory nor recognition theory, here understood as specifying norms of undistorted understanding, can function independently of the other.

[12] For Kristie Dotson, "refusal, intentional or intentional, of an audience to communicatively reciprocate a linguistic exchange owing to pernicious ignorance" (Dotson 2011: 238) amounts to "epistemic violence" insofar as the speaker/listener who is harmed is silenced by another ("testimonial quieting") or is forced to silence herself because of another ("testimonial smothering") (Dotson 2011: 242). The presumption that linguistic reciprocation should be the guiding norm in communicative contexts is true as far as it goes, but when it becomes apparent to one of the interlocutors that communication is "systematically distorted" by cognitive prejudices, defense mechanisms, or a shared background of knowledge and linguistic meaning that one-sidedly misrepresents the experience of that interlocutor or one of the other interlocutors, then, as Habermas observes, the exchange must no longer be regarded by that interlocutor as if it were a genuine communication but rather as a provocative, manipulative, or (self-)deceitful action that masquerades as communication. Under these conditions, a person of color, say, is not obligated to listen or respond to the racist comments of a white person.

[13] Using discourse or recognition theory to diagnose social pathology is a great deal more complicated. Without delving deeply into the ways in which different institutionalized systems of action, or different institutionalized action spheres and their accompanying recognition orders, can clash

Let me begin by briefly adumbrating the contributions of DT to migration ethics as a general field. Discourse ethics draws quasi-cosmopolitan implications from communitarian premises. It reflects a deep social fact about the ethos of mutual accountability informing modern and above all liberal democratic societies. Its reliance on face-to-face deliberation as a mutual check on unreasonable self-interest and cognitive bias provides a needed corrective and complement to mainstream forms of social contractarian reasoning that rely more heavily on personal introspection. That said, in addressing ethical conflicts of widely different scope, both real *and* hypothetical dialogues find equal purchase in this ethics. Practically speaking, normative disagreements that occur within a bounded community can only be legitimately resolved through face-to-face deliberation, even if this process must be dispersed over many "publics" and "condensed" and "filtered" by mass media (Ingram 2019; Ingram and Bar-Tura 2014). Ideally speaking, however, some matters requiring deliberation affect everyone, including future generations. Deliberating on human rights law, for instance, may be spatially and temporally bounded, but the universal scope of any claim advocated on behalf of individuals as instantiations of humanity must refer to an unlimited community, whose possible opinions about ideal justice only personal hypothetical speculation can entertain.

The importance of face-to-face dialogue over personal speculation regarding the justice of borders, immigration policy, and practical application varies considerably. Moral reasoning about meta-political questions concerning appropriate boundaries of deliberation and decision-making and universal human rights principles will rely on personal speculations (ideal theory) that abstract from time and place and imagine possible worlds. Political reasoning about domestic immigration policy will rely on situational discussions taking place in parliament and public arena (nonideal theory). Judicial reasoning in deportation and asylum proceedings will rely on private conversations between judge, defendant, and legal counsel. In some instances, all three types of reasoning may be elicited. Judges, for example, typically recur to policy rationales in applying the law to individual cases; in adjudicating cases that are recalcitrant to mechanical resolution, they are often thrown back on their own moral intuitions.

Discourse ethicists have largely trained their thoughts on border ethics and immigration policy (Habermas 1998; Benhabib 2004; Ingram 2018b; Thompson, 2019). Here I focus on the judicial processing of asylum claims. But before proceeding to my topic, a few general comments about DT and immigration policy are in order. The most striking contribution DT makes to our reasoning about immigration consists in expanding the community of deliberation to include migrants. Most policy debates only take account of domestic opinion. DT demands that policy debates take account of the interests of anyone who is significantly impacted by that policy, so that the debate proceeds "not just from the one-sided perspective of an inhabitant of an affluent region but also from the perspective of immigrants who are seeking

with each other, it suffices to note that forms of unjust cooperation and unjust recognition also often produce forms of distorted communication, false "knowledge," and distorted (self-) understanding, which can decrease a person's capability for exercising agency.

their well-being there; [viz.] a free and dignified existence and not just political asylum" (Habermas 1996: 511). A policy that spells the difference between life and death for some persons obviously impacts their human right to life, which explains why humanitarian arguments for asylum made by desperate economic refugees—and not just political asylum seekers—should carry some weight.

Indeed, the "growing interdependencies of a global society that has become so enmeshed through the capitalist world market," which impose on all of us an "overall political responsibility for safeguarding the planet," also impose *special obligations* on affluent nations to compensate peoples of the developing world for "the uprooting of regional cultures by the incursion of capitalist modernization" (Habermas 1998: 231). Here we see the second contribution that DT makes to our reasoning about immigration. Not only must the claims of immigrants be factored into policy debates, but they must weigh equally. Indeed, they might weigh more than the claims of relatively affluent peoples who are trying to deny them entry into their country. This would especially be the case if the relative differences in standard of living separating these two groups stemmed from past and present injustices of colonialism and imperialism. Giving fair weight to migrants' claims shifts the burden of justification onto affluent nations. Thus, instead of expecting migrants to justify why they should be granted admission as privileged beneficiaries of charity, DT requires that affluent nations justify to migrants why their claims to compensation should not compel some form of restitution. Thus, in discussing immigration quotas,[14] Habermas insists that the needs of migrants count as much (or more) than the economic needs of the host country "in accordance with criteria that are acceptable from the perspective of all parties involved" (232).

DT shifts the burden of justification from needy migrants to affluent peoples by insisting that the weightier claims of migrants for just compensation and humanitarian treatment under the law be factored into policy discussions. For most refugees, however, the only venue where they can press their claims with any hope of receiving an impartial hearing is before an immigration judge.

Adjudication of an appeal may be routine but, but as my sampling of gender asylum cases attests, it often involves judicial discretion in choice, interpretation, and application of statutes, guidelines, and executive orders. Judges rely on open-textured constitutional principles in undertaking this endeavor. The principle mandating equal protection under the law and equal access to due process under the law that often crops up in immigration cases, for instance, is not a rule that narrowly dictates a single correct application but a regulative ideal that requires interpretation in light of the relevant case history and body of law. This principle can impact our understanding of who should carry the burden of proof. It can also impact our

[14] The 1965 US Immigration and Nationality Act ended quotas based on national origin, race, and ancestry that had reserved 70 percent of all slots to the United Kingdom, Germany, and Ireland and very few to Asia and Africa. Unfortunately, the good that came with eliminating a remnant of US racism was counterbalanced by a new set of regional quotas that limited the emission of Mexican work visas to only 20,000, which eventually led to a massive wave of undocumented Mexican migration into the United States.

evolving understanding of what it means to suffer persecution and who can be persecuted. The hermeneutical circle wherein legal precedents, executive priorities, constitutional principles, statutes, and case law mutually interpolate each other can produce new holdings that reverberate throughout the legal system.

DT prescribes a judicial procedure requiring that all relevant perspectives bearing on the most comprehensive description of a case be considered and that all relevant principles (and cases) be weighed in determining which statutory rules and constructions are most applicable. Crucial to this process is a courtroom procedure, which should not be structured simply as an adversarial contest, whereby all sides have equal opportunities to state their cases freely, cross-examine witnesses, and introduce evidence (Habermas 1996: 172; Ingram 2010: 210).

The requirement that a migrant be allowed to state his or her case freely brings into play normative considerations that, while pertinent to DT, also fall within the purview of RT. In keeping with Habermas's characterization of judicial decision-making, judges are properly equipped to apply, not legislate, the law; but applying the law requires choosing which parts of settled law and principled procedure best fit a comprehensive description of the case, which, in turn (and especially within the Anglo-American tradition of case-based law, which Habermas does not consider), can lead to the judicial reinterpretation of settled law and procedure. In this account, judges are institutionally privileged dialogue partners; it is their responsibility, as placeholders of public impartiality, to synthesize the testimony put forward by all litigants into a coherent and comprehensive description of the case. But of course, as Habermas himself observes, the "dialogue" in the court is constrained by an adversarial atmosphere which, in the case of asylum and deportation law, privileges the perspective of the government and places a burden of proof on migrants pleading their cases. Deference to standing law, with its technical distinctions between different grounds for pleading asylum, coupled with severe time constraints, compels judges to describe the case before them in a partial way that does not do justice to the complexities of the asylum seeker's own life circumstances. The Swedish cases recounted above especially attest to this injustice.

2.6 RT: Exposing Epistemic Bias in Asylum Courts

The institutional constraints imposed on immigration courts seem to render a discourse theoretic defense of modern legal systems less convincing than Habermas himself recognizes. DT can be used to criticize clear violations of courtroom procedure that deny clients a fair hearing of their claims, but in its institutional (or legal) form—as distinct from its ethical application—it cannot be used to criticize many of the judge's institutional biases that invariably lead to misunderstanding testimony that has been freely given.[15] In short, it cannot criticize the institutional misrecognition of the migrant as a less than credible witness of his or her own life circumstances. This misrecognition leads to hermeneutical injustice, in which the migrant's perspective is not allowed to adequately inform the description of his or her legal case. This injustice is magnified by the fact that, in most asylum cases, DT should require a reversal of the burden of the proof. Being an eyewitness to one's own injury as an asylum seeker should normally be accorded the same degree of respect as that accorded to a victim of assault testifying in a criminal court. Leaving aside any distortion of memory wrought by the experience of her injury, DT normally attaches the greatest authority to an asylum seeker's eyewitness testimony whenever such testimony finds some independent support and is not contradicted by demonstrable facts. As I argue below, the court should also recognize the special authority of the asylum seeker if he or she happens to be fleeing from circumstances that the court knows have been created in part by the policies of its own government. Conversely, in cases involving children under the custody of adults, the court should help child applicants to asylum become epistemic agents by providing them with legal and social support. As I noted in the case of FK, children who are exposed to FGC, forced marriage, and the like may not have a clear understanding of their oppression. They may express an acceptance of restrictive and harmful gender practices with or without expressing reluctance to undergo them. Their preference for a pathological form of social esteem and social inclusion instead of one that is less harmful, less coercive, and less affirming of patriarchal domination must therefore be questioned as purely adaptive and amenable to change upon assisted reflection.

The kinds of injustices that recognition theory is preeminently equipped to illuminate in the legal processing of asylum claims involve the judicial misrecognition of person's individuality and agency through the lens of gender and cultural stereotypes—injustices that were especially well documented in the Swedish asylum cases I discussed. I have also argued that, besides the personal racial and cultural prejudices of legal authorities, structural biases within the legal system compel judges and lower-level border police to discount the credibility of asylum seekers as authoritative witnesses to, and interpreters of, their situation. This latter

[15] Judges, of course, are obligated to reflect on personal prejudices that prevent them from relating to witnesses as individuals rather than stereotypes. That said, Habermas notes that judicial discourses of application are institutionally bound by precedent legal classifications and traditions of interpretations, so that the scope for any judicial interpretation that radically departs from tradition is severely constrained. By contrast, ethical discourse unbound by institutional constraints requires an unlimited scope for critical reflection on, and radical reinterpretation of, traditional categories and meanings. The distinction between discourse theory in its institutional (legal) and noninstitutional (ethical) forms provides a catalyst for legal reform, as ethical reflection invariably enters into the evolution of law as a progressively inclusive system of rights. Nonetheless, disregarding the exceptional moments of revolutionary legal transformation (as happened, for instance, during the American Civil Rights Movement), the tension internal to law between these distinctive discursive regimes is typically resolved in favor of the institutional regime, so that subaltern experiences of disrespect and misrecognition cannot be sufficiently voiced or acknowledged by the legal system. The kind of injustice referenced here is similar to what Jean-Francois Lyotard designates as a *differend* (Lyotard 1988; Ingram 2018a).

disrespect—which deprives asylum seekers of the esteem due to them as competent witnesses and purveyors of knowledge—is epistemic in nature. It involves a failure to recognize migrants as credible witnesses to their own circumstances, which in turn excludes their knowledge from contributing to the court's knowledge of the case. The growing literature on epistemic injustice that followed in the wake of Miranda Fricker's eponymous book (Fricker 2007) thus provides a useful supplement to RT. Using Gaile Pohlhaus Jr's taxonomies of epistemic injustice, I shall select just some of the most salient features of epistemic injustice that apply to asylum court proceedings. Taken together, they suggest that the testimony offered by asylum seekers, while not irrefutable, should be given the benefit of the doubt unless strongly contradicted by overwhelming evidence. In light of the distorting impact of trauma and, especially, in the case of children who may have developed "adaptive preferences" for restrictive and harmful gender-based forms of treatment, these findings further suggest that courts have a duty to provide asylum seekers with legal and social support capable of promoting the epistemic agency of the clients they represent.

To begin with, epistemic injustices arise from how epistemic institutions, such as courts of law, are structured. They consist of three kinds: (a) refusal to recognize particular knowers as knowers by disrespecting their testimony or by excluding them from epistemic resources, (b) causing epistemic dysfunction by distorting understanding and inquiry, and (c) concealing (a) and (b) behind the façade of institutional authority and epistemic hierarchy (Pohlhaus Jr. 2017: 13).

One of the dangers Pohlhaus mentions regarding (c) is defining the field of knowledge, and therewith the field of epistemic injustice too narrowly. DT can be accused of this injustice to the extent that it defines truth, knowledge, and reality in terms of a procedure of argumentative justification, which privileges competencies related to analytic reasoning, theoretical abstraction, and oral argumentation that find exemplary institutionalization in philosophy departments and courts of law but not in other areas of life. By focusing attention on asymmetries in knowers' opportunities to participate freely in oral argumentation, discourse ethics neglects a very important epistemic field, that of everyday narrative understanding, whose virtuous exercise requires developing listening competencies associated with humility, curiosity, and open-mindedness (Medina 2013: 42).

Pohlhaus mentions several other lenses through which epistemic injustices come into focus. The first concerns the temptation to blend ideal social theory with nonideal social theory in a way that legitimizes extant institutions as if they aspired to realize a just, impartial social contract. Habermas's insistence that the ideal norms of discourse find institutional purchase in democratic institutions, including courts of law, is one example of this.[16] By encouraging judges to believe that they are part of an impartial tribune of higher legal knowledge, the discourse theory of law *unintentionally* encourages their epistemic arrogance, close-mindedness, and

[16] Habermas himself repeatedly underscores the "counterfactual" status of institutionalized discursive norms, whose actual functioning is distorted by power relations.

mechanical lack of creativity and imagination. What José Medina (Medina 2013), with reference to Charles Mills' indictment of the white racial contract underwriting *Herrenvolk* democracy (Mills 1997), describes as an institutional epistemic blindness to subaltern perspectives can just as easily be characterized as a kind of "willful hermeneutical ignorance," to use Pohlhaus's words (Pohlhaus Jr. 2017: 17).

A second lens that is useful for magnifying epistemic injustices concerns the interdependence of epistemic relationships, or the way in which knowledge is a social product built out of the contributions of both consenting and dissenting perspectives. Knowledge presupposes a relationship of social trust. As we saw in Case 3, courts of law notoriously display mistrust toward witnesses they institutionally suspect of being untrustworthy and biased. This disrespect encourages witnesses to doubt their own testimony while also leading them to distrust institutional authority for its own breach of trust (Pohlhaus Jr. 2017: 19). The result is that applicants for asylum are excluded (and sometimes exclude themselves) from contributing to the production of expert knowledge that is relevant to describing their cases.[17]

A third lens through which the layers of epistemic injustice come into view accordingly pinpoints three levels at which knowers are excluded from participating in systems of knowledge. Their credibility can be diminished owing to biases that are not institutionally mandated, such as when the discourse theoretic norm of equitable participation is violated by failure to adhere to due process or when judges let their personal biases cloud their assessment of witnesses' credibility; at a second level, the institutional resources for understanding—such as the technical definitions available to judges in their description of a case as a case of persecution of a certain kind, the expert country-specific data bases available to them, and the time constraints imposed on giving testimony—can, by excluding, invalidating, or ignoring the applicant's sources of understanding, suppress or distort understanding of the applicant's case, thereby resulting in what Fricker dubs a "hermeneutical injustice." Finally, at a third level, the entire epistemic institution might structurally impose testimonial and hermeneutical injustices. In a trivial sense, this appears to be the case with all legal institutions, especially those that rely on adversarial argumentation for filtering legal facts, but it is uniquely applicable to asylum courts, whose default for processing claims reflects a defensive nationalism aimed at upholding and legitimizing restrictive entry in the name of national security and practical expediency.[18]

[17] Schweiger (2019, 56) notes that asylum seekers can exclude themselves for innocuous reasons owing to the distorting impact that their trauma has on their ability to remember the details of their situation accurately.

[18] Alison Wylie observes that persons occupying privileged positions (such as judges) are "invested in not knowing or, indeed, are invested in systematically ignoring and denying" the testimony of those "who are economically dispossessed, politically oppressed [and] socially marginalized" (Wylie 2003, 32). This compounds the systematic ignoring and denying that affects all persons, regardless of background, of testimony deemed by the judge to not respond appropriately (often in simple "yes or no" answers) to questions put forth by officers of the court. Noteworthy in this regard is the fact that adults and persons of privileged backgrounds and educational attainment

A fourth lens for perceiving the systemic epistemic injustice inherent in judicial systems pertains to the unfair distribution of epistemic labor in the production of knowledge. What Pohlhaus calls *agential* injustice occurs when, given the default noted above, migrants pleading their cases are forced to assume a greater burden of proof than the state in arguing their cases. This systemic (or structural) injustice is magnified further by the burden of having to dispute the court's "controlling images" (Collins 2000 69–96; 2017) of asylum seekers as, for example, economic opportunists. As David Coady points out (Coady 2017), although it may seem that credibility and understanding are not scarce resources that can be justly distributed (Fricker 2007, 19–20), the fact of the matter is that credibility and understanding are relative terms. As Fricker concedes, the weight given to the testimony of a black man like Harper Lee's fictional character Tom Robinson in *To Kill a Mockingbird* stands in inverse proportion to the weight accorded the white woman who has accused him of rape (Fricker 2007: 25). In a parallel manner, the credibility of the asylum seeker as expert witness is diminished when confronted by the credibility of the state's expert data banks; the descriptions available to the migrant for understanding his or her experiences compete with the technical legal descriptions available to the court and its fact finders. In this respect, the epistemic labor of the migrant is not only made unreasonably burdensome but often ends up being wasted as "invalid." Finally, the epistemic labor of asylum applicants can be exploited, through coerced testimony, to produce knowledge favorable to the state, as when trafficked women are forced to betray their traffickers in exchange for asylum. Coerced testimony that can be used against the testifier, in turn, leads the testifier to silence herself (what Dotson calls "smothering"), so that mistrust between asylum seekers and asylum boards invariably compels the former to exclude themselves from the latter's construction of their case. Migrants also find themselves caught in a double bind: to the extent that their testimony is credible and understandable according to the court's criteria for rational argumentation, it may be less persuasive to judges (Pohlhaus Jr. 2017: 22). A child who testifies that she consented to be married into her husband's family at the age of 10 because of parental pressure might appear credible and reasonable to a judge who operates with a contextual understanding of oppression; but she still might not persuade the judge that she suffered persecution based on her gender, especially if her husband was a child like herself who was equally vulnerable to parental pressure.

have a greater chance to represent themselves directly before the judge rather than be passively led (or kept silent) by attorneys who are chosen to speak for them.

2.7 Concluding Remarks: Extending DT and RT Beyond Gender-Based Asylum

I have argued that DT and RT uniquely contribute to illuminating the procedural and epistemic injustices suffered by migrants in pleading their cases before asylum courts. Even adversarial courtroom proceedings should abide by discourse theoretic conditions of procedural justice, guaranteeing asylum seekers adequate legal representation, equipped with translators, who have equal opportunities to raise claims and challenge arguments made by government officials. In principle, these proceedings should be conducted in a non-adversarial manner, as an impartial search for the truth regarding the asylum seekers' claims. Furthermore, once it is acknowledged that asylum seekers are victims who have fled coercive (and often violent) circumstances, they should be accorded the same respect as ordinary victims of violence. Although their claims are not impervious to challenge, they should normally be given considerable credence unless clearly undermined upon cross-examination. Once asylum seekers' claims are given the credence they merit, the burden of proof they shoulder should be proportionately lessened and transferred to the immigration boards that are predisposed to rule against them. A just procedure for weighing arguments should therefore permit an extended hearing of the applicant's story that is not perfunctorily truncated by draconian time constraints and interrupted by hostile forms of interrogation.

Because discourse theoretic requirements also mandate that the force of reason (evidence) alone should motivate courtroom deliberation, it is imperative that the applicant's experience, which ought to weigh heavily among the evidentiary sources brought forth as reasons, be properly understood. In this respect, the epistemic virtues that RT enjoins provide a necessary complement to the procedural expectations prescribed by DT. RT requires that judges listen to the applicant's story with empathy and open-mindedness, with full awareness of the law's incapacity to do justice to its complexity and ambiguity. It requires that judges regard the applicant's story as a potential hard case that cannot be easily subsumed under the law as it has hitherto been understood. So construed, the applicant's case represents both a challenge to the justice and integrity of the law and an opportunity for reinterpreting it in a more inclusive way.

The recognitive requirements noted above acquire greater moral urgency when we recall that the testimonial and hermeneutical injustices experienced by women who seek gender-based asylum are typical for all migrants. Personal and structural biases predispose migration judges to deny migrants a fair hearing. The structural biases have their basis in an international legal order that bestows prerogatives on governments and their agents to define who qualifies as a legal refugee. Using national security as a broad pretext for restricting asylum, they refuse admission to all but those who can prove they are fleeing state-mandated political persecution.[19]

[19]20 In late March of 2020, the Trump administration used the Covid-19 pandemic as a national security pretext for suspending the processing of all asylum claims, thereby arguably violating

Persons fleeing forms of domestic violence, economic violence, or climate-related violence do not count as political refugees under this narrow definition, even if the violence they suffer is abetted and condoned by their governments and by the international order.

Recognition theory shows how this conceptual narrowing of who counts as a refugee reflects a vicious hermeneutical circle. The migrant whose testimony is discounted cannot contribute personal knowledge that could revise the very expertise that invalidates her knowledge. To paraphrase Jean-François Lyotard, the migrant's testimony cannot be entered into litigation as a form of dissent that merits argumentative rebuttal because it cannot be recognized as a legitimate truth claim (Lyotard 1988; Ingram 2018a). There is no neutral rule for adjudicating the disagreement between the asylum board's statistical generalizations and classifications that, say, define gender-related violence as an acceptable cultural norm within the migrant's native country and the migrant's unverifiable experience of suffering and resisting her government's acquiescence in the violation of her human rights. Any rule that recognizes the authority of the board's claim—with its structural deference to state sovereignty (both within and without the domestic border of the state it represents)—disqualifies the authority of the migrant's claim to a human right that transcends state sovereignty. One voice must go unrecognized, or at least suffer diminished credibility in relation to the other. And the voice that is effectively silenced will be the less institutionally entrenched voice, which in the case before us is that of the migrant.[20]

To be sure, the injustice committed in silencing the asylum seeker is not inevitable. Sovereignty and human rights need not be conceptualized in stark opposition to each other. Indeed, the Responsibility to Protect doctrine that was adopted by the

international and domestic legal obligations to not repatriate persons who have a credible fear of suffering persecution. Earlier decisions by the Trump administration to deny asylum requests from persons attempting to enter the United States through non-designated ports of entry and persons who have not first applied for asylum in another country while transiting to the United States appear to be legally dubious as well. The Trump administration's Migrant Protection Protocols (MPPs), also known as the Remain in Mexico Program, overturned the legally recognized practice of allowing asylum seekers to enter and remain in the United States while their cases were being processed. Asylum seekers now had to live in squalid and dangerous camps in Mexico while waiting for their names to be placed on a list for eventual processing. Most disturbing from the perspective of DT and RT were the new rules ("Prompt Asylum Case Review" and "Humanitarian Asylum Review Process") implemented for screening asylum seekers by poorly trained officers, who are now instructed to interpret and apply the law governing refugees in the narrowest way possible to applicants that are given barely 24 h to prepare for their interviews, with only a mere possibility of discussing their case over the phone with a lawyer. Together, these policies amount to much more than a gross violation of discursive and recognitive norms; they constitute a grave violation of legal due process.

[20] The UNHCR must also recognize the competing demands of national sovereignty and humanitarian rescue. The protective confinement of displaced populations is politically preferable to their protective resettlement in a foreign country or their repatriation. As in the case of asylum courts, the UNHCR's decision to repatriate is often done without consulting those whose lives are at risk, and even when they are consulted, the testimony of government officials and experts often outweighs the testimony of eyewitnesses on the ground (Ingram 2018b: 141–44; Barnett 2010).

UN in 2005 arguably redefines sovereignty in terms of respect for and protection of human rights. Once respect for and protection of human rights is understood to be the primary rule guiding the interpretation of the Geneva Convention Relating to the Status of Refugees, as the UNHCR recommends that it should be, the restrictive binary categories in terms of which persecution is legally defined must be deconstructed.

Pre-interpreted through the cipher of binary legal categories, the exclusionary logic of the dominant asylum paradigm is challenged by the personal experiences of those who must submit to it. The distinction between private and public domains of life, between domains of life that should be left to individuals to work out among themselves and domains that properly fall under public and political regulation, is questioned by a feminist standpoint that interprets domestic violence as a public harm visited upon women as a politically vulnerable group. The distinction between political persecution and economic oppression is questioned by a critical theory standpoint that interprets government negligence in securing everyone's human right to material security as an act of political discrimination endangering the basic freedoms of the poor.[21] Finally, the distinction between asylum seekers fleeing natural disasters, such as climate change, and asylum seekers fleeing oppressive government is questioned by a globalizing standpoint that interprets climate change, poverty, and political oppression as the shared responsibility of the entire international community.[22]

Thus, the Salvadoran woman seeking asylum in the United States flees from a situation which an inhospitable nature, an impoverished economy, an oppressive society, an unresponsive and hostile government, and an irresponsible global order have conspired together to create. Criminalizing her for bypassing legal ports of entry or punishing her with confinement shows the degree to which the legal system misrecognizes her. If she has voluntarily colluded with a trafficker, she may be misrecognized as a simple accomplice, in total disregard for the coercive bargain and accompanying threats of retaliation leveraged by her trafficker, not to mention the violence compelling her to embark on such a desperate course of action in the first place (Christman 2014; Haynes 2006). Or she may be misrecognized as a passive victim of nonconsensual exploitation in total disregard for her agency. This misrecognition is compounded if she happens to be an unaccompanied minor, in which

---

[21] Urging reconsideration of the usual way in which economic and political categories of government sanctioned endangerment are interpreted does not require abandoning their distinction. Immanent, durable, and severe risk of government persecution (imprisonment, torture, and assassination) targeting specific persons and groups requires urgent action in a way that untargeted and widespread economic suffering caused by government neglect does not. Persons fleeing economic oppression should be afforded temporary protected status but not refugee status (see note 4)

[22] In fact, the Syrian refugee crisis implicates a climatological cause—drought—as contributing to the economic and political causes of the Syrian civil war.

case she will be treated as an incompetent witness without adult standing or as a blameworthy delinquent meriting summary deportation back to her family.[23]

If recognition theory provides a lens for magnifying the epistemic injustices visited upon asylum seekers in the immigration system, it also provides a framework for correcting them. The incorporation of gender-related persecution into refugee law shows that dominant epistemic frameworks informing law are not impervious to change, once alternative resources based on lived experience are given a fair hearing. Accordingly, since 2016, the UNHCR has expanded the definition of a refugee to include anyone living outside their country of residence who faces "serious and indiscriminate threats to life, physical integrity or freedom resulting from generalized violence or events seriously disturbing public order" and other "man-made disasters."[24] Serious and indiscriminate threats to life, physical integrity, or freedom that count as "man-made disasters" include willful neglect of domestic abuse, poverty, and environmental devastation wrought by anthropogenic climate change. Simply reclassifying gender-based asylum seekers as endangered persons who do not merit full protection under the legal definition of refugee, but who nonetheless merit "temporary" or "subsidiary" protection, violates the spirit of the UNHCR's expansive recognition of refugees and misrecognizes the severity, duration, and specificity of their endangerment.

References

Barnett, Michael. 2010. *The International Humanitarian Order*. London: Routledge Press.
Benhabib, Seyla. 2004. *The Rights of Others*. Cambridge: Cambridge University Press.
Buchanan, Allen. 2013. *The Heart of Human Rights*. Oxford: Oxford University Press.
Carens, Joseph. 1995. Aliens and Citizens. In *The Rights of Minority Cultures*, ed. Will Kymlicka, 331–345. Oxford: Oxford University Press.
———. 2013. *The Ethics of Immigration*. Oxford: Oxford University Press.
Christman, John. 2014. Human Rights and Global Wrongs: The Role of Human Rights Discourse in Responses to Trafficking. In *Poverty, Agency, and Human Rights*, ed. D. Meyers, 321–346. New York/Oxford: Oxford University Press.

[23] See the dilemma of misrecognition posed by unaccompanied minors seeking asylum in the United States from Central American gang-related violence (Heidbrink 2013). In defending a prima facie case for prioritizing children for asylum because of their vulnerability (among other reasons), Gottfried Schweiger notes that children are regarded as less reliable witnesses than adults to the circumstances bearing upon their case for asylum (Schweiger 2019: 56–57).

[24] This definition adopts the wording contained in the Organization of African Unity's Convention Governing the Specific Aspects of Refugees in Africa (1974). The Cartagena Declaration on Refugees (1984), adopted in a nonbinding resolution by Mexico and other Latin American countries, builds upon this definition to include "persons who have fled their countries because their lives, safety, and freedom have been threatened by generalized violence, foreign aggression, internal conflicts, massive violation of human rights or other disturbances which have seriously disturbed public order."

Coady, D. 2017. Epistemic Injustice as Distributive Injustice. In *The Routledge Handbook of Epistemic Injustice*, ed. I. Kidd, J. Medina, and G. Pohlhaus Jr., 61–68. London/New York: Routledge.
Collins, Patricia Hill. 2000. *Black Feminist Thought: Knowledge, Consciousness, and the Politics of Empowerment*. 2nd ed. New York: Routledge Press.
———. 2017. Intersectionality and Epistemic Injustice. In *The Routledge Handbook of Epistemic Injustice*, ed. I. Kidd, J. Medina, and G. Pohlhaus Jr. London/New York: Routledge.
Dotson, Kristie. 2011. Tracking Epistemic Violence, Tracking Practices of Silencing. *Hypatia* 26 (2): 236–257.
Fraser, Nancy, and Axel Honneth. 2003. *Redistribution or Recognition: A Political-Philosophical Exchange*. London: Verso Press.
Fricker, Miranda. 2007. *Epistemic Injustice: Power and the Ethics of Knowing*. New York: Oxford University Press.
Günther, Klaus. 1993. *The Sense of Appropriateness: Application Discourses in Morality and Law*. Albany: SUNY Press.
Habermas, Jürgen. 1996. *Between Facts and Norms: Contributions to a Discourse Theory of Law and Democracy*. Trans. William Rehg. Cambridge, MA: MIT Press.
———. 1998. *The Inclusion of the Other: Studies in Political Theory*. Trans. Ciaran Cronin and De Pablo Greiff. Cambridge, MA: MIT Press.
Habermas, Jurgen. 1971. *Knowledge and Human Interests*. Boston: Beacon Press.
Haynes, D.F. 2006. Used, Abused, Arrested and Deported: Extending Immigration Benefits to Protect the Victims of Trafficking and Secure the Prosecution of Traffickers. In *Women's Rights: A Human Rights Quarterly Reader*, ed. B. Lockwood. Baltimore: Johns Hopkins Press.
Heidbrink, L. 2013. Criminal Alien or Humanitarian Refugee: The Social Agency of Migrant Youth. *American Bar Association Children's Legal Rights Journal* 33 (1): 133–190.
Honneth, Axel. 1996. *The Struggle for Recognition: The Moral Grammar of Social Conflicts*. Cambridge, MA: MIT Press.
———. 2007. *Disrespect: The Normative Foundations for Critical Theory*. Cambridge: Polity Press.
———. 2012. *The I in We: Studies in the Theory of Recognition*. Trans. J. Ganahal. Cambridge: Polity Press.
———. 2014. *Freedom's Right: The Social Foundations of Democratic Life*. New York: Columbia University Press.
———. 2015. Freedom, Solidarity, and Democracy: An Interview with Axel Honneth (An Interview Conducted by Morten Raffnsoe-Moller). In *Freedom and Recognition*, ed. J. Jakobsen and O. Lysaker. Leiden: Brill Publishers.
Ingram, David. 2006. *Law: Key Concepts*. London: Continuum/Bloomsbury.
———. 2009. Exceptional Justice? A Discourse-Ethical Contribution to the Immigrant Question. *Critical Horizons* 10 (1): 1–30.
———. 2010. *Habermas: Introduction and Analysis*. Ithaca: Cornell University Press.
———. 2018a. Disputing the Law: Lyotard in Our Time. A Forgotten Critic Bears Witness to Unresolvable Injustices. *Berlin Journal of Critical Theory* 2 (4): 33–54.
———. 2018b. *World Crisis and Human Underdevelopment. A Critical Theory of Poverty, Agency, and Coercion*. Cambridge: Cambridge University Press.
———. 2019. Contesting the Public Sphere: Within and Against Critical Theory. In *The Cambridge History of Modern European Thought. Volume 2. The Twentieth Century*, ed. P. Gordon and W. Breckman, 517–544. Cambridge: Cambridge University Press.
———. 2020a. Recognition and Positive Freedom. In *Positive Liberty: Past, Present, and Future*, ed. John Christman. Cambridge: Cambridge University Press.
———. 2020b. When Microcredit Doesn't Empower Poor Women: Recognition Theory's contribution to the Debate Over Adaptive Preferences. In *Poverty and Recognition*, ed. Gottfried Schweiger. Berlin: Springer.

Ingram, D., and A. Bar-Tura. 2014. The Public Sphere as Site for Emancipation and Enlightenment: A Discourse Theoretic Critique of Digital communication. In *Re-Imagining Public Space: The Frankfurt School in the 21st Ccntury*, ed. D. Boros and J.M. Glass, 65–85. New York: Palgrave.
Lyotard, J.-F. 1988. *The Differend: Phrases in Dispute*. Minneapolis: University of Minnesota Press.
Medina, J. 2013. *The Epistemology of Resistance: Gender and Racial Oppression, Epistemic Injustice, and Resistant Imaginations*. New York: Oxford University Press.
Miller, D. 2016. *Strangers in Our Midst: The Political Philosophy of Immigration*. Cambridge, MA: Harvard University Press.
Mills, C. 1997. *The Racial Contract*. Ithaca: Cornell University Press.
Musalo, K. 2010. A Short History of Gender Asylum in the United States: Resistance and Ambivalence May Very Slowly Be Inching Towards Recognition of Women's Claims. *Refugee Survey Quarterly* 29 (2): 46–63.
Pohlhaus, G., Jr. 2012. Relational Knowing and Epistemic Injustice: Toward a Theory of Willful Hermeneutical Ignorance. *Hypatia* 27 (4): 715–735.
———. 2017. The Varieties of Epistemic Injustice. In *The Routledge Handbook of Epistemic Injustice*, ed. I. Kidd, J. Medina, and G. Pohlhaus Jr., 13–26. London/New York: Routledge.
Rawls, J. 1999. *The Law of Peoples*. Cambridge, MA: Harvard University Press.
Rusin, J., and M. Franke. 2010. Self-Understanding and the Refugee Claimant. *International Journal of the Humanities* 8 (3): 187–198.
Schweiger, G. 2019. Should States Prioritize Child Refugees? *Ethics and Global Politics* 12 (2): 46–61.
Sertler, E. 2018. The Institution of Gender-Based Asylum and Epistemic Injustice: A Structural Limit. *Feminist Philosophy Quarterly* 4 (3): 1–24.
Thompson, D. 2019. *A Discourse Theoretic Contribution to Migration and Sovereignty* (unpublished dissertation).
Walzer, M. 1983. *Spheres of Justice: A Defense of Pluralism and Equality*. New York: Basic Books.
Wikström, H. 2014. Gender, Culture, and Epistemic Injustice: The Institutional Logic in Assessment of Asylum Applications in Sweden. *Nordic Journal of Migration Research* 4 (4): 210–218.
Wylie, A. 2003. Why Standpoint Matters. In *Science and Other Cultures*, ed. S. Harding and R. Figueroa. New York: Routledge.

Chapter 3
Migration and the (Selective) Recognition of Vulnerability: Reflections on Solidarity Between Judith Butler and the Critical Theory

Martin Huth

Abstract In this chapter, I argue that Axel Honneth's theory of recognition provides us with some crucial prerequisites for an ethical analysis of dealing with migration in the Western world. His consideration of the inevitable dependency of individuals on recognition, particularly the respect for rights and the esteem of achievements, is crucial for an evaluation of current practices and the detection of social pathologies. However, there are some pitfalls in this approach that can be illustrated through the application of his theory to migration. First, Honneth assumes a teleology of recognition. He, thus, cannot answer the question *why* differential treatments of various groups emerge (despite of interpreting it as pathological deviation from a normal state); he does not take into account specific power relations; and he cannot target limits of social inclusion. Second, he assumes reciprocity in recognition and is unable to tackle the question of asymmetries between myself and the Other, but also between members of a society and individuals who want to immigrate in this society. Third, I want to challenge the idea that it is the full-blown person or their achievements, which is to be recognized. Instead, I argue that vulnerability should be at the center of recognition, particularly of solidarity. Drawing from Judith Butler, I argue that the structural recognizability of vulnerability helps us to rectify these pitfalls. However, in turn, Butler tends to leave idle the basis for a critical evaluation of current practices. This becomes visible through an analysis of her conceptualization of solidarity. Thus, I propose to complement her approach with the strategy of immanent critique, which has been developed within the critical theory.

Keywords Social pathology · Vulnerability · Dependency · Recognizability · Immanent critique

M. Huth (✉)
Messerli Research Institute, University of Veterinary Medicine, Vienna, Austria

Department of Philosophy, University of Vienna, Vienna, Austria
e-mail: martin.huth@vetmeduni.ac.at

G. Schweiger (ed.), *Migration, Recognition and Critical Theory*, Studies in Global Justice 21, https://doi.org/10.1007/978-3-030-72732-1_3

3.1 Introduction

In what follows, migration forms both the subject of investigation and an illustrative example or a vehicle of explanation. The aim of this chapter is to clarify the relevance of Axel Honneth's recognition theory in the face of migration, but also some pitfalls regarding a decent understanding of ethical issues related to migration. I will argue that particularly his theory of recognition provides us with some important prerequisites for an ethical evaluation of current migration policies in the Western world.[1] Making visible an anthropological, ethically significant human need for respect as rights bearer and for being esteemed as valuable members of society provides us with an important source for a critical investigation of these policies as well as collective attitudes and dispositions. However, Honneth widely fails to clarify the relation between acts and established institutions of recognition on the one hand and structural prerequisites and conditions of these acts (with epistemologically and ethically problematic implications) on the other hand. Therefore, I will suggest to complement his account with Judith Butler's reflections on *recognizability* (Butler 2004a, 2009, 2012), but also with some elements of phenomenological thinking on foreignness, alterity, and responsibility (Waldenfels 1996, 1997; Levinas 1979, 1991). Consequently, the focus shifts from the mutual recognition of subjects and their traits and abilities (Honneth 1996, 2003) to the (potentially asymmetrical) recognition of vulnerability of bodily beings *within the structural conditions and limits of a particular socio-cultural context.* Within such an approach, first, the question emerges whether the subjects of recognition are – and can be – recognized as recognizers. Second, it becomes questionable if the object of recognition consists of personhood and achievements and traits or if, instead, recognition should be primarily focused on vulnerability and dependency. Yet Butler's considerations – even though her focus is the critique of these conditions of recognition – tend to leave idle the epistemological and normative basis for critique of practices that can be conceived of "social pathologies." Thus, I propose to complement her account by turning back to the critical theory and its central concept of *immanent critique*, which has been employed by Theodor W. Adorno (1973) Honneth (2003, 2014), and Titus Stahl (2017).

The phenomenological starting point for the upcoming considerations is the outrage that could emerge in the face of the neglect or exploitation of migrants' particular vulnerability or in the face of structures and strategies that render migrants and their vulnerability invisible (e.g., through the management of the external borders of the European Union or reactionary or ethno-nationalist motivated narratives that derealize and dehumanize asylum seeker by turning them into "economic refugees" and invaders, etc.). Honneth recurrently stresses the importance of the experience of misrecognition for the diagnosis and critique of injustice (or *social pathologies*), though I have to admit that he primarily focuses on negative emotions that emerge

[1] As an Austrian, I will illustrate my considerations mainly with Austrian migration policies in the past years.

on the part of the affected individuals. Particularly, culmination points like the "refugee crisis" in 2015 (and the reaction of some European nation-states who tried to seal their borders as well as the resignification of the opening up of borders for refugees as a profound mistake in the aftermath of this event) or the current conditions in the infamous, overcrowded refugee camp Moria on the island of Lesbos where people – including a considerable number of unaccompanied minors – live under terrible conditions can trigger a dismay that builds the ratio cognoscendi of "social pathologies." This dismay is even augmented if we consider the fact that, while I am writing this chapter, some European governments (for instance, the Austrian) vehemently refuse to receive any unaccompanied minors from this camp arguing that they are in need self-defense against the subversion of the welfare system and the cultural identity – turning vulnerable children into foreign particles.

3.2 Analyzing Social Pathologies: Axel Honneth's Recognition Theory in the Face of Migration

Migration can be conceived as a tough case for liberal theories of justice (e.g., Rawls 2002), which frequently show at least two pitfalls. *First*, they generally reduce persons to independent right bearers and dismantle them of particularities and social relations; they produce equality through reductionism. Thus, the constructed atomistic subject easily turns into an invulnerable, heroic subject whose identity and whose abilities (for instance, the one to independently claim rights) and traits are presupposed. Starting from Hobbes, there is the tendency to separate morality and ethics and to focus on morality (principles that should safeguard the prerequisites for any rational, self-chosen plans, as Rawls puts it; ibid., 49). The genuinely *ethical* question of a good life (and its intersubjective conditions) is left aside (Honneth 2007). Such accounts leave aside the trivial fact that not every individual is an adult with supposedly average capacities, and, what is more, they ignore that dependency and vulnerability do not only occur in early or very late phases of bodily existence but throughout the course of our lives. Moreover, it seems that intersubjectivity and sociality are reduced to derivative phenomena. It is a crucial example that these theories do not take into account that there are hurdles that may deprive individuals – not the least migrants – of opportunities to demand rights and resources if they lack language skills or access to a decent speaker position.[2] In contrast, proceeding from some of Hegel's insights, recognition theories highlight

---

[2]Also Jürgen Habermas' ethics presupposes that there is, at least hypothetically, always an open possibility to be involved in a discourse in which the participants can meet and exchange arguments on an equal footing with the goal and the upright possibility of a consensus or at least a compromise regarding the solution of shared ethical problems (Habermas 1994). He is, thus, barely able to target structural exclusions – misrecognition – of particular individuals and social groups. In contrast, Miranda Fricker's considerations of testimonial injustice, which deprives individuals from being regarded as participants in rational deliberations (2007), provide us with a

the vital significance of recognition for the integrity of the individual and its relation-to-self (Taylor 1994; Honneth 1996). Furthermore, they do not insinuate that we could easily detach persons from their embedding in socio-cultural contexts and their (in some – not the least un-/documented migrants – cases precarious) positions as speakers, knowers, or demanders of rights and social esteem. Social positions clearly determine how individuals are able to participate in public debates or if and how they can turn to institutions.

Second, John Rawls is but one example of theorists who assume that a focus on justice would compel us to think within the Westphalian model. He considers social justice as to be established in nation-states, as he conceives of society as a delineated system of mutual cooperation (2002, 27). Martha Nussbaum's critique in the *Frontiers of Justice* has made visible several pitfalls of this approach. Among others, they include the already mentioned one of a neglect of those who are not able to contribute to such a system of cooperation (due to disability, but we could also think of traumatization, lack of education, and lack of knowledge skill – all of which might be the case in numerous individuals who have been forced to migrate) in terms of mutuality and the one of the exclusion of individuals who are not or not yet considered as citizens (Nussbaum 2006, chapter 5).

Axel Honneth continuously draws from Hegel's early *Jenaer Realphilosophie* and some considerations by George Herbert Mead to stress the existential and normative relevance of recognition for a positive relation-to-self (1996, e.g., 92).[3] Instead of insinuating a self-sufficient subject, he invites us to acknowledge that subjectivity and identity are inevitably couched in and dependent on intersubjectivity. The need for recognition is, thus, an anthropological constant (Honneth 2003, 181) or a "quasi-transcendental interest" (ibid., 174) and, by the same token, forms the basis of an "imperative of mutual recognition" (Honneth 1996, 92). This becomes visible through the famous triad of (a) self-confidence rooted in love as primal form of recognition within the mother-infant relation,[4] which forms the vital basis for, (b) self-respect rooted in having one's rights respected, and (c) self-esteem rooted in solidarity or the esteem for particular capacities and achievements (ibid., 107, 113, 118). "We are, thus, dependent upon receiving recognition from others if we are to understand and value who we are" (McQueen 2017, 206). To put it differently, the moral demand for recognition is rooted in the sort of vulnerability that becomes visible through a lack of recognition. It is important for the upcoming

significant example and a valuable basis for an investigation in those structures of social exclusion that got beyond the scope of rights demanded by full-blown, self-sufficient subjects.

[3] In what follows, I will synoptically summarize some crucial elements of Honneth's recognition theory; it would go beyond the scope of this paper to point out the different shifts and developments in his thinking.

[4] Honneth draws from the psychoanalyst Donald Winnicott to conceptualize love as the primal form of recognition (2003, e.g., 98). Psychoanalysis predominantly detects the basis for healthy relationships in primal relationships and/or primal Others. Yet, the further reflections on migration will of course primarily focus on the other dimensions of recognition (rights and solidarity/esteem).

consideration to note that in Honneth the concept of vulnerability hardly occurs and that it does not play a significant role. Nevertheless, it seems that he acknowledges in the contention of an imperative of recognition that there is an existential dependency on healthy intersubjective relations, which can be exploited or neglected; this presents us with a basic vulnerability of human individuals.

Social pathologies, i.e., the experience of a lack of recognition or misrecognition, undermine the individual's overall integrity and have, thus, normative significance. In the vein of Rousseau, Honneth understands social philosophy's task to diagnose deviations from a healthy normal state of society. Widespread (often institutionalized) forms of deprivations of affective relationships (or even violence), respect, and esteem (or even open indignation) that affect broader populations and groups can be conceived of as social pathologies. In *Freedom's Right*, Honneth asserts: "[A] 'social pathology' indicates any social development that significantly impairs the ability to take part rationally in important forms of social cooperation" (2014, 86).

Drastically, Charles Taylor indicates that misrecognition might inflict a "grievous wound, saddling its victim with a crippling self-hatred" (1994, 26) and that it is "imprisoning someone in a false, distorted, and reduced mode of being" (ibid., 25). Frantz Fanon's famous analysis of racist discrimination is particularly illustrative for the profound and grave effects of social pathologies for the relation-to-self (it is clear that we can transfer these insights to issues related to migration). Drawing from Jean-Paul Sartre, Fanon contends that the misrecognition of the black population by white dominators is even inscribed in the most fundamental layers of the bodily existence – the so-called body schema – of the victims.[5] "In the white world, the man of color encounters difficulties in the development of his bodily schema (...). The body is surrounded by an atmosphere of certain uncertainty (...), the corporeal schema [becomes] crumbled" (Fanon 1994, 83 f.). Hostile atmospheres, (sometimes very subtle) racist discrimination and (sometimes tacit) marginalization, are typical social pathologies that are experienced (not only, but particularly) by migrants. The reification or hypostatization of some traits (ethnic affiliation, color of the skin, mother tongue) to marginalize populations leads to turn them into beings who receive less or nonstandard forms of recognition.

The need for recognition has to be satisfied in order to develop a full-blown personality and to be able to have successful (non-abusive) relationships. Crucially, the experience of social pathologies is not only a theoretical-methodological ratio

[5] In the phenomenological discourse, the body schema and the complementary concept of the body image play a pivotal role for the understanding of existence. Drawing on Shaun Gallagher's classical text on the distinction between body schema and body image, we can associate the habitual body with the body schema as the nonconscious performance of the body. In contrast, the body image is conceived as inconstant intentional object of consciousness (Gallagher 1986: p. 542, 544) with perceptual, cognitive, and emotional aspects (Gallagher 1986: 546). This represents an intentional relation to our body or a self-relation. Crucially, the negative effects of social pathologies relate to the most fundamental dimensions of our being-in-the-world, our basic orientation and capacity of sensemaking. As Fanon asserts in a drastic way, this can produce an existential uncertainty that affects relations to others, the world, and of course oneself.

cognoscendi of injustice but also the mundane starting point for struggles for recognition. Those who are affected by such social pathologies – first and foremost the deprivation of rights and the lack of esteem for capacities and achievements – have a reason and an incentive to demand recognition. Therefore, Honneth challenges Habermas' view that discourse enables a non-agonistic coexistence (Habermas 1994); rather, he contends that *struggles* for recognition are inevitable and have a genuine moral quality.

In Honneth's view, *inclusion* – understood as being or becoming embedded in equal social relationships and as remedy of inside-outside binaries – and *individualization* are the normative goals of acts of recognition and should orient social and political institutions (Honneth 2003, 184–186). The constitution of institutions and their rootedness in interpersonal relationships has a clear direction: Interpersonal forms of recognition get institutionalized; therefore, the normative core of institutions seems to be identified as individuals' needs for recognition (cf. Hirvonen 2016, 29).

Migrants are prototypical figures of individuals who are in a precarious position regarding recognition and are frequently deprived of inclusion and (the means of) individualization. In the name of an alleged self-defense (e.g., against the subversion of welfare systems or of cultural identities), migrants are often deprived of rights (e.g., of asylum, but also of social participation or to have access to the labor market, which would be a prerequisite to gain esteem and solidarity as Honneth conceives it). This can be illustrated with numerous examples such as the refusal of many European countries to participate in an equal allocation of asylum seekers in the European Union, the infamous agreement of the EU with Turkey to prevent migrants to enter the territory of the EU, etc. But the normative dimension cannot be restricted to the detrimental outcomes of misrecognition for the relation-to-self; such social pathologies affect also the relation to others and the social world, as Onni Hirvonen points out: "[T]hose who have fallen to the margins of the society may well become indifferent or hostile towards it" (Hirvonen 2016, 29). If a society fails to realize or live up to its rational possibilities, it does not only deprive particular groups of resources of a good life but impairs its own sustenance. Typical outcomes of the marginalization of migrants (particularly the undocumented) are the development of parallel societies, ghettoization, and pursuits outside of established and legal spheres of labor.

Moreover, we can see that there is the widespread refusal of many countries to provide humanitarian support for migrants (exceeding the scope of rights). This lack becomes particularly visible through the lens of recognition theory and its emphasis of a fundamental need for recognition. An illustrative example is the Austrian government's declination to receive unaccompanied minors from the infamous refugee camp Moria on the island of Lesbos. The justification for this refusal seems to be, again, that the claim for self-defense against the subversion of the welfare system and the cultural identity is a legitimate reason to abandon these children to their fate. Children seem, thus, to be turned into (mere) foreign particles that should not enter the previously unadulterated "body" of particular nation-states or even the EU. However, Honneth's theory of recognition, particularly in the phase of the publication of *Reification* (2008), provides us with a clear normative plea for an *involvement* with other human beings.[6] Humans are to be treated first and foremost according to their characteristics as *human beings* (ibid., 19; again we are confronted with a presuppositional and value-laden concept of the human), the treatment (or neglect) of individuals as mere problem *cases* or *foreign particles* can be unmasked as social pathology. Yet, in Honneth's oeuvre starting from *The Struggle for Recognition*, the question emerges if keeping migrants in the cold might present us with a problem – would it then be a neglect of the rights of these children? Do they really have a legal claim to enter the European Union when they are not considerable as refugees in the proper sense? Alternatively, we could go for a demand for solidarity; however, then the situation becomes even more precarious because Honneth links the concept of solidarity with achievements and shared values – and these children presumably cannot contribute to social cooperation and might not share social values. But the focus on involvement with others provides us with the opportunity to rethink these issues and broaden the horizon of critique of social pathologies. Apparently in the vein of Adorno, Honneth contends that taking another's perspective and involvement are effective remedies against a proliferation of instrumental reason, which tends to reify others or some of their qualities (e.g., if migrants are able to enter a territory, but still remain reduced to cases with a precarious residency status, to their ethnicity, or religion).

However, *The Struggle for Recognition* already provides us with crucial means to analyze fundamental social pathologies of the disrespect of rights and the lack of esteem, which play a significant role when considering the issue of migration (recognition in the form of affective relationships seems to play a minor role in this context). These means allowing us to reveal those social pathologies, which can be detected in the denial of (equal) rights and to the restrictions of opportunities to gain esteem and self-esteem (e.g., opportunities to participate in the labor market, to participate in social spheres that ground solidarity beyond the labor market, etc.), which undermine the possibilities of self-realization or individualization. This might even happen within initially well-intentioned, however tacitly paternalistic, practices and institutions, which are devoted to the management of migration. Crucially, the integration of migrants in specific institutions might also deprive them of such opportunities – and maybe even to start a struggle for recognition. Moreover, Honneth's consideration shed light on the question of individual responsibility insofar as it is not understood as purely individual spontaneity in the Kantian sense but as embedded in social structures (which are in some cases pathological). Social

[6] Basically, Honneth does not proceed from a normative claim but from the inevitability of a genuine praxis that forms the basis for explicit acts of recognition, reciprocity, care, and affirmation of the existence of the other (cf. Butler 2012, 99). However, Honneth's reflections in *Reification* confront us with a significant problem: As Thomas Bedorf (2010) has pointed out convincingly, if recognition is inevitable and takes hold at the primal instance of every intersubjective encounter, then it becomes questionable why we need a moral demand for recognition. These questions clearly go beyond the scope of this paper, but they are crucial and reveal that recognition is not as easily conceivable as a moral principle.

practices and institution can be understood as formative for individual dispositions for perceptions, actions, and affective responses (though Honneth does not thoroughly analyze this constitutive relation between social practices and individual perspectives).

All these considerations only form a raw outline of Honneth's contributions to ethical questions related to migration, but they exemplarily show the strengths of recognition theory for targeting normative issues related to migration. However, the following section will be dedicated to some pitfalls of recognition theory that emerge in the face of moral issues related to migration.

3.3 Some Respects in Which Migration Escapes Honneth's Recognition Theory

In his more recent chapter *Verwilderung des sozialen Konflikts* (Honneth 2013), Axel Honneth complains about an imbrutement of struggles for recognition that thwarts expectations of a general progress toward social inclusion (and individualization). He particularly highlights the erosion of the significance of civil rights as source of recognition. They tend to lose their relevance as symbolic signs of mutual respect but have instead turned into an instrument of individual hedging (ibid., 28). Crucially, Honneth stresses that populations in immigrant countries in the Western world tend to resignify the previous sources of recognition – i.e., rights and esteem/solidarity for abilities and achievements with societal relevance – as means for the defense against claims of others (ibid., 29). Migrants demanding the recognition of their basic rights (the right of asylum and humane treatment in the broadest sense) are then turned into invaders who supposedly violate the rights and entitlements of the people who already live in the respective country. As noted, there is the recurrent assertion that migrants – signified as "economic refugees" – are entering a nation-state to benefit from the welfare system and, thus, deprive those of state subsidies who are supposed to have the primal or the only right to receive them. Therefore, these forms of immunization against "others" subvert the respect for persons as right bearers. What is more, immigrants who are not entitled to work lack the opportunities to gain esteem and, thus, to develop self-esteem. And even if they are allowed to work, they are doing so in often precarious working conditions.[7] Those who try to *enter* the sphere of mutual recognition are excluded, while those who already inhabit a country rely less and less on claiming rights to gain recognition but on property and through distinction from other (marginalized) groups. At first glance, these considerations seem to show recognition theory's potential for the

[7] One former Austrian minister of the interior enacted a provision according to which asylum seekers could work but the municipalities were forced to pay them no more than 1.50€ per hour (while the minimum wage in Austria is approximately 10 Euros). Even if we could assume that this is not a deprivation of right – which we cannot – we still can criticize this proceeding as disrespectful and as expressing a lack of benevolence as well as an ignorance of contributions to the a shared life.

critique of the immunization against and exclusion of migrating populations. Withholding of rights and depriving migrants of opportunities to gain esteem and self-esteem can be identified as social pathologies that are not justifiable, and this critique directly affects policies that are argued for by pretending that this is necessary not to overload the welfare system or – more fundamentally – that there is no obligation to receive migrants but rather an obligation to protect the citizens of a state from migration.[8]

However, I want to argue that in these complaints about an imbrutement of recognition, Honneth betrays some problems of his own theory, which directly affect moral issues related to migration.

First, and most importantly, it seems obvious that Honneth's recognition theory relies on a *teleology* in which an all-inclusive recognition – inclusion and individualization, as he puts it – is the fundamental normative orientation (Bedorf 2010). This is rooted in the formal ethics and the formal anthropology, which form the basis of his analyses. In his earlier writings, he conceives of the need for recognition as an anthropological constant; later, he even assumes that recognition is not only a normative claim but also a *genuine human praxis* (an inevitable primary recognition, which is potentially concealed by tendencies of reification; Honneth 2008). Social issues concerning inclusion and the possibilities for self-realization are, thus, understood as social pathologies that can only be rectified by *more* recognition. Social change is equated with the progressive inclusion of individuals or populations in the sphere of mutual recognition; the mentioned imbrutement of recognition is then the outcome of a *perverted* capitalist (instead of an unadulterated meritocratic) system in which the struggle for recognition and justice is turned in a struggle for self-assertion in a kind of Hobbesian *status naturalis*.

Though I do not at all want to dismiss this analysis *tout court*, this begs at least two (interrelated) questions:

> (a Honneth cannot really explain *why* such forms of misrecognition emerge; he only can point out that there is a social pathology or deviance from an alleged normal state that has to be rectified (though some authors concede that there is a decent etiology of these pathologies I want to contend that this is not really the case; he only can support the detection of deviances). The more we remain clueless, if we try to analyze shifts in traditional structures of recognition which are *not unambiguously progressive from any possible perspective* – Honneth cannot but understand such contingencies as failures to provide recognition. What is more, the consideration of rights seems to be firmly detached from solidarity; therefore, he cannot explain how the respect for migrants as right bearers might come in degrees and how exclusionary tendencies can affect this respect (while a focus on solidarity could also reveal that respect hinges on fraternity or sympathy in the broadest sense).[9]

---

[8] For decades, the buzzword of the "fortress Europe" has been used to criticize European policies in the face of migration. In 2015, during the "refugee crisis," the Austrian minister of the interior contended that we are in need of establishing a fortress Europe for self-defense. The current government of Austria refuses, just like the ones in Poland and Hungary, to receive unaccompanied minors from the refugee camp in Greece.

[9] Timo Jütten convincingly asserts: "Once a class of individuals is characterized as useless and replaceable and therefore not worthy of social esteem, their ability to exact the equal respect that is due to them ascitizens in the form of social rights is undermined too" (Jütten 2019, 91). Though

Furthermore, Honneth conceives of the struggle for esteem as a "horizontal competition of different values" (1996, 122), or it is pathological or ideological. He cannot take into account that, first, competition and solidarity might be mutually exclusive or at least in tension, and, second, that there might be a complex interplay between different dimensions of social positions that predetermine the access to such a competition which is, thus, probably never really horizontal. It is plausible to conceive of the instrumentalization of rights to exclude migrants from the welfare system as a form of misrecognition, yet we have to acknowledge that recognition always takes place within a particular cultural perspective. What is more, recognition – if we understand it as mutual – potentially implies a certain transformation, i.e., a loss or insecurity of the own identity (Butler 2005, 27) as new elements are included in a socio-cultural context. If we are mindful of the potential foreignness or alterity of the previously marginalized or excluded group, recognition is an effort and a venture that potentially jeopardizes own perspectives and identities and is, thus, also limited.[10] Though, basically, it has to be regarded as an ethical problem to deprive individuals or groups of recognition, such processes have to be analyzed as reliant on particular perspectives and identities that cannot be simply abandoned.

Connected to this issue of alterity, it is questionable whether inclusion can be all-inclusive. There is no guarantee that different identities are compatible and can coexist. While Honneth in *Freedom's Right* concedes a tension between inclusion and ethnic and cultural pluralism (2014, 327), Hirvonen opts for the possibility of minimal cohesion (2016, 37), which is plausible for many cases, I want to emphasize that there is the possibility of radical forms of *foreignness* (cf. Waldenfels 1997, 36 f.), which exceed the borders not only of intelligibility but also (and often by the same token) of moral acceptability. But the possibility of radical forms of foreignness thwarts any teleology of recognition and inclusion; otherwise, the cultural identity of an immigration society is not only showing plasticity but porosity or even a veritable collapsibility. To put it pointedly: If we understand recognition as socio-culturally conditioned and perspectival, then an all-inclusive recognition (which would equal Honneth's teleology) would be as oxymoronic as an attention for everyone or everything; an all-inclusive attention ceases to be attention at all. And even if we emphasize democracy as the structure which allows all its members or participants to co-determine how the political sphere is organized (Hirvonen 2016, 29), we have to be mindful that it is always a particular, socio-culturally saturated manifestation of democracy with specific institutions; it is surely not *the* democracy.

Jütten here refers to a text by Honneth and Titus Stahl, I doubt that this insight can be traced back to Honneth's standard consideration of the three forms of recognition.

[10] Let me emphasize that this view does not express a reactionary or identitarian hypostatization of the own cultural identity – or, horribile dictu, even a justification for racism – but builds an attempt to bring to the fore a basic limitedness of recognition due to the limits of any perspective and/or identity, as no identity has endless plasticity. In the upcoming sections, I will use Judith Butler's concept of *recognizability* also to analyze this, in my view, inevitable limitedness.

b) We can contend that Honneth widely fails to acknowledge and analyze that acts of recognition as well as recognition institutions (which are, as noted, only rooted in an unconditioned intersubjectivity) are inevitably saturated with power; his unconditioned teleology does not allow for a conception of recognition as socially mediated and as preceded by structural prerequisites. He tends to "naturalize the idea of recognition by failing to examine more thoroughly the way in which emotions and other aspects of embodied subjectivity are mediated through social relations of power" (Lois McNay cited in McQueen 2015, 50). This becomes particularly visible, when we consider his emphasis on negative experiences as ratio cognoscendi of injustice and as starting point for getting into struggles for recognition. It is restricted to the perspective of the marginalized; the question does not emerge how such affects are socially mediated or even produced.[11] Power is seen as extraneous to families and primal attachments (the sphere of love and self-confidence), legislation (the sphere of rights and self-respect), and the economy (the sphere of esteem and self-esteem). Therefore, on the one hand, he cannot really explain asymmetries and exclusionary tendencies within recognition spheres (which is supposed to be symmetrical in nature) and is unable to decently understand institutional or structural forms of misrecognition (as analyzed convincingly in Iris Marion Young's *Five Faces of Oppression* [1988]) – and possible limits of recognition. On the other hand, he is not able to bring possible ambivalent effects and consequences of recognition to the fore (e.g., adaptive preferences through the subtle subordination under hegemonic structures of recognition by seeking recognition or the development of a "double consciousness" [Young 1988, 286] within which individuals perceive themselves as the non-normal). Even more fundamental, one could draw from Butler's considerations of the (self-)formation of an ethical – or recognizable – subject according to prevailing norms or hegemonic forms of normality (2005, 17). As Paddy McQueen puts it: "We need to analyse how recognition orients us within a given social space and the role that authority and power plays in this" (McQueen 2017, 47).

Second, it is not unproblematic to proceed from the (Hegelian) assumption that recognition is always and inevitably a matter of reciprocity and that the object of recognition is either a full-blown person (rights) or individual achievements against the backdrop of presumably shared values. This assumption tacitly rests on the liberal insinuation of equality of pre-given capacities (e.g., to be an accountable person or to be able to contribute to society's aims). However, I want to stress that entering into struggles for recognition presupposes a process of subjectivation that cannot be taken for granted. As Fanon has put it in drastically: "For Hegel there is reciprocity; here [in the context of colonial domination; M.H.] the master laughs at the consciousness of the slave" (2008, 172; footnote 8). Though it would be questionable

[11] I have been working with asylum seekers from Afghanistan and Syria; some of them felt discriminated because no house with garden had been provided to them by the Austrian welfare system. Their image of the conditions in which the Austrian population lives was obviously distorted. In turn, others develop adaptive preferences and were happy to work for one Euro per hour and did not feel outraged about being underpaid.

to equate slavery and the current conditions of migrants in the Western world, I think that there is one relevant analogy. The spheres of recognition are not always simply accessible for migrants, and in some cases, we are reminded of Orlando Patterson's analysis of the slaves' *social death* (1982); this thwarts Honneth's view who in some passages even assumes that there is a rational social cooperation in which individuals simply complement and complete each other (2014, 86). Migrants who are not entitled to be granted asylum and have no further defined and granted right to stay are illegalized. But also the ones who are not illegalized *sans papiers* are marginalized and left powerless if they have restricted access to the labor market, lack language capacities to address themselves to relevant institutions, and are regarded as outsiders – but not due to malicious decisions. *They are not recognized as recognizers as they are not recognizable as recognizers.*[12]

Generally, addressing is an inevitable presupposition of recognition and also for a struggle for recognition. Particularly with regard to migration and migrants' situation as new arrivals, a bulk of questions emerges: Are they addressed? Can they be addressed? Can they address the other(s)? What is their speaker position and how does it relate to language skills, language codes, etc.?[13] Finally, it is crucial whether and how the mentioned negative experiences can be narrated – or if forms of epistemic injustices, e.g., rooted in the lack of adequate concepts or the lack of trust in the speaker's accountability (Fricker 2007), restrict the possibilities to express these experiences.

Third, related to the question of reciprocity and mutuality, there is another crucial issue. In *Freedom's Right*, Honneth contends that democratic institutions presuppose *and* are supposed to produce and sustain freedom (cf. Honneth 2014, 330). There seems to be an equality that is prior to acts and institutions of recognition, which is not detachable from the one that is or should be constituted by these institutions (Hirvonen 2016, 34, footnote 6). I leave aside the question whether Honneth's considerations are here close to a petitio principii, but I want to emphasize that such an allegation of equality is problematic. Honneth focuses on values like identity, freedom, autonomy, and justice (cf. also McQueen 2017, 208). I am by far not entirely critical of such a focus; however, we find a particularly strong assumption in the emphasis of self-determination and self-realization. While it is hardly contestable that self-determination and self-realization are morally significant, they might

[12] Here we can face sort of an *epistemic injustice* (cf. Fricker 2007) that goes beyond and deeper than testimonial and hermeneutic injustice – even though it is connected to these forms of injustice. While in Fricker individuals are not recognized as knowers (see also Congdon 2018), marginalized or invisible migrants are not even recognizable as full-blown vis-à-vis in an encounter. I assume that this is a particular kind of epistemic injustice, because in more recent texts (2007, 2008), Honneth stresses the deep relation between recognition and cognition (*Anerkennen* and *Erkennen*), and since some individuals are excluded from mutual recognition, they are not perceivable or conceivable as full-fledged members of society. Admittedly, Honneth takes a different perspective as he invites us to acknowledge a primacy of recognition to knowledge and classification; but it would go beyond the scope of this footnote and of the paper to analyze the differences between Honneth's considerations in this respect and the aim of this paper.

[13] Similar questions have been pointed up by Judith Butler (2005) and Jacques Derrida (1992).

also be oppressive because they mirror a particular (liberal) picture that insinuates that individuals strive primarily for autonomy. Moreover, an emphasis on reciprocity and mutuality related to identity as the actual focus on recognition possibly evokes the mentioned problems of the limits of the recognizable (of different identities); particularly solidarity understood as rooted in shared values becomes problematic. Yet if we are to analyze the ethical issues of migrants and their potentially precarious situation, we would be better off to shift the focus from – presumed identities and their acceptability to vulnerability and its "recognizability." On my view, such a change of perspectives brings a more primary form of recognition to the fore, which is the condition of possibility of respect and social esteem. The recognition of vulnerability provides us with the basic humanization and is decisive prerequisite to avoid the derealization of individuals that excludes them from those spheres of recognition, which have been analyzed by Honneth. The recognition of a shared (ontological, yet in particular individuals also situationally realized) vulnerability (Gilson 2014, 7) provides us with a basis for a solidarity that is prior to any achievement and performance. Later, I will draw from Levinas to argue that vulnerability can be conceived of as the primary source of normativity and, as shared vulnerability, as source of solidarity. Crucially, though vulnerability is shared, it does not reinstall mutuality or clear reciprocity as basic characteristic of recognition, because vulnerability is not equally distributed (and migrants build a prototypical example for populations, which are prone to augmented, situational vulnerability). In Henk ten Have's words: "This means that the shared vulnerability that is typical for humans is not merely mutual. Even if we are all vulnerable, one person might be more vulnerable to me than I am to her" (ten Have 2016, 102).

3.4 A Change of Perspectives with Judith Butler

In what follows, I want to point out briefly that Judith Butler's reflections on *recognizability* seem to me to set back recognition theory on its feet – as it seems that in Honneth's version recognition stands respects in some respect on its head. He alleges that we are able to orient our actions and public policies according to inclusion and individualization independently of hegemonic structures and power relations – within a genuine praxis of recognition (2008), Butler points out that *politically charged frames of recognition precede and saturate acts of recognition* (2004a, 4, 58; 2009, 4 f.); thus, recognition is inevitably invested with power, since these frames are to be conceived of as meaning- and value-laden organizational structures (cf. Gilson 2014, 44). Honneth takes the intersubjective, dyadic situation of equals who (should) mutually recognize each other as point of departure. Butler reflects on the embedding of bodily beings in complex social structures that are ingrained in our dispositions of perceiving, affective responses, and schemes of behavior (Huth 2018).

In *Undoing Gender*, Butler highlights the possible discrepancy between gender norms, which make certain identities recognizable, and individual constitutions, which make some people "feel that the terms by which [they are] recognized make life unlivable" (2004a, 4). Therewith, she enables us to acknowledge that inclusion and individualization take place in particular contexts and are shaped by particular norms and particular schemes of behavior. Recognizability forms the normative, but also the epistemological and ontological, horizon within which subjects come to be as subjects (cf. Butler 2005, 17). Its particularity implies that recognition could also be understood as a regulatory practice and is, thus, in nuce ambivalent. Individuals entering a particular socio-cultural context are compelled to "make" themselves intelligible within a particular set of concepts, practices, and images, and they have to comply with certain prerequisites to undergo a subjectivation. Migration, thus, frequently implies that individuals' identities are jeopardized and must become malleable.

In *Frames of War*, she focuses on the even more existential dimension of the recognizability of vulnerability as expressed in the concept of grievability:

> [A]n implicit understanding that the life is grievable, that it would be grieved if it were lost, and that this future anterior is installed as the condition of its life. In ordinary language, grief attends the life that has already been lived, and presupposes that life as having ended. But, according to the future anterior (which is also part of ordinary language), grievability is a condition of a life's emergence and sustenance. (2009, 15)

Schemes of recognition predetermine who will be regarded as a vulnerable subject worthy of recognition (regarding their vulnerability) (2009, 18; 2012, 140). *Recognizability*, thus, determines "who is entered into the category of the human in these times, and who is considered inhuman, monstrous, or subhuman" (Butler 2012, 142) – such as migrants who tend in some instances to be reduced to foreign particles. Recognizability must not be understood as reducible to the possibility of an exercise of rights (ibid., 143) but comes closer to the concept of solidarity that expresses a certain kind of involvement. It builds the essential basis for our "cultural survival" preventing us to fall prey to social death through which the living and dying of the affected individuals becomes derealized (Butler 2004b, 147). Solidarity seems to hinge on the possibility of being categorized in this way. "My wager is that 'recognition' becomes a problem for those who have been excluded from the structures and vocabularies of political representation" (Butler 2012, 140). But even if the categorization does not confront us more or less with a binary of the human and the abjected monstrous, it remains a complex issue: "And for those whose only access to 'recognition' is episodic media exposure or criminalization, we can see that recognition is a vexed problem for the poor and the undocumented" (ibid.).

From *Precarious Life* (2004b) on, Butler draws from Emmanuel Levinas' work, which is dedicated not only to alterity but also the vulnerability of bodily beings. He is particularly concerned with the – in his view unconditioned – fact that the Other's very vulnerability addresses me and imposes an obligation on me (1979, 1991; see below). Thus, vulnerability is to be regarded as the root of normativity. In contrast, Butler emphasizes that the recognition of vulnerability is socially mediated and even socially enabled. Therefore, she distinguishes between *precariousness* and *precarity*:

> The more or less existential conception of 'precariousness' is thus linked with a more specifically political notion of 'precarity'. And it is the differential allocation of precarity that, in my view, forms the point of departure for both a rethinking of bodily ontology and for progressive or left politics in ways that continue to exceed and traverse the categories of identity. (Butler 2009, 3)

Frames of recognition inevitably condition not only our intentional decisions, e.g., through language, but already our *perception* of individuals (2009, 29). Perceptions, affective responses, and (dispositions for) acts of recognition are saturated with social (political) frames of intelligibility. In the vein of Michel Foucault, Butler contends that a particular regime of truth decides what will or will not be a recognizable form of being (2005, 21), i.e., a vulnerable human who addresses me in a normatively relevant way. Those who appear to belong to a "suspect ethnic group" (2009, 26) are perceived as less grievable and are, thus, less intelligible as vulnerable than other individuals or even abjected as inhuman or monstrous.

Thus, Butler enables us, first, to shift the emphasis from the recognition of identities and full-blown persons (as right bearers and as contributors to social aims) to the recognition of vulnerability. Second, vulnerability becomes conceivable as basis for normativity. Third, recognition is conditioned by politically charged frames, which saturate our perception, our initial affective responses, and our dispositions to constitute acts of recognition. Crucially, these frames determine the intelligibility of lives as (vulnerable) lives, i.e., as sharing vulnerability with a socio-cultural "us." In this view, solidarity is not rooted in a meritocracy but in the avowal of the common vulnerability.

3.5 Migration and the Selective Recognition of Vulnerabilities: Reflections Between Judith Butler and the Critical Theory

Proceeding from this shift of perspective through Butler's work, I want to outline some elements of a theory of the recognition of vulnerability, which seems to be particularly suitable to tackle ethical issues related to migration. Furthermore, I will also focus on the question of *critique* of current conditions. At this point, it might seem that in Butler's view, the frames of recognition (described as a *historical a priori*; 2009, 6) could even fully determine the recognition of migrants as vulnerable individuals; critique would then be futile or impossible as allegedly responsible subjects would have to be regarded as mere products of a subjectivation dependent on contingent frames of a particular. To be sure, Butler would of course never defend such a view, yet it seems that critique remains a critical issue in her analysis of recognizability.

In the vein of Emmanuel Levinas, we can contend that vulnerability is the source of normativity. He invites us to comprehend the *face* of the Other as an excess over the mere appearance of the very Other (Levinas 1979, 23, 194ff., 1991, 91). So, the concept of the face does not describe (in a positivist manner) a visible or touchable part of another one's body but rather represents a deep structure of the experience of an encounter with the Other. Using the concept of *nakedness*, Levinas expresses two different issues: First, the Other is naked as she/he is beyond any horizon of understanding; her/his appearance even thwarts any socio-cultural context, institutions, or images, and the Other is unconditioned (Levinas 1979, 74). Thus, radical forms of alterity become thinkable. Second, as naked body, the Other is vulnerable and mortal (Levinas 1991, e.g., 88). By virtue of this vulnerability, the Other addresses me and raises a moral claim. I cannot but respond to this claim – as even ignorance or violence is a response to this vulnerability. This is the root of responsibility, since I cannot but decide how to respond to the Other. Therefore, the primal scene of an encounter with the Other presents me with an address that is prior to any political consideration of asylum policies, but also legal claims for asylum or the allocation of esteem (though the plurality of Others compels me to weigh different needs and moral demands, i.e., to allocate care, respect, and solidarity according to respective criteria).

However, Butler highlights – in a way against Levinas – that this being addressed takes place within a particular socio-cultural context, i.e., with a specific structure of recognizability. As noted, recognizability precedes acts of recognition and predetermines institutions and common practices. Therefore, vulnerability is always related to some frame of intelligibility and norms of responsibility. Against the backdrop of Butler's conceptual distinction between existential precariousness and politically induced precarity, this is expressed as follows:

> So as soon as the existential claim is articulated in its specificity, it ceases to be existential. And since it must be articulated in its specificity, it was never existential. In this sense, precarity is indissociable from that dimension of politics that addresses the organization and protection of bodily needs. Precarity exposes our sociality, the fragile and necessary dimensions of our interdependency. Whether explicitly stated or not, every political effort to manage populations involves a tactical distribution of precarity, more often than not articulated through an unequal distribution of precarity, one that depends on dominant norms regarding whose life is grievable and worth protecting and whose life is ungrievable, or marginally or episodically grievable and so, in that sense, already lost in part or in whole, and thus less worthy of protection and sustenance. (Butler 2012, 148)

As noted, recognizability is efficacious already on the level of perception and imbues our affective responses. The recognizability of vulnerability, which forms the basis of a humanization and is a possible remedy of the derealization of humans, renders a specific vulnerability particularly visible or conceals it through abjecting narratives, tacit exclusionary practices, or structural/discursive invisibility (constituting "good" or "bad," "needy," or "dangerous" foreigners). The structures of recognition constitute or contribute to abjection could turn refugees into invaders or parasites who undermine the welfare system ("economic refugees") and who take away the women from the autochthonous men (and intend to undermine our culture and ethnicity). Thus, some of these discourses and practices evoke a profound de-vulnerabilization of migrants undermining those forms of solidarity that are rooted in shared vulnerability. This seems to be a matter of the historical situation, the contemporary forms of representation, the limits of the sayable,[14] and, generally, the nature of those frames that bring vulnerability into particular forms of visibility. During the refugee wave in 2015 in Europe, the description emerged that we would have to face hordes of randy young men, ready to use (sexual) violence, entering our tranquil lifeworld. This generated a hypervisibility of a distorted picture of refugees. At the same time, a structural invisibility emerged, i.e., a lack of representation, the absence of migrants in the public discourse and the media (at least the absence of their own voices), but to some degree also in the public space (as detention often was – and is – a common measure to treat migrants without an asylum status). Both the distorting hypervisibility and the invisibility contribute to the marginalization, the structural exclusion, and, thus, the misrecognition of migrants.

However, I want to argue that the recognizability of vulnerability (as well as identities, rights, and achievements) is not only contingent but also *inevitably* selective and exclusive (Waldenfels 1996); as noted, the idea of an all-inclusive recognition is oxymoronic. Recognizability is not only invested with power and contingent but also *necessarily perspectival*. These considerations boil down to the insight that every constitution of a cultural identity equiprimordially produces an inside and an outside, thus, forms of foreignness. Drawing from Levinas, the German phenomenologist Bernhard Waldenfels has analyzed the possibility of *radical foreignness*, which implies a non-includability of particular behaviors and customs or, at worst, the abjection of populations.

Migrants may well be beyond the line due to their cultural affiliation; this is not only the case in examples such as the abjection of the Arab population in the United States in the aftermath of 9/11, which Butler uses to illustrate her theory of recognizability (Butler 2009). Solidarity, which is not only passive tolerance but also felt concern for the other (cf. Honneth 1996, 129), is then undermined by a politics of emotion that affects individuals beneath the level of higher-order intentionality. However, if we are mindful of the fact that any cultural context constitutes and, thus, excludes particular forms of foreignness, the attention should shift from such somehow initiated and controlled politics of emotions to the inevitability of exclusion of foreignness. As noted, Butler indicates that individuals, but also communities, are transformed through recognizing, which implies sort of a loss (of the previously given identity) in the process of recognition (Butler 2005, 27). A consequence – which has not been drawn by Butler – is that any socio-cultural field of recognition (by which I am partly dispossessed as a recognizer) *must* exclude radically unrecognizable groups and individuals because of insurmountable limits of liberality and

[14] The mentioned former Austrian minister of the interior decreed to fix signs with the inscription "emigration center" ("Ausreisezentrum") at the entrance of a huge immigration center. Such cynicism betrays hostility against foreigners and racism, but it seems clear that this would not have been sayable and doable in another historical context in Austria, e.g., as refugees from Bosnia (many of them Muslim) immigrated and racism well played a role in the debates and the social atmospheres, but such acts of political abjection did not occur. Moreover, Erinn Gilson convincingly points out that the denial of vulnerability underlies other types of ignorance, for instance, racial oppression (2014, 75ff.).

flexibility; there is neither an endless plasticity of identities nor of the tolerable. This has significant implications for migration. First, there seems to be no guarantee for migrants to be able to identify with a set of available norms, even if they are accepted and recognized as vulnerable, and, thus, dependent on care, nor is there a guarantee that there is a chance to influence and shift these norms to become recognizable through a struggle for recognition. Second, there is no guarantee that different identities and vulnerabilities can coexist without ongoing friction (while Honneth insinuates that at some point any struggle for recognition can lead to inclusion and individualization). This is not meant as an argument against ethnical and cultural pluralism but to increase the attentiveness for those forms of exclusion (e.g., of behaviors and customs that are considered as violent or criminal) that are tacitly considered as "normal" and to increase the awareness of the limitedness of solidarity and tolerance, which might entail cases of hardship.

Finally, I want to turn to the ethical consequences of the proposed shift of perspective concerning the recognition of vulnerability and solidarity based on the avowal of shared vulnerability. Recognizability cannot be traced back to intentional decisions and policy-making. Butler emphasizes that (generalizing) norms and the terms by which I confer recognition are not mine alone; there is an inevitable dispossession at the heart of recognition (Butler 2005, 26). As noted, the frames of recognition are tacitly powerful and imbue our perceptions and affective responses prior to our intentional decisions. Therefore, it would be an absurd simplification and responsibilization to blame particular individuals or groups for such practices, though it is clear that some nationalists purposefully stir the opinion against migrants. Thus, the question becomes pressing how a critical relation to frames of recognition is possible. I assume that it is possible only through a distance to our habitual dispositions to perceive and conceive of migrants (and other often marginalized, yet vulnerable, groups, for instance, homeless or addicts) whose vulnerability is often disavowed. Such a distance seems only possible by virtue of a process of critical distancing. In turn, this begs the question what enables us to be critical.

Butler and Athena Athanasiou suggest in their book *Dispossessio*n to consider *solidarity* as vehicle of critique, as power to dismantle social conventions and foreclosures that render some lives and desires impossible (2013, 185). Solidarity is conceived as a form of sympathy or empathy that opens us to vulnerabilities that exceed those frames that tacitly organize our perceptions and affective responses.[15] However, it is not an unambiguous or unambivalent as it is inextricably enmeshed with power. They contend: "Solidarity is unavoidably intervowen in the normative

---

[15] In a similar vein, in *Frames of War*, Butler uses the concept of *apprehension*, which denotes the possibility of apprehending lives beyond the structures of recognition: "We can apprehend, for instance, that something is not recognized by recognition. Indeed, that apprehension can become the basis for a critique of norms of recognition. The fact is we do not simply have recourse to single and discrete norms of recognition, but to more general conditions, historically articulated and enforced, of 'recognizability.' If we ask how recognizability is constituted, we have through the very question taken up a perspective suggesting that these fields are variably and historically constituted, no matter how a priori their function as conditions of appearance" (2009, 5).

violence inherent in the ways we come to imagine and recognize a viable life in accordance with given prerequisites of intelligibility. At the same time, though, it somehow offers a space for exposing and exceeding such prescribed limits" (185).

Now it is utmost important how we understand this "somehow" to make sense of these reflections, and this is the point where I think we shall turn back to critical theory and the concept of recognition. Basically, Butler conceives of critique as a social practice, which exceeds the sphere of the exercise of rights (Butler 2012, 143). She stresses that it consists of the establishment of a *distance* from any naturalized version of the schemes of recognition and conceives it as an intervention "that exposes the implications that such schemes have for the differential prospects of living and dying for various populations" (ibid., 141). First and foremost, the naturalization of ethnicity or authenticity (of an allegedly autochthonous population) forms the basis for the exclusion or refugee refusal, which is often linked but not necessarily overlapping with blatant racism. Yet Butler is adamant about the availability of such exclusionary tendencies. "The practice of critique then expresses the limits of the historical scheme of things, the epistemological and ontological horizon within which subjects come to be at all" (Butler 2005, 17).

Using an argument borrowed from Walter Benjamin (and, actually, also from Ferdinand de Saussure and Jacques Derrida), she indicates that the circulability of frames and norms, i.e., their actualization in variegated contexts and situations, implies that they break away from themselves (2009, 9). Thus, the constituted distance to the normative schemes provides us with the possibility to perceive, empathize with, and comprehend others in alternative ways. However, on my view, it remains unclear how such a gap could *ethically orient* these alternative ways and how we are to understand a vulnerability, which is inevitably framed *and* exceeds these frames. There seems to be no normative quality in the mere possibility to perceive and conceive others in previously nonestablished ways. How is it possible to direct our attention particularly to the invisibility of some vulnerable populations or specific forms of vulnerability (as some kinds of vulnerability of migrants might not be conceivable in the dominant framework of a particular society) to include them into the schemes of intelligibility and the connected norms? How is it also possible to make visible the inevitable violence of any frame of recognition that always enacts a certain generalization that turns individuals into typical cases?[16]

[16]The other is singular, yet norms are impersonal and indifferent (Butler 2005, 24; Adorno 1973, 173 f.). Adorno's criticism of the administered world (cf. ibid., 20) problematizes this neglect of singularity in the process of categorization. At worst, vulnerability becomes naturalized, which may well lead to paternalistic relationships and institutions. Moreover, this compels people to *enact their vulnerability* in particular ways so that they can expect specific, expectable responses. This could lead to the development of adaptive preferences, in which individuals become a distorted picture of their own needs (here we can see again the importance of the insight that recognition is constitutive for any relation-to-self). Needs, which are to be understood as culturally imbued, become bent according to the immigration culture. Moreover, this might also lead to cultural imperialism as described by Iris Marion Young: She points out that members of cultural minorities could find themselves compelled to detect their own particularities – in this case vulnerabilities – as abnormal and might even hide them (1988, 285ff.). But these are often tacit and

To gain some ground in this aspect, I suggest to turn back to the critical theory and to take up the concept of *immanent critique*. On my view, immanent critique relies on two conditions of possibility: First, acts of recognition are not to be considered *merely* as parts of the mechanisms of power (as McQueen indeed seems to insinuate; 2017). They are related to frames saturated with power, yet there is a gap between the general frames and their situational actualization (which resembles Butler's idea of the circulability of norms). Second, we are immersed in a *multidimensional* normative infrastructure that enables us to criticize practices from an already socially approved standpoint (Stahl 2017, 3). While no external standpoint exists from which we can (dogmatically) make use of purely rational principles to evaluate hegemonic practices and structures, we can proceed from indisputable feelings of moral repulsion (Adorno 1973, 286) that reveal *ex negativo* the "promises" that are implicit in this very infrastructure (see also Horkheimer and Adorno 2002, 202). Honneth reflects on the possibilities of immanent critique in various texts; his basic assumption seems – a specific interpretation of Horkheimer's and Adorno's claim – that there is a *surplus of validity* in norms of recognition (cf. 2003, 186) that goes beyond the manifest practices and institutions and particularly beyond explicitly acknowledged dimensions of norms. This surplus together with a hermeneutical interpretation of these very norms' meaning forms the basis for critical evaluation of respective practices in our lifeworld.

However, I find Titus Stahl's approach of immanent critique particularly convincing. In his view, *particular* experiences of the exploitation of vulnerability from the perspective of marginalized social groups create a point of departure from which structural and collectively habituated practices can be criticized; again, these experiences builds the ratio cognoscendi of social pathologies. Crucially, this makes clear that social practices and their notion do not build an integral whole. The particularity of their perspective predestines them to detect and express an otherwise implicit conflict in social practice. In the vein of Robert Brandom, Stahl contends: "[C]ritical judgments are justified by making explicit a conflict between commitments immanent to a practice and that practice's actual form, a conflict that reveals itself through experiences of injustice" (Stahl 2017, 3). The basis is not solely the surplus of more or less explicit (and in part implicit) norms that has to become accessible. In Stahl's view, immanent critique means making explicit ethical reasons that are already *embodied in our practice* (not as an abstract surplus of validity of norms – though I do not want to dismiss Honneth's account of immanent critique *tout court*) but that become only explicable through recalcitrant moral experiences and "which (…) are prevented from making explicit by a distorting conceptual framework" (ibid., 8). Notably, the particularity of the experience and the perspective of those who suffer from injustice implies that any attempt to make explicit the implicit norms is sort of a request (in sense of Kant's *Ansinnen*), an imposition to reconsider norms and transform practices that might not go without

hardly noticeable events beneath the surface of the normal practices and institutions of a particular society and its dominant framework that – due to this *inconspicuous* normality – cannot be easily addressed as ethical issues.

friction with other perspectives (e.g., if the foreignness of migrants is qualified differently by various social groups).

Immanent critique becomes, thus, a vehicle "to articulate what precludes ethical response, how the sense of responsibility that might grow out of awareness of common vulnerability is truncated, how the actions and attitudes that might fulfill responsibility are inhibited, and how the ways in which we live channel our energies away from ethical responsiveness" (Gilson 2014, 61). The normative basis remains an emphasis of vulnerability and a solidarity that is rooted in the recognition and avowal of the shared vulnerability of bodily beings; this solidarity is potentially asymmetrical, because it does not target vulnerability per se, but proceeds from different manifestations and degrees of vulnerability. Migration is but one example of the constitution of a potentially increased, situational vulnerability that transfers individuals in a comparatively weak position. Yet the social multiperspectivity and the pluridimensionality of overlapping practices and connected frames that organize our perception and notion of vulnerability potentially allow to recognize previously unrecognized forms of vulnerability and groups previously unrecognized as vulnerable. Whenever common practices and particular experiences show mutual tension, then the ethical possibility of an alternative form of recognition emerges. However, in spite of this plasticity of schemes of recognition, there is always a limitedness of the includable and acceptable, and the most pernicious way to deal with this limitedness is its disavowal.

3.6 Conclusion

I have argued that Honneth's account of recognition provides us with some crucial prerequisites for reflecting on the ethical implications of migration. His emphasis of the inevitable dependency of individuals on recognition and the analysis particularly of the respect for rights and the esteem of achievements are significant means to detect social pathologies. However, first, Honneth assumes an unconditioned teleology of recognition. Therefore, we miss an explanation why differential treatments of various groups emerge; moreover, he fails to take into account power relations; and he cannot target limits of social inclusion. Second, he assumes reciprocity in recognition and is unable to tackle the question why some individuals might be not in the position to be recognized as recognizers, which applies not the least to migrants who are unfortunately conceived of as invaders. Third, he shifted the focus from personhood and achievements to vulnerability as the center of recognition, particularly as basis for solidarity. Drawing from Judith Butler, but also from Levinas' considerations of vulnerability as source of normativity, I have argued that the recognizability of vulnerability helps us to rectify these pitfalls. Within this approach, it is possible to tackle the question of power in recognition, but also to be sensitive for *inevitable* structural exclusions of forms of vulnerability and of particular groups from being recognized as vulnerable since any particular sociocultural framework constitutes a certain perspectivity. However, in turn, Butler in some respects leaves idle the basis for a critical evaluation of current practices. Thus, I have proposed to complement her approach with the concept of immanent critique as has been established within the critical theory to find such a basis of what normatively orients possible critique. Yet this critique might be a radical, but not a total, one since any cultural context has its limits of inclusion as well as multilayered practices which are not to be entirely unified.

References

Adorno, Theodor W. 1973. *Negative Dialectics*. Trans. E.B. Ashton. London: Routledge.

Bedorf, Thomas. 2010. *Verkennende Anerkennung*. Suhrkamp: Frankfurt am Main.

Butler, Judith. 2004a. *Undoing Gender*. New York: Routledge.

———. 2004b. *Precarious Life: The Powers of Mourning and Violence*. New York: Verso.

———. 2005. *Giving an Account of Oneself*. New York: Fordham University Press.

———. 2009. *Frames of War: When is Life Grievable*. New York: Verso.

———. 2012. Precarious Life, Vulnerability, and the Ethics of Cohabitation. *Journal of Speculative Philosophy* 26 (2): 134–151.

Congdon, Matthew. 2018. "Knower" as an Ethical Concept: From Epistemic Agency to Mutual Recognition. *Feminist Philosophy Quarterly* 2018: 4. (4).

Derrida, Jacques. 1992. Force of Law: The 'Mystical' Foundation of Authority. In *Deconstruction and The Possibility of Justice*, ed. Cornell Drucilla, Michel Rosenfeld, and David Gray Carlson, 3–67. New York: Routledge.

Fanon, Frantz. 1994. *Black Skins, White Masks*. London: Pluto Press

Fricker, Miranda. 2007. *Epistemic Injustice: Power and the Ethics of Knowing*. Oxford: Oxford University Press.

Gallagher, Shaun. 1986. Lived Body and Environment. *In Research in Phenomenology* 16: 139.

Gilson, Erinn C. 2014. The Ethics of Vulnerability. *A Feminist Analysis of Social Life an Practice*. London and New York: Rutledge.

Habermas, Jürgen. 1994. *Justification and Application. Remarks on Discourse Ethics*. Trans. C. Cronin. Cambridge: Polity Press.

Hirvonen, Onni. 2016. Democratic institutions and recognition of individual identities. *Thesis Eleven* 134 (1): 28–41.

Honneth, Axel. 1996. *The Struggle for Recognition. The Moral Grammar of Social Conflicts*. Trans. J. Anderson. Cambridge: The MIT Press.

———. 2003. *Contributions to Redistribution or Recognition: A Political-Philosophical Exchange*. London: Verso.

———. 2007. *Disrespect. The Normative Foundations of Critical Theory*. Cambridge: Polity Press.

———. 2008. *Reification. A New Look on an Old Idea*. Ed. By Martin Jay. Oxford: Oxford University Press.

———. 2013. Verwilderung des sozialen Konflikts. Anerkennungskämpfe zu Beginn des 21. Jahrhunderts. In *Strukturwandel der Anerkennung. Paradoxien sozialer Integration in der Gegenwart*, ed. Axel Honneth, Ophelia Lindemann, and Stephan Voswinkel, 17–39. Frankfurt am Main: Campus.

———. 2014. *Freedom's Right: The Social Foundations of Democratic Life*. New York: Colombia University.

Horkheimer, Max, and Theodor W. Adorno. 2002. *Dialectics of Enlightenment. Philosophical Fragments*. Trans. E. Jephcott. Stanford: Stanford University Press.

Huth, Martin. 2018. Incorporated Recognizability. A Handshake Between Maurice Merleau-Ponty and Judith Butler. *Acta Structuralica* 2 (Special Issue): 119–145.

Jütten, Timo. 2019. The Theory of Recognition in the Frankfurt School. In *The Routledge Companion to the Frankfurt School*, ed. Peter Gordon, Espen Hammer, and Axel Honneth, 82–94. New York: Routledge.
Levinas, Emmanuel. 1979. *Totality and Infinity. An Essay on Exteriority*. Trans. A. Lingis. The Hague/Boston/London: Martinus Nijhoff Publishers.
———. 1991. *Otherwise than Being or Beyond Essence*. Trans. A. Lingis. Dordrecht: Kluwer Academic Publishers.
McQueen, Patrick. 2015. Honneth, Butler and the Ambivalent Effects of Recognition. *Res Publica* 21: 43–60.
———. 2017. The Promise and the Problem of Recognition Theory. In *The Persistence of Critical Theory*, ed. Gabriel Ricci, 205–221. London/New York: Routledge.
Nussbaum, Martha C. 2006. Frontiers of Justice. *Disability, Nationality, Species Membership*. Cambridge: The Belknap Press of Harvard University Press.
Patterson, Orlando. 1982. *Slavery and Social Death: A Comparative Study*. Cambridge: Harvard University Press.
Rawls, John. 2002. *A Theory of Justice*. Cambridge: The Belknap Press of Harvard University Press.
Stahl, Titus. 2017. Immanent Critique and Particular Moral Experience. *Critical Horizons*. https://doi.org/10.1080/14409917.2017.1376939.
Taylor, Charles. 1994. *Multuiculturalism and the Politics of Recognition*. Princeton: Princeton University Press.
ten Have, Henk. 2016. *Vulnerability. Challenging Bioethics*. London/New York: Routledge.
Waldenfels, Bernhard. 1996. *Order in Twilight*. Ohio: Ohio University Press.
———. 1997. *Topographie des Fremden. Studien zur Phänomenologie des Fremden* Band 1. Frankfurt am Main: Suhrkamp.
Young, Iris Marion. 1988. Five Faces of Oppression. *The Philosophical Forum* XIX (4): 270–290.

Chapter 4
Transnationalizing Recognition: A New Grammar for an Old Problem

Gonçalo Marcelo

Abstract This chapter analyzes migration through the angle of what I propose to call a critical hermeneutical theory of transnational recognition. Forced migrants are often denizens in a space of political indetermination, leaving their countries for humanitarian reasons but sometimes not even recognized as valid subjects of rights, given the fact of their political exclusion. As Nancy Fraser (Scales of justice. Reimagining political space in a globalizing world. Columbia University Press, New York, 2010) has argued, these transnational problems mark the limits of a justice envisaged in Westphalian terms and push us to think justice in a post-Wesphalian framework. Within this backdrop, this chapter develops an alternative framework to help shed some light on migration problems and to understand the claims put forward by migrants. Drawing from critical theories of transnational justice such as Fraser's (Scales of justice. Reimagining political space in a globalizing world. Columbia University Press, New York, 2010) and Forst's (Metaphilosophy 32(1/2): 160–179, 2001), I apply this transnationalizing move to recognition theory and spell out the new grammar of recognition that comes to light when one thinks of social actors claiming for due recognition across borders. One of the key aspects is that this theory needs to be hermeneutical because unlike cosmopolitan theories that mainly insist in the need to welcome asylum seekers on the basis of respect for their human rights, I claim that one needs to go deeper and resort to thicker descriptions of the identities, communal ties, and cultural background of these people in order to

My research benefited from the postdoctoral scholarship granted by the Foundation for Science and Technology, FCT, I.P. (SFRH/BPD/102949/2014), and the contract signed under the D.L. 57/2016 "norma transitória." This research is also financed by national funds through the Foundation for Science and Technology, FCT, I.P., in the framework of the CECH-UC project: UIDB/00196/2020. A first version of the paper was presented in the workshop *Recognition, Migration and Critical Theory*, organized by Gottfried Schweiger and which was held at the Centre for Ethics and Poverty Research (University of Salzburg). I thank the participants of the workshop for their comments and mostly Gottfried for the many suggestions on the first drafts of the paper, which helped me to clarify many points in my claims – without them, it would have been a different paper.

G. Marcelo (✉)
CECH, Universidade de Coimbra, Coimbra, Portugal

Católica Porto Business School, Porto, Portugal

G. Schweiger (ed.), *Migration, Recognition and Critical Theory*, Studies in Global Justice 21, https://doi.org/10.1007/978-3-030-72732-1_4

understand their motives and what they can actually bring to enrich host societies. In order to flesh this out, some intersubjective settings are especially well suited such as narrative research with migrants themselves. The chapter thus spells out the tasks for this theory, from the hermeneutical understanding of recognition claims and their normative assessment to the political and institutional implications for host societies of seriously tackling migration from a recognition-theoretical perspective.

Keywords Recognition · Migration · Identity · Transnational justice

4.1 Introduction

The topics of recognition and migration have received a fair share of attention both in philosophy and the social sciences in recent years. However, and judging from most literature in those fields, they would seem, at first glance, rather disconnected. Recognition reemerged in the late 1980s and early 1990s mostly from a Left-Hegelian standpoint. It spawned multiple debates, with an emphasis on two different, yet intersecting, perspectives: (1) a sweeping philosophical anthropology aimed at grasping the evolution of societies through the struggles of social movements against moral injuries and toward "moral progress" in the critical theory of Axel Honneth (1995) or (2) in the debates connected with the "communitarian" critique to liberalism and associated with how to develop multicultural societies in which each culture could have a valid claim to see its value recognized, and therefore be granted protection, in the works of Charles Taylor (1994). In the wake of these debates, recognition also gradually emerged as a key aspect to take into account in the whole debate on the theories of justice – that has of course taken a huge role in Anglo-Saxon political philosophy ever since the 1970s – not only with Honneth (2014) but also in the works of Nancy Fraser (2010), Rainer Forst (2001), and many other critical theorists, to the point of certainly becoming one of the most central aspects of third-generation critical theory, and a focal point also for fourth-generation critical theorists, not to mention how it eventually extended to become a paradigm for analysis in the social sciences.

Migration, on the other hand, has for a long time been devoted theoretical attention in the context of global or cosmopolitan theories of justice, for instance, in the works of Pogge (1992), and it should rightly hold a privileged place in the theories of justice for which the primary context of justice (as Forst puts it) is the whole world (see Beitz 1999) rather than domestic polities. And in this context, many have made the case for open borders from a human rights-based perspective (see Carens 2013) or in the context of a critique of the domination processes taking place within state borders in polities de facto ruled by self-serving elites (Kukathas 2012). These debates have of course been a locus of tensions between "statists" and "globalists" to determine whether or not domestic polities have a morally justified right to exclude access to political membership to those coming from outside and also to therefore exclude them from access to redistributive mechanisms and decision-making processes, in the name of a priority given to those who are already members of that political community. And while this debate has been going on also for some decades, the truth is that, at least in the European context, it became much more poignant from 2015 onward, given the increase in the flow of forced migrants coming to European shores from the Middle East after the Syrian crisis. This is a debate that certainly became much more complicated and gained more traction with the unfortunate scapegoating of migrants promoted by right-wing exclusionary populists in Europe (and elsewhere), thus leading to a complicated entanglement between the so-called refugee crisis – which is of course only a crisis due to the political backlash that followed it, given the numbers of refugees coming to Europe are not that significant – and the crisis of liberal democracy itself. Consequently, migration became one of the most widely discussed issues in the public sphere and also one of the most significant political challenges of the past decade.

Given this backdrop, I was drawn to the challenge put forward by Gottfried Schweiger, which can, in a first approach, be stated in very general terms as follows: can recognition theory at least partially illuminate what is at stake in migratory phenomena? In order to better grasp at least some of the implications of this general inquiry, it might help to break it down in a set of more concrete and limited questions, such as (1) does a recognition-theoretical framework allow us to give an account of the experiences forced migrants[1] go through and (2) based on the reconstruction of migrants' claims, can this recognition-theoretical perspective unpack some of the duties that host societies owe these migrants, for instance, in what comes down to hospitality and solidarity?. My wager concerning the first question is that indeed recognition theory can give an account of the significance of migratory experiences and that this, in turn, might allow us to better unpack the duties involved in question 2. Answering the first question involves the adoption a hermeneutical approach to social reality, while the second involves a normative analysis,

[1]Here I am willingly using the terminology "forced migrants" even though this category might involve further scrutiny, insofar as it might ultimately be difficult to utterly determine who was actually forced to migrate. I am using it because I believe that the distinction that recently came into fashion between refugees/asylum seekers and "economic migrants," and which is often used to deny entry to the latter, is itself very problematic. The distinction is based on the attribution of potentially valid reasons for the flight of the former, such as fear of persecution due to war or membership of a particular social group, whereas the latter are said to only seek material improvement of their livelihood. But by referring to forced migrants, I want to emphasize that the element of coercion is more widespread than usually acknowledged. Following a conflict, those who flee famine are also wanting to improve their material conditions, but they flee for sheer need. Further than that, many of those who suffer from "multiple domination" (Forst 2001) in the sense that I will allude to below, in peripheral countries, are often faced with dire circumstances that they feel forced to escape, for their own or their families' sake, and this is completely different from, for instance, highly specialized workers from affluent countries who benefit from globalization who migrate to seek better salaries, and in which the element of coercion is completely absent. Forced migrants are thus many more than those host countries, with the EU being a case in point, are willing to consider refugees.

and I will be claiming that the recognition theory at stake here must be both hermeneutical and normative, which brings us to question 3: should such a theory be successful, what kind of recognition theory would this be? For instance, could it spell out recognition claims only as they would apply within the closed borders of a domestic polity under the strict application of the duties of solidarity toward fellow citizens? And here my claim is that such a framework would be insufficient. Indeed, in order for recognition theory to really be able to tackle migration in a sufficient manner, this would require devising some sort of a "transnational theory of recognition,"[2] one that would be able to develop, to use Honneth's words, a "new grammar" for an "old problem."[3] This is, of course, a grammar of recognition involving not only the reconstruction of the "moral injuries" affecting migrants – and doing this by interpreting their experiences, which has to be done hermeneutically and dialogically – but also spelling out the normative claims inbuilt in these problems and how they should help us reshape institutions and policies in order to tackle the "old problem" of migration.

As such, and in order to address this set of questions and challenges, this chapter intends to spell out some of the features of this, so to speak, "cross-border theory of recognition." A transnational recognition will of course share many traits with so-called transnational theories of justice such as those put forward by Nancy Fraser (2010) and Rainer Forst (2001), and I will use them as a starting point. However, given that recognition theory can – and indeed it does, for instance, in Axel Honneth's works – be more encompassing than a "theory of justice" properly speaking, insofar as the "moral grammar" is rooted in a philosophical anthropology attentive to thick aspects of agents' identities, a transnational recognition will have as a starting point a level that is prior to normative theoretical discussions on justice.

Therefore, in the first section of the chapter, I start by mentioning the insufficiency of currently dominant theories of justice (both global/cosmopolitan and statist) to sufficiently grasp what is at stake in migratory phenomena. My starting point will be the way "transnational" theories of justice stemming from the critical theory tradition, such as Forst's and Fraser's, respond to the inadequacies of the Westphalian framework, given that migrations are a paradigmatic example of such an inadequacy. Then, having taken stock of the ways in which this move of "transnationalizing" justice is needed, I analyze how such a move could be applied to recognition theory in order to tackle the problem of migration. Thus, in the second section, I

[2] Even though I am pushing for this agenda in my own terms, I am of course not the first to use recognition theory to tackle justice problems that are thought of in global or transnational terms. I am here borrowing some inspiration from theories of transnational justice but others before me have, for instance, pushed for "globalizing Honneth" (see, for instance, Heins 2008, Schweiger 2012), or tried to prove the significance of recognition theory for global politics (Hayden and Schick 2016). As for using the theory of recognition to think the integration of immigrants, see Goksel 2017.

[3] I am borrowing the notion of the "moral grammar of social conflicts" from Honneth's *Struggle for Recognition* (Honneth 1995) and applying it to the "old problem of migration" much like Honneth aims to revive the "old idea" of reification by taking a "new look" at it (Honneth 2008). We can thus see the Honnethian inspiration both in the title of this short essay and in some of its key claims.

argue that this theory will certainly be a critical theory, but also a hermeneutical theory, and start unpacking the ways in which the hermeneutical method can help to spell out the several forms of misrecognition that forced migrants suffer from, as well as bring to light their claims. Finally, and before my brief conclusion, the third section of this chapter tackles the topic of how this grammar of recognition can impact policies and institutions of hopefully hospitable and multicultural societies, against the backdrop of the populist backlash.

4.2 Migration and the Inadequacy of the Westphalian Framework: Framing the Transnational Subject of Justice

Migrants are, by definition, people whose lives are transported to contexts in which they are not primarily subjects of rights. For a long time, it was taken for granted that domestic polities were the primary context of justice and that, as such, the ties of membership to a particular state made it not only de facto possible but also morally justifiable to prioritize fellow citizens in redistributive mechanisms as well as to refuse membership status to foreigners or even to deny them entry. This, of course, never applied equally to everyone: those denied entry were more often than not only poor migrants, even though they might be claiming refugee status and fleeing their home country for humanitarian reasons; but "visa gold" and other privileged schemes to allow access to those who have capital always proved that it was usually the downtrodden that were excluded. Be that as it may, the right of political communities to choose whether or not to welcome migrants and to veto those that they did not want to be allowed access, as well as the level of political membership attained – and, ultimately, to send back all those who were refused entry – was taken for granted in what Fraser (2010, p. 12) calls the "Keynesian-Westphalian framework."

To be sure, this framework applied even when thinking about international justice. As Forst makes clear, international justice "takes political communities organized into states to be the main agents of justice" (Forst 2001, p. 160). Therefore, from this standpoint, which takes for granted the principle of sovereignty behind closed borders, it suffices "to regulate the relations between states in a fair way" (ibid.). Global (or cosmopolitan) justice, on the other hand, aims to "regulate the relations between all human beings in the world and ensure their individual well-being" (ibid.) and thus takes the whole world as the primary context of justice, and individual members as its main focus. Therefore, from a globalist point of view, the priority should be to reach a global context of justice that is well-ordered, whereas for statists, this priority is reversed: domestic justice being a "thick" (communal) context of justice, the "thin" context of global justice must be sorted out in the interplay between sovereign states fully legitimated in their decisions within their borders.

Now, in what comes down to migrants' rights, one cannot deny the good intentions of the several formulations of global or cosmopolitan justice. But these are met with several problems. From a nonideal standpoint, they of course have to face the difficulties of *Realpolitik*: in what comes down to border control and membership rights, sovereign states still call the shots. It is clear that international law still lacks the means to effectively enforce migrants' rights, and in what concerns forced migrants even the more compelling human rights-based cases for open borders run against the simple lack of political will of many domestic or even supranational polities (such as the EU) to welcome forced migrants. But even at the strictly theoretical level, they seem to fall short. Indeed, with their mere emphasis on individual rights (more often than not from a liberal standpoint) and formulation at a purely normative level often with a deep deficit of critique, it seems to me that these theories are unable to give a sufficient account of migration phenomena.

So, on the one hand, we find a hardheaded insistence on the Westphalian framework, which is simply incapable – or perhaps sometimes unwilling – to tackle problems that are taking place cross-borders, at a transnational level. In my view, they fall short both in ideal and practical terms, given they adopt a context and frame of justice that are not appropriate to solve these issues. And on the other hand, we find theories that make a move in the right direction and try to picture these problems on a larger scale but that simply do not have the means to fully grasp what is at stake in these problems, mainly because, methodologically, they are not equipped to delve in the thick contexts of migration experiences – something that necessitates a hermeneutic approach, as I will try to show below. Against this backdrop, a third option recently appeared, that of critical theories of transnational justice and whose focus partially overlaps with theories of global justice but with a very different emphasis, stemming from the critical standpoint. But before we turn to the traits of this different strand of theories of justice, it should be noted exactly what is at stake here: determining who is the "subject of rights." To be sure, (usually global/cosmopolitan) human rights-based theories do try to solve this question too, emphasizing that human beings qua human beings are the primary subjects of rights. But I find that the question is more fully grasped when one takes the critical standpoint and delves on existing insufficiencies of the currently dominating framework, and in this Nancy Fraser's contribution proves to be useful.

Nancy Fraser notes that within the dominant Westphalian framework, discussions on justice were for long dominated by the priority of the "what" of justice over the "who": it felt self-evident that the individual subjects to whom justice should be served were the citizens of a given state (Fraser 2010, p. 13). However, it is obvious that decisions taken within a given state often affect the lives of those who live outside it. In the case of "illegal immigration," for instance, when asylum seekers are denied refugee status, what happens is that there is a deep gap instituted between citizens who have a status as full rights bearers and "denizens" who dwell in spaces in which their lives seem to not really matter. Fraser argues that this raises a second-order, meta-level question of justice, which is how to settle the right frame of justice: e.g., to whom does justice apply. And this leads her to add a third layer to her formerly bidimensional theory of justice.

Nancy Fraser had long argued that justice is bidimensional: it needs to address concerns of economic redistribution and what she calls "cultural recognition." But insofar as there are significant obstacles to participatory parity to those who do not have access to political representation precisely because, like migrants, they are not full standing members of a given political community, she later came to argue that justice must become three-dimensional, adding the political dimension of representation to the two other dimensions (ibid., p. 15). We can see that the criterion here is that of participation, and of pinpointing the obstacles that prevent some people from fully participating as peers. These obstacles might come in the form of distributive injustice and status inequality/misrecognition (ibid., p. 16) but also, significantly, in the form of political misrepresentation/misframing. By this, Fraser basically means two different things: misrepresentation "occurs when political boundaries and/or decision rules function wrongly to deny some people the possibility of participating on a par with others in social interaction – including, but not only, in political arenas" (ibid., p. 18) It is a first-level injustice, de facto denying some people proper access to justice. But misframing runs deeper: it "arises when the community's boundaries are drawn in such a way as to wrongly exclude some people from the chance to participate at all in its authorized contests over justice" (ibid., p. 19), and here we should stress the impossibility of participating *at all* because this means that this is a constitutive exclusion. Fraser characterizes this as a meta-injustice because when people are affected by it, they are "denied the chance to press first-order justice claims in a given political community" (ibid). And even more serious, Fraser adds, is when people are excluded from membership in *any* political community because when this happens, people lose, in Arendt's words, the "right to have rights" and Fraser equates this with some sort of "political death" (ibid. pp. 19–20).

Now, this exclusion from political participation is of course a major source of injustice because it denies people who suffer from various forms of injustice access to ways of challenging these injustices. And this is, in my view, a definitive blow to the attempt to think justice only within the Westphalian frame. As Fraser puts, it "is a powerful instrument of injustice, which gerrymanders political space at the expense of the poor and despised" (ibid., p. 21). To be sure, ideally, justice could be served if, within the Westphalian frame, everyone would have citizen status and states would provide citizens with a very extensive set of rights which would be close to an ideal of participatory parity (among citizens). However, this is nowhere the case, and forced migrations provide plenty of examples of the exclusion addressed here. But the second issue that Fraser raises is equally important, because access to frame-setting is of course unequally distributed and this impacts the "who" of justice. In fact, not only domestic polities but also transnational elites have the capacity to influence or even determine the frame (ibid., p. 26) within which claims are disputed. But those who suffer from misrepresentation or misframing do not have the power to do it. So, in a nutshell, excluded from parity of participation, they are denied a voice. And they are also excluded from deliberations and decisions concerning not only their claims but also their inclusion as valid subjects of right, i.e., from being people to whom justice applies. Excluded from being the "who" of justice, they are also, eo ipso, barred from having a say in the "what" of justice (which includes decisions on solidarity mechanisms). Consequently, as Fraser argues, "no redistribution or recognition without representation" (ibid., p. 27). One could of course argue that issues of redistribution, recognition, and even representation are more easily solved at a state level rather than at a (as of yet) utopian cosmopolitan governance. However, it is the de facto failure in the attribution of due representation to those who fall outside of the border, and the unequal power to determine the frame itself that makes all this architecture of justice to ultimately fail when thought of merely at the state level.

This pushes Fraser to reflect on the development of transnational mechanisms to remove these obstacles to participatory parity, including "transnationalizing the public sphere" and applying the "all-affected principle," and I will come back to these proposals in the third section. But before I analyze the political and institutional implications of this transnationalizing move, it is important to mention a few more details of what this means for a critical theory of justice. And here the perspective of Rainer Forst adds meaningful methodological and substantive issues to the debate.

Forst clarifies that a critical theory of transnational justice must take as a starting point the "fact of multiple domination" (Forst 2001, p. 167). This means that rather than "ideal" theories of justice which start with a picture of accomplished justice and engage in a purely normative debate, a critical theory of transnational justice starts by looking at different instances of the absence of justice. As Clive Barnett (2017) has recently argued, this priority of injustice is a striking feature of approaches to justice and democracy coming from the critical theory tradition, and it is a good vantage point to tackle many of the blind spots of more traditional theories of justice, insofar as it is neither purely normative nor purely empirical, and instead, resorting to both normative and empirical elements with an emphasis on the critique of the already existing reality, it aims at changing that reality with a progressive view. Forst recognizes that the world as a whole can be considered a context of justice but that to see it as being marked by unqualified relations of "cooperation" or "interdependence" would be rather naïve, given that a closer look reveals a context of relations of force and domination (Forst 2001, p. 166). This is seen in poorer regions and countries that are "forced into a subordinate economic and political position" (ibid.) but also in the people affected by these power structures: "Shifting perspective to that of the dominated, then, reveals that theirs is a situation of *multiple domination*: most often they are dominated by their own (hardly legitimate) governments, elites, or warlords, which in turn are both working together and are (at least partly) dominated by global actors" (ibid.).

This leads Forst to conclude that a conception of justice must address these situations of multiple domination at various levels (local, national, international, and global) because they are connected through the injustices they bring about. What is thus called for is a comprehensive analysis of the multiple and widespread phenomena of injustice and of their roots in the unequal distribution of power. Forst then develops this theory into a theory of justification – every social relation must be able to withstand critique and be justifiable; otherwise, they are unjust, and critique must be addressed to "those institutions, rules or practices which either pretend to be justified without being so or appear to be beyond justification in terms of being either natural or unchangeable. In both respects, ideology critique is necessary" (ibid., p. 168). Forst further argues that this type of theory of justice must include an analysis of the historical genesis and current state of social relations (including the inequalities and power asymmetries they contain), as well as a critique of false justifications and a theorizing of the practices of justification themselves. One important aspect of Forst's take, when used to tackle migration issues, is that the critique of multiple domination goes hand in hand with a practice of justification that should also apply to migrants. Justification (as argued by Forst) and representation (as claimed by Fraser) are thus two central elements to do justice to people who are in this situation.

Furthermore, it seems to me that this transnationalizing move as applied to critical theories of justice (and democracy) is a step in the right direction in that it both criticizes the insufficiency and injustices brought about by the Westphalian framework and avoids the pitfalls of naïve global theories of justice. As can be seen, these transnational theories of justice tend to encompass recognition claims as some sort of regional part of their broader pictures of justice – something akin to a "sphere of justice" in Walzer's sense. I do not wish to dispute this assessment, as I do not believe that all justice claims or all forms of injustice can be reduced to matters of recognition or the lack thereof. However, I do want to argue that in many ways, and from its own perspective, recognition also encompasses justice, insofar as suffering from injustice is only one of the forms in which lack of recognition can express itself. And I want to further argue that in order to sort out what is at stake in migratory experiences and the plea of forced migrants who are in fact victims of what Forst calls multiple domination, or who suffer from what Fraser names misrepresentation and misframing, a transnational critical theory of *recognition* is called for. This implies applying to recognition theory the same transnationalizing move adopted in the theories of transnational justice I am alluding to in this section and, mutatis mutandis, to similarly start from the negative experiences of those affected. In this chapter, I am of course only sketching some traits of what such a theory could look like.

In the next section, I start spelling out how this grammar of transnational recognition could be grasped, especially when assessing the experience of migration and the claims of forced migrants. I argue that this grammar needs to be unpacked hermeneutically, because a hermeneutical methodology is fundamental to give migrants a voice. Therefore, the following section is mostly theoretical, in that it lays out the fundamentals of a critical theory of transnational recognition as well as the several levels in which it must operate.

4.3 Toward a Critical Hermeneutical Theory of Transnational Recognition

One of the long-standing debates in critical theory concerns the role of the critical theorist: what are we, qua critical theorists, supposed to be doing when we analyze social reality? Are we aiming to change it, as seemed to be the case of first-generation critical theorists? And if so, how? This type of questioning is more complex than what it could seem at first glance because it ultimately involves a whole set of social epistemology assumptions. If, for instance, we assume that most social actors are necessarily caught in the web of distorting ideology and that only some sort of social science (in the strong sense) is able to see beyond that distortion, in virtue of an "epistemological break" of some kind, then the theorist really has a strong role to fulfill: unmask and pave the way for liberation. However, this image does seem a little far-fetched today, because it presupposes some sort of special type of reasoning to which only a very few enlightened theorists would have access to. Contrary to this elitist view of social theory, a recent strand in critical theory has tended to frame it as much more open to democratic dialogue, rather than being enclosed in "monological theory." But this also has, of course, an assumption: that of a much more universal sharing of the use of reason and of the capacity to grasp meaning and to provide reasons (justifications) for actions.

Coming back to the debate on the "who" of justice: assuming that there would be fully functioning institutionalized ways to translate justice claims by social actors on democratic decision-making, we could also assume, at least from an ideal standpoint, that all those who belonged to a given polity and thus had a voice could express their claims and, were them to be fair, to see them adjudicated by institutions. However, as we have seen in the last section, in the Westphalian framing of justice, many are excluded from this state of affairs, and forced migrants are a case in point. Left without a voice, without representation, they have no say – or at any rate very little say – in matters that affect their daily lives. When they are subject to multiple domination in their home countries and decide to flee them, then they are often met with a denial of hospitality in the countries of destiny. Left in a space of indetermination, they might be able to identify their ailments – or maybe sometimes not even that, given that they are often exposed to trauma during their trips and might suffer from "hidden injuries" that are often hard to express – but even if they do, there are significant obstacles to their "parity of participation." Indeed, they are frequently scapegoated by the populist right and suffer from a deep estrangement, insofar as they are uprooted from their culture and coming to a new society, different from their own, and oftentimes lacking in access to solidarity mechanisms and democratic participation, when not having to face persecution and oppression. Faced with this problem, what is then a critical theory of recognition do?

As I argued above, a new grammar of recognition unfolds here. Fleeing multiple domination, making a perilous journey at the risk of their own lives – a case in point being the refugees going to Europe by traversing the Mediterranean, as so many did in recent years – and then being confronted with a lack of recognition on many different levels are "moral injuries," to borrow Honneth's framework, that need to be spelled out. And in this, some sort of "normative reconstruction" will at some point be needed. But I want to argue that this process of reconstruction needs to run deeper and that it needs to be hermeneutic through and through. Allow me to explicate the tasks of this theory in a little more detail.

Like the critical theories of justice and democracy I alluded to above, I would like to argue that the starting point of a theory of transnational recognition has to be the cases of denial of recognition, and mostly disrespect, facing social actors – and in this case, forced migrants. If justice is better grasped through identifying and reflecting upon cases of injustice, the same goes, mutatis mutandis, for recognition. In what manners do migrants suffer from misrecognition? These forms are varied and widespread, and this critical theorist here can only report so many (as we've seen, there's only so much "theory" can do). But some examples can be provided.

First, they might see their story be denied credibility, as when migrants seeking asylum/refugee status see their plight downplayed or get tagged as "economic migrants" (as if escaping extreme poverty was not a strong enough reason to move) to deny them permission to stay in host countries. Second, they might see their status as valid partners in interaction denied by being encapsulated in false and denigratory narratives by right-wing exclusionary populists. Everyone knows the anecdote of the "Schrödinger immigrant" who is too lazy to work while at the same time stealing someone's job. But the truth is that these stereotypes – no matter how contradictory or delusional they might be – are demeaning and harmful. Forced migrants are variously pictured as potential terrorists, free riders that come to take advantage of the welfare state, or barbarians pictured as so exotic and uncivilized that they are below the threshold of humanity and thus deserve rejection in the name of allegedly superior values and customs. Third, and in virtue of the Westphalian framework that was discussed in Sect. 1, they are often also excluded from counting as a valid subject of rights and thus formally barred from solidarity mechanisms or to participate in the decision-making processes[4] that directly affect their lives – and,

[4] There are of course arguments for leaving decisions to those who are already part of the membership of a given bounded polity, including arguments that justify limits to immigration as being an integral part of the rights of the citizens of a given state. The most notable defender of such a political philosophy of limited immigration is perhaps David Miller (see Miller 2005 and 2016). Such an attempt at justification often brings in communitarian or cultural arguments: e.g., a society might have an interest in protecting a given language or their public culture so as to ensure "cultural continuity over time" (Miller 2005, p. 200) or point to undesired political outcomes of mass migratory movements, such as a political backlash, or the distortion of that cultural continuity. To be sure, I am not making here a claim for a full open border policy, nor I am denying existing political communities the right to take a part in the decision-making processes concerning who is or is not allowed to stay in these communities or benefit from the web of social institutions and benefits that exist there. Ideally, the freedom of movement principle should lead to a world in which current restraints to migration should not exist – but from a nonideal standpoint, we should of course grapple with the situation we have right now, which includes striking a balance between the interests of current political memberships within existing states and the interests and needs of migrants. What I am arguing here though is that due to the moral irrelevance of someone being born in a given country and not in another one, and the moral demand to take into account the rights and

as such, even if they were to make justice claims, chances are that no one would hear them; and even if they were to be heard, many existing institutions would still leave them out.

These are just cursory examples, but they start to give us a picture of the varied ways in which forced migrants can suffer from misrecognition.[5] It becomes apparent that they might suffer from two types of misframing: political misframing – which is also a form of misrecognition – in the sense defined by Nancy Fraser and also what we could call narrative misframing. This second case is more elusive, but with direct political effects. Indeed, if forced migrants, as denizens, often seem themselves denied rights of political participation, and if, facing linguistic and other barriers, they are de facto voiceless, then they have no control over their own narrative. That is, the way in which the potential host society will look at them will not really depend on them, but on the political narratives that will be told by third parties about them either by political actors or by the media or other civil society actors such as NGOs, care workers, and so forth, and that then will likely spread through social media and other outlets. And here the danger is, especially – but not exclusively – in "illiberal" societies, that the denigratory and unfair narratives put across a distorted story/picture of these migrants and therefore contribute to stir rage and rejection against them. Consequently, narrative misframing will serve to allegedly justify political misframing and to reinforce the exclusionary mechanisms to the point of naturalizing them: after all, the narrative goes, who would want to welcome such people?

And it is against this backdrop that a critical hermeneutical theory of transnational recognition can be of use. I argued above that this theory needs to be hermeneutic through and through. But why should this be so? In order to understand it, I propose that we recall what is hermeneutics as a method for social science. In 1987, Michael Walzer (1987) argued that there are three paths for social criticism:

---

interests of forced migrants, as well as the fact that currently the interests of migrants are seldom – or not at all, depending on the polity we are referring to – taken into account, taking membership as the sole criterion in a decision-making process involving migration (thus including citizens and excluding denizens) is wrong. And this is why, in my view, Fraser is right in claiming that we need to overcome the Westphalian paradigm. I also take issue with Miller's distinction between refugees and economic migrants (Miller 2016), which is the basis for his distinction between those migrants (refugees) who should be let in, for humanitarian reasons, and those whom polities can rightfully reject or let him in on the basis of their (morally justified) criteria. But in the postscript of *Strangers in Our Midst* (Miller 2016, pp. 166–173), Miller himself recognizes that in the wake of the 2015 refugee crisis, such a distinction seems to collapse, insofar as the overwhelming majority of them seemed to count as "survival migrants" (a notion he draws from Betts), i.e., persons who flee their countries due to existential threats which cannot, in a foreseeable near future, have a solution within that same home country. It seems to me that this a strong enough reason to blur this neat distinction between refugees and "economic migrants" and make a plea for the application of the all-affected principle in the case of forced migrants.

[5] These examples also fall into the category of injustices. But as I show later in this section, framing them as specific cases of misrecognition illuminates the reasons why their interests and needs should be taken into account – therefore, removing the obstacles to due recognition can also be the correction of existing injustices.

discovery, invention, and interpretation. And while discovery is not very credible as a method – given that it would presuppose a second-order social world, e.g., a metaphysical reality from which one would derive, that is, discover, the norms and values that should guide us in our own world – invention actually became the dominant method in the wake of Rawls. Concerning the method of invention, enough has been already said in this chapter about the shortcomings of ideal theory when dealing with the problem of injustice and migration. Interpretation, on the other hand, seems to be the most appropriate method for social theory – and also the one having more affinities with critical theory, especially from Honneth onward – given that it takes real societies as they are, in their historicity, contingency, and complexity, and draws the values and norms from already existing practices, institutions, and cultural artifacts, aiming to reinterpret and assess them anew. And here one might note the similarities with the notion of immanent critique.

Now, it goes without saying that there are many varieties of "hermeneutics," and here I am of course not alluding to Heideggerian hermeneutics of facticity, or to textual hermeneutics properly speaking. Rather, I am drawing on "critical hermeneutics" – an expression coined by Ricœur (1981) and used by John B. Thompson (1981) and others – as a method applied to social reality and that aims both to take stock of that same reality in an interpretive fashion and to be "liberating," i.e., transformative. Some authors, like Vattimo and Zabala (2011), make of this progressive, emancipatory tendency a constitutive feature of hermeneutics, as a "thought of the weak" aiming to change social reality. Furthermore, even within hermeneutics, different interpretive methods are to be found, and that range from historical, literary, or cultural inquiries to psychological/psychoanalytical investigations or phenomenological descriptions. Hermeneutics has a broad toolbox to analyze social reality (Marcelo 2012), and we can put it to use to analyze different phenomena, whether this is populism (Marcelo 2019) or, in this case, migratory phenomena.

So this theory of transnational recognition is hermeneutic in nature. But how can this help solve the problem of political and narrative misframing alluded to above? It would do so, or at least I am arguing here that it would, because recognizing migrants as specific human beings worthy of rights and esteem requires understanding them at a non-superficial level and, if possible, in their own terms. But when they lack the hermeneutic resources to project their self-understanding narratively,[6] the mediation brought about by critical hermeneutics can kick in. It seems to me that duly recognizing migrants involves a thick reconstruction of their identities, communal ties, and cultural backgrounds.[7] And this strikes me as a very important task,

[6] Even though I am not analyzing this aspect in detail here – it must be done elsewhere – this aspect is of course close to what Miranda Fricker (2007) calls "hermeneutical injustice."

[7] This is a recognition of their life story as a life story that deserves credibility (such as to avoid what Fricker calls "testimonial injustice") and it is ultimately tied to several other levels of recognition. For instance, it is related to the recognition of esteem: insofar as every one of us has a life story that is unique, it is the recognition of that uniqueness (and potential specific value of that life) that is also at stake. To be sure, some levels of recognition do not need such a thick grasp in order for the moral demand for recognition to be justified: for instance, in the sphere of respect, recogni-

mainly because it allows us to understand the motives for their migration – often the "multiple domination" by warlords, corrupt government or elites, systematic exploitation by broken economic structures, etc. And only this thick understanding can bring about a fairer and clearer framing of these people in their singularity and their worth.

Now, to be clear, this can be done in a number of ways.[8] A striking positive example is to be found in Behrouz Boochani's narrative *No Friend but the Mountains* (2018), translated and co-written by Omid Tofighian. Boochani is a Kurdish migrant who fled Iran in the hope of emigrating to Australia and, after almost dying in the sea, saw himself trapped in Manus Island, an illegal detention center in Papua New Guinea where the Australian government kept unwanted migrants. Boochani wrote the book, which eventually received numerous awards and granted him international recognition, by sending text messages to Omid Tofighian, who then translated them to English and edited and discussed the text with Boochani, thus becoming, more than a translator, almost a co-author. The book is poetic in tone, is autobiographical, and mixes the description of what Boochani calls the "Kyriarchal system" of the prison – a term used to convey the domination exerted by the prison itself in the prisoners, almost as if it had an agency which it used to confuse and demean those who inhabited it, by making an explicit use of rules that seemed arbitrary and ever-changing, but in reality only functioned to assert its unfounded authority – with a phenomenological account of Boochani's inner experience as read from his own cultural background and life experience. It conveys the story of his journey and his suffering from a unique viewpoint. And this of course would have been impossible without Tofighian's collaboration; Tofighian who, as a hermeneutician – interpreting Boochani's messages and helping him to assemble the text – and as a critical theorist helping Boochani craft a new language to denounce the oppressive character of the prison, helped to bring this narrative to light in Boochani's own terms. This, of course, helped bring about an almost unique act of recognition because the specific way in which Boochani – with the help of Tofighian – was able to express himself (artistically, philosophically, politically) not only allowed him to, so to speak, own his own story but also gave visibility to the plight of forced migrants captured in Manus Island, exposed the unacceptable acts of the Australian government, and granted him recognition of his artistic and philosophical value. This is, of course, a notable example. But it is also an exception.

tion is (or should be) granted to each in every human being vis-à-vis a shared humanity and a shared attribution of equal rights. But when recognition of specific traits is at stake, it seems to me that this type of recognition through narrative means is involved.

[8] Here I am only hinting at a few of them. One might ask what is the purpose of such a thick reconstruction. I would argue that its main aim is that of making the host society more keenly aware of who migrants really are, what they have been through, and what they can bring to that society – and in this researchers only act as mediators. But it can further be argued that sometimes the techniques that help bring to light these thick aspect's of one's identity can also be of help for making migrants themselves get a better sense of their own identity, as is argued by Lechner and Renault (2018).

Another example can be found in the work of biographical research done by social scientists. In Coimbra, Elsa Lechner and Letícia Renault (2018) have conducted what they call "biographical workshops" with migrants from different origins. Adopting a Ricœurian standpoint, also inspired by social scientists such as Johann Michel (2012) and the work on storytelling and hospitality by Richard Kearney, they wager, on the one hand, that identities have a narrative fabric that can better be captured in an intersubjective, dialogic manner. On the other hand, they are keenly aware of the power dynamics that traverse the social world, and in which migrants are often ascribed a passive role, without a chance, as we've seen, to frame themselves narratively. Moreover, they note that very often the experience of migration produces a disruption in migrants' autobiographies (Lechner and Renault 2018, p. 7): there is a before and an after the migration experience and this can be a source of solitude and sorrow (ibid.) or even trauma. Therefore, migrants can be at pains to actually reconstitute what has really happened to them and to integrate that experience in their own life story. So what Lechner and Renault do, in their role as social scientists, is to provide migrants with the means to develop reflective self-awareness and foster mutual recognition. They provide them with an intersubjective setting in which their personal experiences, their views on migration, and their experiences with the host community, as well as views on the problems affecting migrants (e.g., debating topics such as racism, discrimination, gender, work, and so forth), are put forward, the results being spontaneously shared oral and written pieces of biographical work.

Lechner and Renault report that participants in these workshops come to grasp their experience more fully and in a new light: they deploy new meanings on what has happened and "become aware of the collective dimension of their situation (their rights, vulnerability, possibilities for civic participation" (ibid.)). Moreover, this engagement in multicultural dialogue fosters empathy and mutual hospitality in what Lecher and Renault willingly describe as a process of mutual recognition that also aims at social transformation by taking stock of the larger historical and political structures (ibid., p. 9). Rather than being passively described, migrants become co-producers of knowledge about their own lives, experiences, and the overall social significance that they have. In a nutshell: migrants can become used to not having a voice, and they in fact usually don't have one, for the reasons explained above. But what these social scientists do is to provide hermeneutic mediation for that voice to be given to them: they help migrants become aware that they do have agency, that they too can at least tell their own story, in their own terms. By itself, this of course is not sufficient to level the playing field in terms of the distribution of power. But I think it is a good example of a narrative reframing as a process of mutual recognition.

These two examples show one of the ways in which social scientists or critical theorists can help unpack the grammar of recognition by means of a hermeneutic mediation that puts migrants center stage. And these can of course be of help not only in reconstituting the instantiations of misrecognition but also in overcoming one of them, that of narrative misframing. However, a grammar of transnational recognition should also make a further move because it should aim to unpack in positive terms the claims for recognition that can be put forward by migrants. This grammar should of course be thought of beyond the Westphalian paradigm[9] and aim to correct the situation of misrecognition at several levels and could perhaps even be thought of as happening in different spheres of recognition. In this framework, I would mention the following: (1) recognition of civic rights and of a right to democratic participation; (2) recognition of their life story and experience of migration; (3) recognition of their vulnerability, and of their right to have access to solidarity mechanisms/economic redistribution; and (4) recognition of access to opportunities to be worthy of esteem (in the spheres of work, culture, etc.). It is impossible, in this short piece, to fully explicate what is at stake in all of these spheres. The narrative/symbolic level was already addressed in this section, and in the next one, I will briefly address the political and institutional implications of the framework I am proposing here. But something must also be said concerning the level of an "ethics of recognition."

Here, by "ethics of recognition," I mean more than just the normative assessment of the recognition claims put forward in each sphere, even though this theory of recognition must also encompass that task. Often the emphasis on the social aspect of recognition, especially as seen from the standpoint of a demand of reciprocity, overlooks the voluntary acts that gladly grant recognition rather than just demanding it/struggling for it for oneself. However, as I argued before (Marcelo 2011), some authors like Ricœur or Levinas do develop aspects of an ethics of recognition that can be seen as "inspirational" or "edifying." Without delving too much on the details of that sort of ethics here, I would like to argue that in the context of this theory of recognition, such an ethics would involve fostering the act of welcoming the stranger (for which we can find resources in Levinassian, Derridean, or Ricœurian ethics) or, as Kearney (Marcelo 2017) puts it, of making the impossible wager of going from hostility to hospitality. Kearney himself argues that this can be done by exchanging stories, in what is itself an act of mutual recognition, as hinted at above. And in the case of migrants, this forces the host society to reflect on how open to cultural plurality it wants to be, as I will analyze in the next section.

Finally, one last task falls upon this theory, which is, precisely, to normatively assess the recognition claims stemming from social actors and, in this case, from migrants. They need to have their voices heard and see their everyday pleas connected to justice and recognition principles. But certainly not every claim will be valid, and this brings us back to the procedures of justification argued for by Forst, Boltanski and Thévenot (2006), and others. And theory does have a role to play in this. Because even though "normative reconstruction" is of course "reconstruction" and thus hermeneutically starts from the analysis of social reality, it is also "normative": and to spell out and assess these normative elements, the theoretical background is also needed. And this is a level in which theory and democratic dialogue complement each other. Again, this is not to say that the "theorist" needs to have the

[9] To be clear, from the migrants' viewpoint, acceptance into a given state might be exactly what she/he is looking for. However, the grammar of recognition unfolding here does have to take into account those who do not belong yet; it needs to develop, so to speak, across borders.

last word and decide what is valid, which narratives are true, etc. As I argued, there is no superior viewpoint that the critic could claim to have. Be that as it may, the theoretical background helps in connecting narratives to claims and claims to principles and can also be of help in making a mediation between all these elements to the political and institutional domains that affect the lives of both citizens and denizens. And this would all be rather abstract, should we not be ready to discuss precisely the political and institutional implications of these problems. Let us briefly discuss them.

4.4 Welcoming the Stranger and Recognizing Diversity

From what has been said in the preceding section, we can conclude that the theory I am proposing here would extend the spheres of recognition beyond those of purely intersubjective recognition and institutional concern only for fellow citizens. This recognition would extend beyond borders and consider also those who are not part of the same community or domestic polity.

But beyond this, theory also needs to reflect on the kinds of institutions, policies, and political and social changes needed to do justice to these migrants and denizens. Contrary to other cases of misrecognition, migrants do not necessarily suffer from invisibility (Honneth, 2011) but rather from excess visibility in a distorted way, in that they are scapegoated and misframed in the ways already discussed. But reflection on how to tackle these problems needs to go further.

Nancy Fraser (2010), for instance, argues for what she calls a transformative politics of framing, and also for transnationalizing the public sphere, procedures which are meant to apply the "all-affected principle" – e.g., all those significantly affected by a decision should be able to have a say in it, that is, they should see the obstacles to their participatory parity removed – and this could eventually impact the democratic decision-making of domestic polities. But this would need further spelling out, because it involves hard matters of sovereignty and collectively shared responsibility. Today, as Fraser recognizes, public spheres are indeed already transnationalized. But in political terms, this is true not only for progressive social movements but also for the populist right. Indeed, in what concerns the public sphere, state borders mean very little in the digital age. And if, for instance, indignation with racism and police brutality can rapidly spread to fuel protests across the world pushing for an anti-racism agenda, as happened in 2020 following George Floyd's death – like it had happened before with the Occupy Wall Street and the *Indignados* movement following the 2008 economic crisis – the anti-immigrant sentiments and the spread of fake news and other similar tactics have also been striking features of the so-called nationalist international. This means that the symbolic struggle for hegemony is nowadays not only narrative but also digital. And also that the institutional frame-setting is partially dependent on the narrative frame-setting for its legitimacy, which makes the task of unmasking false and demeaning narratives all the more important.

I argued before (Marcelo 2019) that political proposals (including so-called "populist" ones, either from the right or the left) should be assessed according to the concrete policies they put forward and by the ideals they promote. Therefore, right-wing exclusionary populist proposals could be rejected on the basis of their undemocratic nature, while so-called left-wing progressive proposals could be accepted if they respected the democratic ideals by which we live. And this sort of assessment can also be useful when dealing with narratives about migration, such as was analyzed in the preceding section.[10] This can have important political and institutional implications because, ultimately, political frame-setting draws its legitimacy from the very values that are the heart of a given polity. And so the main question here is how open and welcoming do we – we, who are already members and can therefore take part in the formal and informal deliberations about our common future – want our societies to be? And I believe that this forces us to examine once again the model of multicultural or intercultural societies.

Multiculturalism, as we know, was deemed defunct in the last couple of decades, following the 9/11 terrorist attacks and the scapegoating of Muslim immigrants. But, as I hinted at in the preceding section, the grammar of transnational recognition includes the possibility of seeing worth recognized in different aspects. And at least one of these involves the recognition of value in cultural-specific terms. And this is exactly what Taylor (1994) was aiming at when he described multiculturalism as involving some sort of hermeneutic fusion of horizons: to understand another culture and to at least give it the opportunity to have value recognized involves a process of deep understanding and of grasping the standards of evaluation proper to that very culture. And so I believe that rethinking migration through the lens of a transnational theory of recognition also means reviving the debate on the merits of a multicultural society. Certainly, in nonideal terms, this model will always have its challenges too. But in the terms of a politics of recognition, I think this forces us to devise the details of a non-homogenizing inclusion, and this does bring us closer to the multicultural model. As societies, we have to understand how welcoming different cultures can enrich, rather than obliterate, our own culture.[11]

This hospitality is thus simultaneously political and cultural/symbolic. It involves some degree of "linguistic hospitality," to borrow Ricœur's words which have been further developed by Kearney and also, as mentioned before, a stronger narrative participation of those formerly excluded, and which sometimes can form what Fraser calls "subaltern counterpublics." If these narratives have the upper hand against false and demeaning exclusionary narratives, it is more likely that the

[10] To be clear, this is an assessment taking place in the political public sphere and in which potentially everyone should be able to participate – theorists, citizens, and, of course, those formally excluded from membership, provided they are given a platform to do so.

[11] This is of course my own assessment, and different societies might wish for different things, including, of course, maintaining some level of "cultural homogeneity." I cannot fully develop this argument here, but I do think we can argue that hospitality is in itself a strong ethical trait and that recognition of diversity and respect for it – as well as recognition of the potential value that every civilized culture can develop – is a mark of a decent society and of "moral progress."

political conditions for welcoming and progressive societies are met. And these, in turn, are also likely to push for reforms at the institutional level, for instance, creating stronger social security nets and further extending solidarity mechanisms. Welcoming the stranger and recognizing the value of diversity can thus be the condition of possibility for keeping our societies open and plural.

Within the European Union, this will probably mean, at some point, a stronger transfer of competences from the national to the supranational level, thus reinforcing the shift of framing from a Westphalian to a post-Westphalian political setting, and therefore making it impossible for member states to refuse welcoming refugees. Within the global context of justice, this would probably involve giving teeth to international human rights law, making it possible to enforce regulations that today are mere idealizations. It is not my goal here to fully spell out the political, legal, and institutional implications of these moves. But they should be further discussed in the future.

4.5 Conclusion

Throughout this chapter, I argued for the development of a critical hermeneutical theory of transnational recognition and tried to spell out what this theory looks like, what its tasks would amount to, and also some of its ethical, political, and institutional implications. I also mentioned before that this is not a task for "monological theory" alone but for participatory dialogue in a democratic fashion. Given that this issue directly addresses migrants, this should involve migrants themselves albeit, sometimes, through hermeneutic mediation as I argued above – but not only migrants, because the whole point of intersubjective, dialogical frame-setting is that different people should be involved.

It goes without saying that this theory deals with issues that are of the utmost social importance and that when discussing the tasks of this theory the analytical and normative levels are somewhat entangled. Normatively reconstructing the claims of migrants, siding with hospitality in the political public sphere, and unmasking right-wing populist exclusionary narratives is, at the same time, reviving the emancipatory and transformative potential that has for a long time been a staple of the critical theory tradition. Today, as in the past, nothing less is required from us.

References

Barnett, C. 2017. *The Priority of Injustice: Locating Democracy in Critical Theory*. Athens: The University of Georgia Press.

Beitz, C. 1999. *Political Theory and International Relations*. Princeton: Princeton University Press.

Boltanski, L., and L. Thévenot. 2006. *On Justification: Economies of Worth*. Princeton: Princeton University Press.

Boochani, B. 2018. *No Friend But the Mountains. Writing from Manus Prison*. Sydney: Picador.

Carens, J. 2013. *The Ethics of Immigration*. Oxford: Oxford University Press.
Forst, R. 2001. Towards a Critical Theory of Transnational Justice. *Metaphilosophy* 32 (1/2): 160–179.
Fraser, N. 2010. *Scales of Justice. Reimagining Political Space in a Globalizing World*. New York: Columbia University Press.
Fricker, M. 2007. *Epistemic Injustice. Power and the Ethics of Knowing*. Oxford: Oxford University Press.
Goksel, G. 2017. *Integration of Immigrants and the Theory of Recognition: Just Integration*. London: Palgrave Macmillan.
Hayden, P., and K. Schick, eds. 2016. *Recognition and Global Politics: Critical Encounters Between State and World*. Manchester: Manchester University Press.
Heins, V. 2008. Realizing Honneth: Redistribution, Recognition, and Global Justice. *Journal of Global Ethics* 4 (2): 141–153. https://doi.org/10.1080/17449620802194025.
Honneth, A. 1995. *The Struggle for Recognition. The Moral Grammar of Social Conflicts*. Cambridge: MIT Press.
———. 2008. *Reification. A New Look at an Old Idea*. Oxford: Oxford University Press.
———. 2011. 'Invisibility': On the Epistemology of Recognition. *The Aristotelian Society, Supplementary Volume* 75 (1): 111–126.
———. 2014. *Freedom's Right. The Social Foundations of Democratic Life*. Cambridge: Polity Press.
Kukathas, C. 2012. Why Open Borders. *Ethical Perspectives* 19 (4): 650–675.
Lechner, E., and L. Renault. 2018. Migration Experiences and Narrative Identities: Viewing Alterity from Biographical Research. *Critical Hermeneutics* 2 (2018): 1–25.
Marcelo, G. 2011. Paul Ricœur and the Utopia of Mutual Recognition. *Études Ricœuriennes/ Ricœur Studies* 2 (1): 110–133.
———. 2012. Making Sense of the Social. *Hermeneutics and Social Philosophy, Études Ricœuriennes/Ricœur Studies* 3 (1): 67–85.
———. 2017. Narrative and Recognition in the Flesh. An Interview with Richard Kearney. *Philosophy & Social Criticism* 43 (8): 777–792.
———. 2019. Towards a Critical Hermeneutics of Populism. *Critical Hermeneutics*, Special Issue, pp. 59–84.
Michel, J. 2012. *Sociologie du Soi. Essai d'herméneutique appliquée*. Rennes: Presses Universitaires de Rennes.
Miller, D. 2005. Immigration: The Case for Limits. In *Contemporary Debates in Applied Ethics*, ed. A.I. Cohen and C.H. Wellman, 193–206. Oxford: Blackwell.
———. 2016. *Strangers in Our Midst. The Political Philosophy of Immigration*. Cambridge, MA and London: Harvard University Press.
Pogge, T. 1992. Cosmopolitanism and Sovereignty. *Ethics* 103: 48–75.
Ricœur, P. 1981. In *Hermeneutics and the Human Sciences*, ed. J.B. Thompson. Cambridge: Cambridge University Press.
Schweiger, G. 2012. Globalizing Recognition: Global Justice and the Dialectic of Recognition. *Public Reason* 4 (1–2): 78–91.
Taylor, C. 1994. *Multiculturalism: Examining the Politics of Recognition*, ed. Amy Gutmann. Princeton: Princeton University Press.
Thompson, J.B. 1981. *Critical Hermeneutics. A Study in the Thought of Paul Ricœur and Jürgen Habermas*. Cambridge: Cambridge University Press.
Vattimo, G., and S. Zabala. 2011. *Hermeneutic Communism. From Heidegger to Marx*. New York: Columbia University Press.
Walzer, M. 1987. *Interpretation and Social Criticism*. Cambridge, MA: Harvard University Press.

Chapter 5
Transnational Struggle for Recognition: Axel Honneth on the Embodied Dignity of Stateless Persons

Odin Lysaker

Abstract In the last 15 years, scholars have increasingly applied Axel Honneth's recognition theory to global issues such as justice, poverty, solidarity, peace, cosmopolitanism, and climate change. UNHCR estimates that there are currently 12 million stateless people worldwide. Their citizenship rights and human rights are mis-recognized. As humans, their human rights and human dignity should be protected. However, this requires being a citizen of a state, which, by definition, excludes stateless people. Although representing a fairly large group among today's irregular migrants, within the above Honneth scholarship, statelessness is seldom investigated. And if being explored, the primary focus is rights-based recognition within states. In contrast, therefore, I reconstruct what I perceive as a Honnethian idea of a "transnational struggle for recognition" of stateless people. First, I reframe the concrete universalism of the early Honneth's philosophical anthropology, which includes stateless persons. It is concrete by being grounded in subjects' bodily experiences, and it is universal by understanding recognition as a human condition. Second, building on his philosophical anthropology, I argue that embodied, vulnerable humans are motivated by mis-recognition. Stateless peoples' transnational struggle for recognition is driven, then, by the lack of love, respect, and esteem. In some instances, however, statelessness even generates "refugee patients," where stateless persons may be subjected to bodily experiencing, e.g., extreme traumatization due to the limbo situation often caused by statelessness. By being non-recognized, and not merely mis-recognized, stateless persons may be hindered from struggling for recognition. Finally, I emphasize the relevance of the entire Honnethian framework in the case of statelessness based on the suffering caused by the experience of mis-recognition as well as non-recognition. This approach involves applying all three forms of recognition: respect, love, and esteem. I conclude that this framing invokes what I conceptualize as a "transnational

O. Lysaker (✉)
Department of Religion, Philosophy, and History, University of Agder, Kristiansand, Norway
e-mail: odin.lysaker@uia.no

G. Schweiger (ed.), *Migration, Recognition and Critical Theory*, Studies in Global Justice 21, https://doi.org/10.1007/978-3-030-72732-1_5

recognitive demand" that stateless people be recognized—as humans—in order to live a full life with dignity.

Keywords Axel Honneth · Recognition · Statelessness · Transnational struggle · Embodied dignity

5.1 Introduction[1]

During the last approximately 15 years, scholars have increasingly applied Axel Honneth's recognition theory to a global scale (e.g., Hacke 2005; Fossum 2005; Linklater 2006; Weber 2007; Lysaker 2008; Heins 2008; Seglow 2009; Hayden 2012; Lysaker 2013; Thompson 2013; Lysaker 2020c). Notably, his idea of how the mis-recognition of bodily vulnerable humans can motivate them to struggle for recognition struggle has been a focal point in many of these approaches. Within the framework of, e.g., global ethics and international political theory, the abovementioned Honneth scholars have contributed to the discourse on such issues as justice, poverty, health, and human rights together with solidarity, peace, cosmopolitanism, and climate change.

Although difficult to count and contested figures, the Office of the United Nations High Commissioner for Refugees (UNHCR) estimates that there are currently 12 million stateless persons worldwide (UN 2018). According to the United Nations (UN), a stateless individual is defined as "a person who is not considered as a national by any State under the operation of its law" (UN 1954, 6). Consequently, stateless peoples' citizenship rights, along with human rights and human dignity, are not ensured. Because they are human beings, their dignity and rights should, ideally, be protected; however, in reality, this requires being a citizen of a state, which by definition, excludes stateless persons. In cases such as extreme traumatization from torture during war, statelessness can even produce "refugee patients" (Lysaker 2020a). They can be, then, subjected to non-recognition, in addition to mis-recognition, which might undermine the resilience they need to struggle for recognition (Lysaker 2020a).

Despite the fact that globally representing a reasonably large group among today's refugees and other irregular migrants as well as subjected to non-recognition and the above "statelessness problem" (Arendt 1951, 289), the aforementioned Honneth scholars have seldom investigated statelessness. Moreover, if statelessness is examined, it is primarily done in the context of state-based recognition as rights. Consequently, only one—i.e., rights as a subcategory of respect-based recognition—of the three recognition forms (i.e., love, respect, and esteem) of Honneth's theory is considered.

[1] I am deeply grateful to Patrick Hayden and the editor Gottfried Schweiger for their insightful comments and suggestions on earlier drafts of this chapter.

In contrast, in this chapter, I reconstruct what I perceive as the Honnethian idea of a "transnational struggle for recognition." A recognition struggle is transnational in the sense that the struggle takes place outside states and/or by people without citizenship. My aim is to contribute to the above discourse within global ethics and international political theory by shedding light on today's transnational flows of stateless persons. In doing that, I develop a framework that could ensure full human lives with dignity for stateless people.

In a first step, I reframe what I view as the concrete universalism of Honneth's philosophical anthropology, which I argue includes stateless persons. It is concrete by being grounded in subjects' bodily experiences of mis-recognition and/or non-recognition, and it is universal by understanding the vital need for recognition as ontologically a human condition. In the second step, building on my reconstruction of his anthropological core, I introduce Honneth's multidimensional recognition theory, which includes the three recognition forms of love, respect, and esteem. I further address Honneth's account of how bodily, vulnerable, and dependent humans are motivated by mis-recognition. Here, I hold that stateless persons' transnational struggle for recognition is driven by the lack of love, respect, and/or esteem within his multidimensional framework. In the third, and last, step, in line with such mis-recognition experiences—and in contrast to what I understand as Kelly Staples' (2012) reductionist Honneth reading in the case of statelessness—I underscore the relevance of the entire Honnethian framework. Subsequently, I go beyond merely applying rights in terms of a subcategory of respect-based recognition to include recognition as love and esteem. I also elaborate on the earlier introduced connection between non-recognition and stateless people as refugee patients. I conclude that Honneth's multidimensional framing articulates what I conceptualize as a "transnational recognitive demand" of stateless people—as humans—to live a full life above the moral threshold of embodied dignity.

5.2 The Anthropological Core

5.2.1 Unchanging Preconditions

In his 1988 book *Social Action and Human Nature*, co-authored with Hans Joas and initially published in German as early as in 1980, Honneth puts forward his philosophical anthropology, that is, a philosophical theory about human nature. In doing so, Honneth wishes to achieve "a general concept of the human form of life" (Honneth and Joas 1988, 22). Honneth aims to identify the preconditions of human action and interaction. He approaches his philosophical-anthropological idea of certain unchanging preconditions of the human life-form in the following way:

> [Philosophical] [a]nthropology must not be understood as the theory of constants of human cultures persisting through history, or of an inalienable substance of human nature, but rather as an inquiry into *the unchanging preconditions of human changeableness*. (Honneth and Joas 1988, 7, emphasis added)

Within this context, Honneth explains what he takes to be the intersubjective nature of human action and interaction. He underscores that "natural *invariability* [might] help explain *universal* features of species-specific human historicity and plurality" (Honneth and Joas 1988, 7, italics added). What interests Honneth is not identifying something fixed or limited. Rather, he reconstructs the "*invariant conditions* of human historicity" (Honneth and Joas 1988, 7, emphasis mine). Furthermore, Honneth wants to capture "the *normatively* orientated determination of the natural basis of specifically human sociality" (Honneth and Joas 1988, 26, italics mine). He explores a wide range of features characterizing the unchanging preconditions of human changeableness, such as embodiment, sensuousness, basic needs, and socialization processes along with verbal and nonverbal communication. He also links his idea of humans' unchanging preconditions to social emancipation (Honneth and Joas 1988, 85; Deranty 2009, 22).

Later, in his 1995 book *The Struggle for Recognition: The Moral Grammar of Social Conflicts*, originally published in German in 1992, Honneth seemingly bases his idea of recognition on a similar philosophical anthropology as he did in 1980. More than 10 years after, at the beginning of the 1990s, the original anthropological core of the unchanging preconditions is now articulated as "enabling conditions" of recognition (Honneth 1995, 173):

> The [structural elements of the formal] concept of 'ethical life' is now meant to include the *entirety* of intersubjective *conditions* that can be shown to serve as *necessary preconditions* for individual self-realization [through recognition]. (Honneth 1995, 173, emphasis added)

In connection to this, Honneth (1995, 1, italics added) explains that his "intersubjectivist concept of the person" requires "the possibility of an undistorted relation to oneself." This "proves to be *dependent on [all] three forms* of recognition," namely, love, respect, and esteem.

Taken together, the enabling and necessary preconditions of recognition on this most basic, ontological level, on the one hand, and the dependency and the entirety of the three recognition forms, on the other hand, indicate that Honneth's recognition theory is anthropologically grounded. That is because both these elements resonate with the early Honneth's story about the unchanging preconditions of human changeableness. The first element of necessary preconditions echoes the aspect of unchanging preconditions, whereas the latter element of recognition as love, respect, and esteem corresponds to the aspect of human changeableness. Also, viewed from the fundamental level of recognition as love, humans are dependent on others' recognition. Due to being philosophical-anthropological characterized by our embodied vulnerability, humans depend on others' recognitive protection against mis-recognition. Although I find it important when Honneth here speaks about the entirety of all the three recognition forms, I leave this matter for now and return to it in my discussion of Staples' Honnethian approach to statelessness in the second last part of this chapter.

As in my analysis, Danielle Petherbridge (2013, 8, emphasis mine) underscores that for Honneth, "*all* social philosophical research *requires* a form of *[philosophical-]anthropological* reconstruction." This anthropological dimension refers to "certain constants" or "enduring conditions" (Petherbridge 2013, 8, italics added). Petherbridge (2013, 13, italics mine) further explains that the "*normative* foundation of recognition is *grounded anthropologically.*" Thereby, these "anthropological structures of recognition are intended to provide a *context-transcending* claim to validity that is *universally applicable* regardless of historical or socio-cultural context" (Petherbridge 2011, 14, emphasis added). Parallel to Petherbridge, Christopher Zurn (2000, 115) captures this holistic interpretation of Honneth in terms of a "structural interconnection" between all the three stages of individual identity development, all the three recognition forms needed on each level, and all the three social organization forms required to achieve an undistorted, positive, and healthy self-realization. Zurn argues:

> For Honneth, it is important to realize that *[all] these three forms* of relation-to-self – self-confidence, self-respect and self-esteem – [and their corresponding recognition forms, i.e., love, respect, and esteem] – are *ontogenetically fulfilled in a developmental hierarchy with a directional logic*. (Zurn 2000, 117, italics added; see also Honneth 1995, 141; Lysaker 2015)

Zurn's viewpoint implies that to become fully recognized as humans, individuals need to develop an undistorted relationship to themselves and others within an institutional framing involving all the above recognition forms. Building on that, since this development is contingent upon an ontogenetical hierarchy, I propose, by way of Zurn, that we cannot have the one recognition form fulfilled without having the other two realized as well. If so, this is in line with the above quote in Honneth regarding the interplay between the ontologically necessary preconditions, on the one hand, and the dependent entirety of the recognition forms, on the other.

In my view, Honneth's philosophical anthropology is a key to understanding his thought. As the above quotes suggest, due to his original idea of the unchanging preconditions, this anthropological core runs like a red thread through his writing, at least between 1980 and 1995. The reason why I believe Honneth's ideas make up a philosophical anthropology is that this field explores the most fundamental building blocks of human existence. Thus, to investigate other issues—such as statelessness—we first need to formulate the unchanging preconditions of human changeableness explicitly. If not, methodologically, one either implicitly takes certain philosophical-anthropological conceptions and premises for granted or superficially rejects such anthropological exploration altogether. We, therefore, first need to get a clearer picture of the unchanging preconditions if we are to explore Honneth's perspective on issues such as statelessness, which is my main aim in this part of the chapter.

To investigate the human conditions, I build further on my own development of the concept of "existential preconditions" (e.g., Lysaker 2020a; Lysaker 2013;

Lysaker 2008). I reformulate Honneth's anthropological core through preconditions that are non-choosable, i.e., conditions that are embodied and shared by every human subject. Ontologically, humans' existential preconditions exist prior to and are therefore more foundational than other everyday aspects of the human life-form.

I argue that vulnerability and dependency are the most basic among our existential preconditions (Lysaker 2020a; see also Honneth 1995, 131; Anderson and Honneth 2005, 130). Vulnerability refers to our capacity to be bodily harmed by others, through, for instance, mis-recognition, injury, and dehumanization (Hayden 2016, 102–103). To be born and exist as a vulnerable human being, we are constantly and throughout our entire lifespan, in various ways and to differing degrees, subjected to our vulnerability (e.g., illness, disease, disability, aging, and mortality). Humans are, therefore, existentially preconditioned to depend upon others (in terms of, e.g., love, respect, and esteem) to protect themselves against harm. If humans do not safeguard each other concerning this interplay between our basic vulnerability and dependency, we might be injured. Morally, then, our embodied vulnerability presupposes protection against such harm by being recognized by others. I return to this issue in the last part of my chapter regarding my Honnethian idea of embodied dignity.

5.2.2 Concrete Universalism

In addition to the added value of being an anthropological stepping-stone, in my reading of Honneth, the above-explicated idea of existential preconditions serves as the normative justification for his recognition theory. I believe that Honneth justifies his theory based on what I conceive as a "concrete universalism" (Lysaker 2020b; Lysaker 2013, 130; see also Hayden 2016, 102).

The notion of concrete universalism refers to the combination of what is often assumed to be two separate categories: universalism and particularism. The justification is concrete by being situated within the particular, diverse, and unique everyday contexts, experiences, and practices of, e.g., Honnethian recognition. It is also universal as it relates to human persons by virtue of the ontologically shared human condition of embodied vulnerability and dependency; these preconditions invoke a vital human need for recognition (Ikäheimo 2009) and thus for all humans to be protected against mis-recognition and thereby ensuring their dignity.

Concrete universalism aims to justify normative claims in a way that balances universalism and particularism. Universalism often justifies normative claims independently of time and space, whereas particularism usually justifies normative claims time- and space-dependently. From my viewpoint, concrete universalism to a greater extent considers both universalism and particularism—not relying on the one or the other—of the diverse and complex reality within which our moral evaluations and performance take place, which not necessarily can be categorized as either universal or particular.

In his book *The Struggle for Recognition*, in my interpretation, Honneth introduces such concrete universalism under the heading "formal conception of ethical life":

> If the idea of a 'struggle for recognition' is to be viewed as a critical framework for interpreting the processes by which societies develop, there needs to be, by way of completing the model, a *theoretical justification for the normative point of view* from which these processes can be guided. In order to describe the history of social struggles as moving in a certain direction, one must appeal *hypothetically to a provisional end-state*, from the perspective of which it would be possible to classify and *evaluate* particular events. (Honneth 1995, 171, italics mine)

By establishing a justificatory middle way, Honneth wishes to avoid what he perceives as the pitfalls of both universalism (e.g., Kantian morality and autonomy) and particularism (e.g., communitarian ethical life and substantive values). To escape these two unsatisfying extremes, he nonetheless draws on both:

> The line of argument that we have been following in the reconstruction of the model of recognition, however, points to a position that does *not seem to fit clearly into either of these two alternatives*. Our approach departs from the Kantian tradition in that it is concerned not solely with the moral autonomy of human beings but also with the conditions for their self-realization in general. Hence, morality, understood as the point of view of universal respect, becomes *one of several* protective measures that *serve the general purpose* of enabling a good life. But in contrast to those movements that distance themselves from Kant, this concept of the good should not be conceived as the expression of substantive values that constitute the ethos of a concrete tradition-based community. Rather, it has to do with the *structural elements of ethical life*, which, from the *general* point of view of the communicative enabling of self-realization, can be *normatively extracted* from the plurality of all particular forms of life. (Honneth 1995, 172–173, emphasis added)

Honneth further explains his idea of a golden mean as follows:

> [I]nsofar as we have developed it as a *normative* concept, our recognition-theoretic approach stands *in the middle* between a moral theory going back to Kant, on the one hand, and communitarian ethics, on the other. It shares with the former the interest in the most general norms possible, norms which are understood as conditions for specific possibilities; it shares with the latter, however, the orientation towards human self-realization as an end. (Honneth 1995, 173, italics added)

To balance the two extremes of Kantian morality and communitarian ethics, Honneth then suggests the following justificatory strategy:

> The desired characterizations must, then, be *formal or abstract enough* not to raise the suspicion of representing merely the deposits of concrete interpretations of the good life; on the other hand, they must also have *sufficient substantive content* to be of more help than Kantian references to individual autonomy in discovering the conditions for self-realization. (Honneth 1995, 173, emphasis mine)

This Honnethian middle way resonates with my concept of concrete universalism. Honneth's justification is concrete by being sufficiently substantive through the three recognition forms (i.e., love, respect, and esteem), and it is universal by being formal enough to build on the unchanging preconditions of embodied, vulnerable, and dependent subjects.

This concrete universalism seemingly takes advantage of what I previously explicated as the anthropological core of Honneth's recognition theory. Here, I think of how both the concrete and universal aspects of his normative justification resonate with my conceptualization of existential preconditions. To illustrate, the precondition of vulnerability is concrete as humans bodily experience it in unique and diverse ways in everyday life. At the same time, this precondition is universal by being shared by each human subject.

For this reason, Honneth's formal conception of an ethical life involves both the dependent entirety of all the three recognition forms and the necessary, existential preconditions of recognition as a vital human need. On this note, Honneth (2007, 138, emphasis added) sums up his concrete universalism like this: the "moral point of view has to encompass not just one, *but three* independent modes of recognition." In the next part of my chapter, I introduce these three recognition forms of his theory in more detail. Still, I wish to close this first part of the chapter by underscoring that Honneth's normative justification includes all the recognition forms, that is, love, respect, and esteem. I will return to this matter and its impacts concerning stateless persons later.

5.3 Recognition Struggles

5.3.1 A Dependent Entirety

In Honneth, we find traces of his idea of the struggle for recognition at least as early as in the 1980s. In the 1988 afterword of the second German edition of his book *The Critique of Power: Reflective Stages in a Critical Social Theory*, originally published in German in 1985, Honneth (1991, xviii) refers to "Hegel's idea of a 'struggle for recognition'." Also, in his 1989 Habilitationsschrift titled *Kampf um Anerkennung: Ein Theorieprogramm im Anschluss an Hegel und Mead*, Honneth (1992, 200, footnote 2), he articulates the idea, which he later developed into a mature recognition theory. Interestingly, during the same period of the 1980s and the 1990s, Honneth develops his philosophical anthropology and his recognition theory in parallel, along with their intersections. I will return to this issue later in my chapter.

Within a multidimensional framework, Honneth articulates his recognition theory through three different, but still interconnected, categories of recognition. As Honneth (2002, 506) later sums up, recognition designates "a conceptual species comprising three subspecies." Furthermore, he draws attention to three dialectical levels: first, the conditions of individual self-realization (or, practical relation-to-self and practical identity-formation) (Honneth 1995, 92, 128–129, 144); second, the development of relationships to others (being oneself in another and being at home in the other) (Honneth 1995, 9, 105; Deranty 2009, 291); and third, certain institutions (or, spheres) of social reproduction (Honneth 1995, 93). As part of this complex framing, Honneth outlines three ontogenetical recognition forms, which correspond to three kinds of mis-recognition (or, harm, injury, and injustice) (Honneth 1995, 131, 138; Honneth 2007, 133–137).

Love is the first and most basic category among the three recognition forms. In Honneth, love and care refer to "primary relationships insofar as they (...) are constituted by strong emotional attachments among a small number of people" (Honneth 1995, 95). Recognition as love is gained by a "highly complex process, in which both participants acquire, through practice, the capacity for the shared experience of emotions and perceptions" (Honneth 1995, 97). Therefore, recognitive love "presupposes liking and attraction, which are out of individuals' control" (Honneth 1995, 107). On top of that, Honneth (1995, 99) situates recognition as love within the relationship between children and their parents: "the utter helplessness [i.e., embodied vulnerability and dependency] of the infant, who is unable to articulate his or her physical and emotional needs communicatively." Here, recognition requires that the infant—along with each human person as a vulnerable subject—"trusts the loved person to maintain his or her affection, even when one's own attention is withdrawn" (Honneth 1995, 104). Consequently, through such love-based recognition, humans develop individual self-realization which Honneth (1995, 107; see also 129) defines as a "basic (...) self-confidence." This individual development takes place within close relationships in various social realms, mainly in the private sphere, which include, e.g., parent-child, family, friendship, and love relationships (Honneth 1995, 95). In contrast, on this basic level, Honneth (1995, 143; see also 131) refers to mis-recognition as "the violation of the body." Rape, torture, and parental abuse and neglect are some examples of how our bodily integrity and dignity may be violated.

Respect is the second recognition form. This category covers a wide range of recognition types—in contrast to what I later present in this chapter as Staples' reductionist reading of Honneth regarding statelessness. For example, Honneth (1995, 125) refers to "equal legal protection." One aspect of such protection is being "bearers of rights," namely, "when we know (...) what various normative obligations we must keep vis-à-vis others" (Honneth 1995, 108). To Honneth (1995, 116–117), the overarching notion of rights designates a wide spectrum of classes of rights, such as welfare rights (Honneth 1995, 116–117) along with citizenship or basic civil rights (Honneth 1995, 110, 116). Moreover, respect-based recognition generates duties through which we "can be sure that certain of our claims will be met" (Honneth 1995, 108). Additionally, Honneth relates this second type of recognition to "moral respect" (Honneth 1995, 112, 113) and the development of "moral responsibility" (Honneth 1995, 119). On top of that, recognition as respect involves universal human rights (Honneth 1995, 116, 119, 125, 132, 158) and human dignity (Honneth 1995, 125). In terms of individual self-realization, recognition as respect develops what Honneth (1995, 118) termed as "self-respect," meaning that individuals should become "capable of autonomously making reasonable decisions about moral norms" (Honneth 1995, 109). Concerning the social realms where respect-based recognition takes place, Honneth (1995, 110, 116–117) explains that such recognition partly occurs within various state-based institutions, ranging from the welfare state (i.e., welfare rights) to the constitutional democratic state and the public sphere (i.e., citizenship rights and discursive will-formation). Also, as I return to in the last part of this chapter, recognition as respect involves *transnational* institutions (i.e., human rights and human dignity) (Lysaker 2013; Lysaker 2008). On this note, Honneth (1995, 119, italics added) underscores that "*only* with the establishment of *universal human rights* [and thus their moral grounding in human dignity] [can] self-respect (…) assume the character associated with talk of moral responsibility as the respect-worthy core of a person." In the instance of respect-based recognition, Honneth (1995, 129) designates mis-recognition as, e.g., "exclusion" and the "denial of rights."

Esteem is the final category of recognition in Honneth's theory. This recognition form is defined as "a form of social esteem that allows [individuals] to relate positively to their concrete traits and abilities" (Honneth 1995, 121; see also 129). Also, Honneth (1995, 121) speaks about esteem-based solidarity like this: "To the extent to which every member of a society is in a position to esteem himself or herself, one can speak of a state of societal solidarity." He uses the notion of "achievement" to describe this third dimension of his recognition theory (Honneth 1995, 125). Recognition as esteem is, then, a way to "to view one another in light of values that allow the abilities and traits of the other to appear significant for shared praxis" (Honneth 1995, 129). This presupposes an "essential openness to interpretation of every societal value-horizon" (Honneth 1995, 129) concerning the "degree of pluralization of the socially defined value-horizon" (Honneth 1995, 122). Therefore, when considering societal "value pluralism," Honneth (1995, 125; see also 126) underlines that such value-based esteem is "inevitably accompanied by the opening of societal value-ideas for differing forms of personal self-realization." Furthermore, he argues that societal disagreement produced by such value pluralism can play a positive, democratic role (Honneth 1995, 126f.; see also Lysaker 2017). Regarding self-realization, esteem-based recognition is crucial for individuals developing "self-esteem" and "feeling of self-worth" (Honneth 1995, 129). Such recognition takes place within the social realm which Honneth (1995, 129) identifies as "communities of value." These spheres might be located whenever and wherever individuals' traits, abilities, and achievements are positively valued by others. In contrast, therefore, when esteem-based recognition is withdrawn or withheld, it constructs mis-recognition through "denigration" and "insult" to human "ways of life" (Honneth 1995, 129, 131, 134).

5.3.2 Invariant Love

Now, after having explored all the three recognition forms in Honneth's recognition theory, I link this explanation to the discourse around recognition as love. Here, one of the main philosophical-anthropological issues is the nature of love. In Honneth's writing, one can find at least two different accounts of the concept of love, namely, an invariant (or, ahistorical) and a variant (or, historical) account, respectively. In my view, it is crucial to explicate these two approaches to love. Although Honneth might have undertaken a turn regarding his conceptualization of love, we can demonstrate that he still based his recognition theory on a particular philosophical anthropology. When applying Honneth's recognition theory to concrete cases—e.g., statelessness—we should address the issue of love and its anthropological core, since it is the foundation of his theory.

In his 1992 book *The Struggle for Recognition*, Honneth explicates that both recognition as respect and recognition as esteem are produced by and become visible only under historical processes, change, and progress. In contrast, however, Honneth approaches recognition as love like this:

> [A] mode of basic self-confidence [that] represents *the basic prerequisite for every type* of self-realization in the sense that it allows individuals to attain, for the first time, the inner freedom that enables them to articulate their own needs. Accordingly, the experience of love, *whatever historical form* it takes, represents *the innermost core of all forms of life* that qualify as 'ethical' (...). Because it does *not* admit of the potential for normative development, the integration of love into the intersubjective network of a post-traditional form of ethical life *does not change its fundamental character*. (Honneth 1995, 176, emphasis mine)

Honneth further elucidates his concept of love as follows:

> Since such forms of basic psychological self-confidence carry emotional *preconditions* that follow a largely *invariant* logic (...), this experience (...) *cannot* simply vary with the historical period or the cultural frame of reference. (Honneth 1995, 133, italics added)

Consequently, he defines love as "the *most basic* form of recognition" (Honneth 1995, 162, italics mine) and "the *structural core* of all ethical life" (Honneth 1995, 107, emphasis added), which is why recognition as love is "*prior to every other* form of reciprocal recognition" (Honneth 1995, 107, emphasis mine). Love, therefore, conditions "all later forms of affectional bonds" (Honneth 1995, 97) and "all further attitudes" (Honneth 1995, 107).

Interestingly, Honneth (1995, 108; see also 38, 163, 164; Lysaker 2017, 2–3) links love to politics in the following way: "it is only this symbiotically nourished bond, which emerges through mutually desired demarcation, that produces the degree of basic individual self-confidence indispensable for autonomous participation in public life." Therefore, since all humans are preconditioned by vulnerability and thus to a lesser or greater extent might be harmed by misrecognition throughout their lives, humans depend on the recognition of others as protection against such harm. When acting and interacting democratically—as vulnerable and dependent creatures—both children and adults are, therefore, characterized by Honneth through their vital need for recognition as love.

As Petherbridge (2015, 169, italics added) underscores, the idea of recognition as invariant love "constructs the entire recognition model in *fundamentally anthropological* terms." Jean-Philippe Deranty (2009, 287, emphasis mine) adds to this that the "*first* sphere of recognition [as invariant love] is *first* not only in a *[onto]genetic*, but also in a *logical* sense." The reason is that love-based recognition "relates to the establishment of the *most basic condition*s of subjective agency" (Deranty 2009, 287, italics mine).

In my reading, Honneth here moves beyond the traditional divide between the private and the public found in liberal thought. According to the above-introduced anthropological core along with the concrete universalism of his recognition theory, Honneth frames all the three recognition forms (i.e., love, respect, and esteem) as a mutually dependent entirety along with the necessary enabling preconditions of recognition. This implies that there exists a mutual dependency between the so-called private (i.e., recognition as love) and the public (i.e., recognition as respect and esteem). I return to this issue in the third part of this chapter while discussing Staples' connection between Honnethian recognition and statelessness.

In 2002, Honneth redefined his concept of love as variant. This turn regarding his articulation of recognitive love is explicated as follows: "the differentiation of various kinds of recognition *not* as an ahistorical given *but rather* as the *result of a directional process*" (Honneth 2002, 511, emphasis added). From my viewpoint, Honneth here exchanges his original invariant and ahistorical concept of love with a variant and historical one. Building on that move, all the three recognition forms in Honneth's theory are now accounted as the outcome of a historical process, change, and progress. To recall, in the period from 1980 and until the mid-1990s, Honneth's thought was based on what I termed as an anthropological core. Then, he put forward the idea of certain unchanging preconditions of human changeableness, which seemingly goes hand in hand with his notion of recognition as invariant love. In 2002, however, Honneth seemingly turns around and rejects the very core of his initial anthropology.

In 2008, in the book titled *Reification: A New Look at an Old Idea*—initially published in German in 2005 and therefore came out only a few years after his 2002 rejection of the anthropological core of recognition as invariant love—Honneth surprisingly appears to recover an anthropological core similar to his initial entry (Lysaker 2013, 60ff.). Honneth (2008, 37, my italics; see also Lysaker 2015, 150–152) explains that he now deals with recognition "in its *most elementary* form," which appears to exist on an even more basic ontological level than invariant love:

> We are (…) dealing with a *more elementary* form of recognition than the one that I have dealt with in my previous treatments of the issue. (…) As a result, I now assume that this '*existential' mode* of recognition provides a *foundation for all other*, more substantial forms of recognition [i.e., love, respect, and esteem] in which the affirmation of other persons' specific characteristics is at issue. (Honneth 2008, 90, footnote 70, italics added)

Honneth (2008, 152, emphasis mine) further explains that "[t]he implication for the *structure* of my own theory of recognition is that I must *insert a stage* of recognition *before* the previously discussed forms, one that represents a kind of *transcendental condition.*" Later, Honneth comments upon the above statements as follows:

> [T]he idea of bringing in a *fundamental form* of recognition (…) is simply the result of finding out what the whole idea of reification of others means, and I think it means basically to ignore the *personhood* of others as *human beings*. It is a *very fundamental* kind of negative reaction to others. And then, if you are clear about this, you realize that a *basic form* of recognition is *constitutive for every* kind of societal reproduction and integration, namely that we recognize each other on a *fundamental level as human beings*. So, the introduction of a *more elementary form* of recognition is meant to solve this problem.

> *Before* I can recognise a person in a specific way (...) I must recognise *the human being in this person*. Therefore, with this *fundamental recognition* I mean an awareness of *the humanity of human beings*. I believe that we are dealing with something like a *transcendental necessity*: *before* we, in a society, can even begin to differentiate between different forms of recognition, we must *recognize each other as human beings, as persons*. (Honneth cited in Jakobsen and Lysaker 2010, 165, italics mine; see also Lysaker 2015, 150–152)

So, on this very fundamental level, Honneth introduces a recognition form that is transcendentally constitutive of and necessary for all the other three and thus less fundamental recognition forms, namely love, respect, and esteem. This more elementary recognition form is located before—i.e., ontologically prior to and therefore appearing as the foundation of—the less basic and more substantial categories belonging to his recognition theory. Also, on this most fundamental level, we recognize the personhood of each other. According to Honneth, this fundamental stage of recognition defines the idea of humanity and what it requires to become a human being.

To me, this description of the most basic recognition form echoes my earlier analysis of the anthropological core of the early development of Honneth's recognition theory. Nonetheless, does this solve the problem raised by Honneth? He seemingly runs into a contradiction in terms: On the one hand, he holds that all the three substantial recognition forms (i.e., love, respect, and esteem) are produced by historical and societal processes and are therefore non-transcendental; on the other hand, however, Honneth argues that this recognitive meta-form is transcendental and more fundamental than the former categories. The question, then, is whether his recognition theory—at least in its later version—can simultaneously consist of elements that are basic and nonbasic, formal and substantial, as well as transcendental and non-transcendental.

5.3.3 Mis-recognition and Non-recognition

I believe that it matters which of the accounts of love-based recognition we find in the current version of Honneth's recognition theory when applying it to statelessness. If love is still conceptualized as invariant, it can constitute a concrete universalism that includes stateless people. If the notion of love is defined as variant, it seems to force Honneth to adopt a historical context within which struggles for recognitive love take place. If so, it is unclear whether such a move still includes all humans—such as stateless people—outside what can be accused as representing Honneth's methodological nationalism (Staples 2012; Lysaker 2013). Still, it can be argued that since Honneth currently operates with the idea of the most elementary and existential recognition form—which is identified prior to the substantial recognition forms—we can find an anthropological core as well as a normative foundation that will include even stateless persons.

In line with this issue, we can differentiate between two forms of recognitive negation (Lysaker 2020a, 96–97; Lysaker 2013; see also Hayden 2016): first, mis-recognition, which negates the three substantial recognition forms of love, respect, and esteem, and second, *non*-recognition, which negates the most basic, existential recognition form, recognition as a human person along with humanity and dignity that command. I return to this distinction and its relevance to the matter of statelessness in the next part of the chapter.

Another way to deal with the issue about love-based recognition is to draw on my above-introduced idea of concrete universalism, where the later Honneth could argue that his combination model does not imply a contradiction in terms. This justificatory strategy, he could claim, bridges the gap between the often-assumed division between the universal and the concrete, or, in the present analysis, bridges the gap between the transcendental and non-transcendental. Nevertheless, I argue that such a reply ostensibly calls for the same kind of anthropological core that I introduced above.

5.4 Transnational Recognitive Demand

Today, there are approximately 12 million stateless persons globally (UN 2018). Due to their lack of citizenship and thus lack of state-based membership, they are subjected to powerlessness and exclusion. These people become stateless by being denied access to the Honnethian dimension of respect-based recognition covering rights. I am especially thinking of citizenship rights together with human rights and human dignity.

Stateless persons experience what Hannah Arendt (1951, 289) defines as the "statelessness problem": the discrepancy between the universal ideal, on the one hand, of being a human and thus having the moral, legal, and political status of human dignity and human rights and, on the other, the reality that citizenship is required to protect the millions of stateless persons against mis-recognition. So, although international frameworks like the UN 1954 *Convention Relating to the Status of Stateless Persons* and the UN 1961 *Convention on the Reduction of Statelessness* along with the UN 1948 *Universal Declaration of Human Rights* are supposed to protect every human being—including stateless people—the opposite is closer to the truth.

Therefore, in this part of my chapter—by drawing on my previous exploration of the anthropological core and concrete universalism of Honneth's recognition theory—I discuss the Arendtian statelessness problem. In doing that, I argue that we should adopt the entire multidimensional framework of Honneth's theory, involving all the three recognition forms of love, respect, and esteem.

Although stateless people lack a state-based membership and citizenship rights, within the Honnethian framework, statelessness involves more than only one (i.e., rights) among the several subcategories of the second recognition form of respect. In addition, stateless persons—as human beings—also need to be recognized through love and esteem, which requires stable social environments and resilience.

Stateless people, however, are often subjected to long-lasting and intractable exile, limbo, or displacement (Lysaker 2020a, 90); they are by the UNHCR defined as among the world's "most vulnerable" groups (Lysaker 2020a, 82). On top of that, stateless persons might become "refugee patients" (Lysaker 2020a, 83–84). Then, their dignity and self-worth are negated. Stateless persons as refugee patients are also subjected to different kinds of losses, e.g., the loss of familial or cultural belonging and the loss of their bodily capabilities due to traumatization and refuge, along with the loss of time needed for self-realization. Finally, refugee patients lack resilience while attempting to live through traumas and other types of significant stress during extended time periods. If stateless persons become refugee patients, it further decreases their possibilities of fulfilling the vital human need for recognition as love, respect, and esteem compared to citizens within more stable surroundings.

Here, I think it is worthwhile drawing attention to my above differentiation between mis-recognition and non-recognition. In the case of stateless people becoming refugee patients, I believe it is crucial to examine whether they are experiencing mis-recognition or non-recognition. In the former case, despite their reduced capacities, e.g., regarding health or resilience, they can still be motivated to struggle for recognition. In the latter case, however, the consequences of non-recognition—e.g., through absolute poverty, extreme traumatization, or torture—disable them, at least temporarily, to struggle for recognition.

As I see it, this distinction between mis-recognition and non-recognition resonates with Honneth's two-leveled idea of recognition. To recall, on the less basic and substantial level, he speaks about love, respect, and esteem, whereas on the most foundational and existential level, Honneth refers to humanity and thus, I believe, to humans' inherent dignity. I argue that there is only a gradual instead of an absolute distinction between mis-recognition and non-recognition. Thus, the question of whether people subjected to non-recognition might struggle for recognition or not is an experiential and empirical question instead of a merely conceptual one. The same goes for cases where persons are exposed to mis-recognition over an extended time, which can hinder their recognition struggle.

Due to the principally different degrees of injury at these two levels, I think that the difference between mis-recognition and non-recognition within the framework of Honneth's comprehensive theory is helpful to more concretely identify the various ways in which the vital human need of recognition is at stake in the case of, inter alia, stateless peoples' transnational recognition struggle. I return to this matter in the last part of this chapter in my proposal for a recognitive idea of embodied dignity.

Applying the anthropological core of Honneth's framework, I suggest that the motivation of recognition struggles might occur even on the transnational level of statelessness. I use Honneth's (1995, 138–139; see also Lysaker 2020a) perception of vulnerable subjects' bodily experiences of mis-recognition and how such experiences can motivate their struggle for recognition. Against this backdrop, I put forward the idea of a *transnational recognitive demand*. This ethical demand is based on the minimal requirement concerning my extension of Honneth's theory—which recognitively includes and safeguards the lives of stateless individuals—people who have never had citizenship to begin with or have lost and have not yet regained their citizenship.

I argue that by adopting the entire Honnethian recognition theory—including its anthropological core and concrete universalism—we will be better equipped to tackle the above-described Arendtian statelessness problem as well as non-recognition concerning stateless people becoming refugee patients. Furthermore, in contrast to what I perceive as Kelly Staples' (2012) reductionist interpretation of Honneth, I argue that this statelessness problem is neither a blind spot nor something that is reproduced by his recognition theory. Rather, the transnational recognitive demand sets a moral threshold for the embodied dignity of all humans.

In my view, Staples is thus far the scholar who has contributed with the most elaborate adoption of Honneth's recognition theory applied to statelessness. Adopting her analysis to the fields of global ethics and international political theory, she provides new insights to the discourse around the possibility of extending Honneth's thought to a transnational scale.

However, contrary to her own aim, Staples' critique of Honneth's recognition theory regarding statelessness reproduces the problem that she claims to identify in his theory and wishes to solve. Here, by not applying all the three recognition forms of love, respect, and esteem in the broadest sense of these categories (Ikäheimo 2017), Staples overlooks how the mutuality of the entire Honnethian framework can be extended to the most far-reaching—say, transnational—range of contexts. Subsequently, I find Staples' Honneth reading reductionist since it is based on both a misinterpretation and a false premise. I am thinking specifically of how Staples (2012, 95) almost only draws narrowly on what she designates "legal recognition." According to this interpretation, there is a tight linkage between respect and rights. In result, legal rights are the only means by which respect-based recognition might be articulated. Equally important, Staples merely briefly mentions recognition as love and esteem.

In the following, I address Staples' reductionist reading through what I take to be her four most essential objections against Honneth. I label them as the problem of dualism, private/public, inclusion/exclusion, and autonomy, respectively.

5.4.1 The Dualism Problem

The dualism problem is directed against what Staples (2012, 97–98, 101–102) holds to be Honneth's "ontological trap." Although Staples in the quote below relates this presumed trap to the divide between the private and the public sphere associated with liberal political thought, in the present subsection, I simply focus on the dualism problem. I return, however, to the private/public problem in the next subsection of this chapter. Staples formulates the dualism problem as follows:

> There seem to be valid theoretical reasons to take steps to avoid the 'ontological trap' that equates recognition with personhood in the strong sense by setting up an unnecessary opposition between the private and the political. We have seen so far that the pre-theoretical

> experience of disrespect helps Honneth avoid the self-defeating conclusion that relations of respect are *necessary* pre-requisites of personhood. Even though he retains a clear distinction between the private and the political, his Hegelian understanding of recognition enables him to avoid conceiving of personhood as a zero-sum property. However, I want to suggest that the normative dimension of his theory recreates the constraints of that trap insofar as it leads Honneth towards an insufficiently critical appreciation of respect. (Staples 2012, 97–98, original italics; see also 96)

I agree with Staples that we should avoid what she calls an ontological trap and which I designate as the dualism problem.

In contrast to Staples, however, I hold that regarding Honneth's equation of recognition with personhood, there exists no such trap or dualism. Instead, the above-explicated anthropological core of his recognition theory involves a mutually dependent entirety between all the three recognition forms (i.e., love, respect, and esteem). Also, this mutually dependent entirety is grounded in what Honneth refers to as the necessarily enabling preconditions of the different recognition forms. Consequently, the Honnethian framework is based on an ontological holism rather than an ontological dualism.

Both Honneth's early and later conceptualizations of recognition demonstrate his ontological holism, because they both consist of an anthropological core that presupposes at least two crucial things: First, this core designates an ontological basis upon which recognition as both invariant love and the most elementary recognition form existing prior to the substantial three recognition forms of love, respect, and esteem is founded. Second, the remaining recognition forms—i.e., recognition as respect and esteem—are ontologically grounded in what Honneth defines as a transcendental condition of his theory. So, an ontological holism characterizes both Honneth's earlier and later articulations of his recognition idea.

Thus, I claim that Staples leads herself to a misguided criticism of Honneth by overlooking this ontological holism. Consequently, Staples reproduces the ontological dualism that she attempts to escape through her unconvincing interpretation. By rejecting Staples' reductionist reading and upholding Honneth's holism, I believe that the above idea of a transnational recognitive demand of stateless persons is valid.

5.4.2 *The Private/Public Problem*

As mentioned, Staples links the dualism problem to the private/public problem: "[Honneth's] theory [is] (...) prone to being caught in an 'ontological trap' resulting from a *too-strong public/private opposition.*" Staples even claims that Honneth "retains a *clear* distinction between the private and the political." Confusingly, then, since Staples never explores this view further, she additionally holds that Honneth's "Hegelian understanding of recognition enables him to avoid" the private/public problem (Staples 2012, 98, emphasis added).

If Staples' Honneth reading is correct, I support her problematizing of the often-assumed divide between the private and the public within the liberal political tradition. I am not, however, convinced by her critique. In fact, I think that Honneth's theory does not represent this private/public problem in the first place.

As explained earlier by Zurn, Honneth's recognition theory—including all its three substantial recognition forms—is ontogenetically fulfilled through a developmental hierarchy with a directional logic. Based on my interpretation of Honneth's recognition theory as consisting of the anthropological core and mutually dependent entirety, together with the link between love in the private and politics in the public, it is unlikely that he separates the private from the public. Rather, to develop, e.g., one's self-respect through respect-based recognition and thus the capability of, e.g., democratic participation in the public sphere, humans' prior need for recognition as love and hence one's development of self-confidence in the private domain must be fulfilled. It is, therefore, important to safeguard the private domain to ensure public action and interaction.

In my view, Staples ignores this Honnethian private/public interplay due to her reductionist reading of Honneth's theory by overlooking its mutually dependent entirety. By identifying the private/public interplay, we come closer to articulating the above idea of a transnational recognitive demand of stateless persons, a demand that depends on this interplay in terms of a moral threshold for embodied human dignity. I explain this connection in more detail in part five of this chapter.

5.4.3 The Inclusion/Exclusion Problem

In her approach to the Arendtian statelessness problem from the viewpoint of Honneth's recognition theory, Staples (2012, 97) explains the inclusion/exclusion problem like this: "there is potentially something productive in the move away from a dualistic ontology of (…) *exclusion/inclusion*." From Staples' (2012, 99, my italics) standpoint, stateless people are exposed to "the discriminatory sources and cumulative exclusions of *state mis-recognition*." Accordingly, the "mis-recognition of stateless persons is perhaps the clearest evidence of the arbitrary and exclusionary power of the state" (Staples 2012, 102). Moreover, to "the extent that legal recognition can have disciplinary effects, there is good reason to be wary of *essentialising the state* as the appropriate arrangement for Honneth's normative ideal of human flourishing" (Staples 2012, 103, italics added). Staples (2012, 102) also criticizes what she takes to be Honneth's "normative connections between the state and the idealised" ability to gain recognition as respect. She holds, therefore, that Honneth remains "wedded to a conception of the state as the source of legal rights." Thus, Honneth seems "destined to remain caught in the ontological trap which grants states full power over self-respect."

Against the horizon of the before-explained statelessness problem of Arendt, I think Staples is correct in pointing out the fact that states may illegitimately produce power and exclusion. Nonetheless, in my opinion, there are at least two problems regarding this part of her critique of Honneth. First, as I showed in the subsection about the dualism problem, there is no dualistic ontology in Honneth. Consequently, it is unpersuasive when Staples bases the inclusion/exclusion problem on what she holds to be an ontological dualism in his thought.

Second, as I explicated in the part of this chapter dealing with Honneth's multidimensional recognition theory, the second recognition form of respect covers a broad spectrum of individual self-realization, developmental relationships, institutional arrangements, and mis-recognition (Ikäheimo 2009). These deontic features include universal human dignity and human rights, which is relevant to the transnational struggle for recognition of stateless persons. Here, we should recollect the mutually dependent entirety of the recognition forms, which infers that we cannot judge Honneth's theory simply by exploring one out of the three recognition categories. In contrast to Staples, we cannot choose merely one subcategory of Honneth's second recognition form, such as the legal rights of the state. My point is not that legal rights are an unimportant part of the second recognition category. Rather, I emphasize our need to adopt the full Honnethian recognitive picture. Staples does the opposite by basing the inclusion/exclusion problem on a false premise, namely, the second recognition form narrowly interpreted as merely legal rights within states.

Based on a false premise, Staples reproduces the inclusion/exclusion problem, which she creates herself. In contrast to her straw man argument, then, I hold that a fuller understanding of Honnethian inclusion can, in fact, open the pathway for articulating the idea of a transnational recognitive demand of stateless people.

5.4.4 The Autonomy Problem

Staples formulates the autonomy problem this way: "[Honneth] tries to ground the possibility of full personhood in the potential of practical autonomous judgements about just and unjust social and political relations." Subsequently, Honnethian autonomy is "optimistic that our capacity for decision-making makes the greater achievement of full human flourishing possible" (Staples 2012, 99). According to Staples (2012, 100), however, Honneth's autonomy concept "restricts the range of justifiable social relationships." This is due to the "*strand of Kantian universalism* in his theory of recognition" (Staples 2012, 100, emphasis mine). Certainly, Staples here merely talks about a strand of Kantianism, that is, an element instead of the entire picture of Honneth's philosophy. Still, she primarily links this element to the state. Through her lens, the state seems understood as a somewhat Kantian, universal institution. This is at least the case when Staples on the issue of statelessness argues that Honneth is too optimistic concerning "the extent that *the state* is judged to be a practical *condition* of human flourishing" and "rational political arrangement" (Staples 2012, 99, italics added).

I find Staples' critique unsatisfying for at least two reasons. First, the concept of autonomy—at least the Kantian account of autonomy which Staples holds to be present in Honneth—does not play as significant a role in Honneth as she proposes. As clarified, due to the mutually dependent entirety of Honneth's multidimensional recognition theory, autonomy is only one of the many subcategories of the second recognition form, namely, recognition as respect. Linking his theory too closely to autonomy without clarifying this concept in more detail and doing so within a rather narrow Kantian context of the state does not resonate with Honneth's framework. In fact, from the standpoint of the anthropological core of his recognition theory—together with its focus on the precondition of humans' embodied vulnerability and dependency—Honneth' holistic approach explicitly challenges this very same Kantian idea and its liberal-individualistic picture of independent humans (Anderson and Honneth 2005, 128).

My argument here is further supported by what I explained above as Honneth's concrete universalism. As pointed out, in contrast to Staples' reading, this justificatory strategy aims to balance Kantian universalism and communitarian particularism. Due to his formal conception of ethical life, Honneth wants to avoid an overemphasis on either side.

Second, contrary to Staples' interpretation, Honneth defines his concept of autonomy as "recognitive." Against the horizon of the anthropological core of his recognition theory—which is conditioned by humans' embodied vulnerability and dependency—Honneth approaches the notion of autonomy as "relational" and "full" in terms of "the real and effective capacity to develop and pursue one's own conception of a worthwhile life" (Anderson and Honneth 2005, 130). Building on this, Honneth states that:

> [The] conception of autonomy (…) goes by various names – relational, social, intersubjective, situated, or recognitional – but can be summarized in the claim that 'Autonomy is a *capacity* that exists *only* in the context of *social relations* that *support* it and *only* in *conjunction* with the *internal sense* of being autonomous.' (Anderson and Honneth 2005, 129, italics mine)

To realize this recognitive ideal of autonomy, then, we need to consider individuals' "autonomy-related vulnerabilities." Such vulnerabilities—i.e., various threats to a relational and full autonomy—should, therefore, be "reduced to an acceptable minimum." This highlights "the ways in which individuals' autonomy can be diminished or impaired through damage to the social relations that support autonomy" (Anderson and Honneth 2005, 127). The realization of recognitive autonomy requires, then, "social justice" in terms of the "material and institutional circumstances" of autonomy (Anderson and Honneth 2005, 130).

Based on the before-introduced ontogenetically developmental hierarchy of the recognition forms of love, respect, and esteem, Honneth further explicates recognitive autonomy like this:

> [A]n impressive accomplishment that, *on the path from helpless infancy to mature autonomy*, we come to be able to trust our own feelings and intuitions, to stand up for what we believe in, and to consider our projects and accomplishments worthwhile. We *cannot* travel this path alone, and we are *vulnerable at each* step of the way to autonomy-undermining injustices – not only to interference or material deprivation, but also to the

> disruptions in the social nexus that are necessary for autonomy. (Anderson and Honneth 2005, 130, italics added)

Due to humans' unchanging preconditions, we are bodily vulnerable to others' misrecognition and non-recognition. As a result, we depend on others to protect our vulnerability through recognition or recognitive autonomy (Anderson and Honneth 2005, 130; Lysaker 2020a).

Staples' autonomy problem, therefore, appears to be incoherent. I here hope to have made clearer how and why Honneth's idea of recognitive autonomy does not echo the Kantian strand she ascribes to him. By resolving her criticism, I believe that we are getting closer to a Honnethian justificatory strategy that articulates the recognitive demand of stateless peoples' recognition struggle on a transnational scale.

5.5 Embodied Dignity

In this last part of my chapter, I wish to assemble the pieces from the previous parts to demonstrate what I perceive of as a Honnethian idea of *embodied dignity* (Lysaker 2020a, 87–88; Lysaker 2015). My aim here is to further articulate the before-introduced transnational recognitive demand of stateless people.

In the first part, I explicated the anthropological core and the concrete universalism of Honneth's recognition theory. According to these foundational elements of his thought, humans share an embodied vulnerability and dependency. In the second part, I introduced the three recognition forms within the context of the mutually dependent entirety of his multidimensional recognition theory and the necessary preconditions of the three different recognition forms of love, respect, and esteem. I also emphasized that the anthropological core can be understood both through recognition as invariant love and the most elementary recognition form of human dignity. In both cases, I claim that since the unchanging preconditions (i.e., vulnerability and dependency) are shared by every human, they normatively ground Honneth's concrete universalism.

To live a fully human life with inherent dignity and bodily integrity requires recognition—a recognition either based on the mutually dependent entirety (i.e., recognition as invariant love prior to respect and esteem or the substantial recognition forms of love, respect, and esteem) or the most elementary recognition form (i.e., human existence, humanity, and dignity).

These various recognitive approaches to dignity and integrity springs out of a "moral injury" (Honneth 2007, 133–137; see also Lysaker 2020a, 87–89; Hayden 2016, 101). To be morally injured means experiencing harm against one's embodied integrity and inherent dignity. Furthermore, moral injuries violate humans' intuition about justice. As noted, such injuring refers to emotionally experiencing the suffering that injustice causes. Such experiences can motivate injured persons to regain their violated dignity through struggles for recognition. I think Jay M. Bernstein (2005, 313) sums up Honneth's recognitive notion of embodied dignity precisely, when he writes that "the dignity of the person just is what comes to be through the forms of recognition through which the intact, self-moving body comes to be." Thus, "the dignity of the self is the reflective articulation of the moral integrity of the body."

Since Honneth's framework consists of what I earlier explained as the two levels of recognition—i.e., the existential and the substantial—I hold that although Bernstein apparently does not mention the concept of non-recognition, his reading of Honneth's idea of integrity and dignity seemingly resonates with this most basic level of humanity. To illustrate, stateless persons who become refugee patients can be subjected not merely to mis-recognition, but non-recognition as well. To have their embodied dignity recognized necessitates existentially avoiding recognitive negation of these people's bodily integrity and human dignity. If not, they will become nonhumans, and such dehumanization can undermine their very motivation to struggle for recognition, even on the substantial level.

Patrick Hayden, too, supports my argument around Honnethian embodied dignity. Within the discourse of global ethics and international political theory, Hayden (2012, 575) claims that Honneth's framework is "particularly apparent when considering (…) the day-to-day struggles of those seeking recognition of their human status on the simultaneously local-national-global terrain of contemporary politics." Hayden (2012, 576) views statelessness as a "the paradigm case" concerning the transnational relevance of Honnethian recognition. Interestingly, Hayden (2012, 577, italics added) underscores how Honneth's recognitive outlook identifies "a *deeper* understanding of the place of *embodiment* in the mediation between physical being and sociopolitical existence." This viewpoint is highly appropriate given that stateless persons and other globally marginalized people are "*unheard* and *unseen* by others who do *not recognise* the claimant as *sufficiently human.*" To flourish fully as a human with embodied dignity, then, "a sociopolitical condition that comes from being recognised by another *as human*" should be satisfied (Hayden 2012, 576, emphasis mine). In Hayden's (2012, 576, italics mine) eyes, Honneth's recognition theory is significant on this transnational level of statelessness since it suggests why such excluded groups recognitively struggle to recover their embodied dignity after being morally injured: "If mutual recognition is the *prerequisite* for becoming *fully human*, then there is a *shared human interest* in attempting to create and recreate socio-political arrangements that *extend* recognition to *all.*" Hayden (2012, 576, emphasis added; see also Lysaker 2020a) further explains that recognition struggles are struggles "for *due* recognition of *all* as equal and distinctive, yet *vulnerable* persons." This "necessarily entails critique of those prevailing conditions that foster *asymmetric* relations of *mis-recognition*, and of social, economic, and political *inequalities* that *violate dignity* and human rights." Subsequently, such capacities "condition both the *agency* and the *vulnerability* of human beings as *embodied* persons" (Hayden 2012, 577, italics added).

By extending Honneth's recognition theory to the transnational level of, e.g., statelessness, as both Hayden and I do, I hold that stateless people—as humans who all equally share the unchanging preconditions of vulnerability and dependency—should be able to live a full human life with embodied dignity. In terms of the moral threshold of the transnational recognitive demand, therefore, stateless persons, hereunder refugee patients, should live a full life without both mis-recognition and non-recognition.

5.6 Conclusion

In this chapter, I have explored the role of Honneth's recognition theory as applied to stateless persons within the discourse of global ethics and international political theory. I have demonstrated how his ideas can be extended to recognition struggles on a transnational scale using what I coined as Honneth's anthropological core and concrete universalism. Moreover, I explicated the mutually dependent entirety of his multidimensional recognition theory. I also showed the relevance of the distinction between mis-recognition and non-recognition through the case of stateless people who become refugee patients due to, e.g., extreme traumatization.

In my analysis, I criticized what I perceive as Staples' reductionist Honneth reading, which reproduces the problems she wishes to sidestep in the question of statelessness. In contrast to Staples, I showed how Honneth's recognition theory, in fact, is much more robust and applicable to the situation of the currently around 12 million stateless people worldwide. In addition, I further developed Bernstein's and Hayden's articulation of a Honnethian notion of dignity related to humans' embodied vulnerability and dependency. In my view, Hayden's line of argument comes much closer to an adequate application of Honneth's recognition theory applied to stateless persons.

I also claimed that our shared unchanging preconditions of vulnerability and dependency invoke an ethical demand that we all be fully, transnationally recognized as humans with embodied dignity. Stateless people—as humans—should, therefore, have their embodied dignity safeguarded, even on the scale of transnational struggles for recognition.

References

Anderson, Joel, and Axel Honneth. 2005. Autonomy, Vulnerability, Recognition, and Justice. In *Autonomy and the Challenges of Liberalism: New Essays*, ed. John Philip Christman and Joel Anderson, 127–149. Cambridge: Cambridge University Press.

Arendt, Hannah. 1951. *The Origins of Totalitarianism*. New York: Schocken Books.

Bernstein, Jay M. 2005. Suffering Injustice: Misrecognition as Moral Injury in Critical Theory. *International Journal of Philosophical Studies* 13 (3): 303–324.

Deranty, Jean-Philippe. 2009. *Beyond Communication: A Critical Study of Axel Honneth's Social Philosophy*. Leiden: Brill.

Fossum, John Erik. 2005. Conceptualizing the EU's Social Constituency. *European Journal of Social Theory* 8 (2): 123–147.

Hacke, Jürgen. 2005. The Frankfurt School and International Relations: On the Centrality of Recognition. *Review of International Studies* 31 (1): 181–194.

Hayden, Patrick. 2012. The Human Right to Health and the Struggle for Recognition. *Review of International Studies* 38 (3): 569–588.
———. 2016. Lost Worlds: Evil, Genocide, and the Limits of Recognition. In *Recognition and Global Politics: Critical Encounters Between State and World*, ed. Patrick Hayden, and Kate Schick, 101–120. Manchester: Manchester University Press.
Heins, Volker M. 2008. Realizing Honneth: Redistribution, Recognition, and Global Justice. *Journal of Global Ethics* 4 (2): 141–153.
Honneth, Axel. 1991/1985. *The Critique of Power: Reflective Stages in a Critical Social Theory*. Cambridge: MIT Press.
———. 1992/1990. Integrity and Disrespect: Principles of a Conception of Morality Based on the Theory of Recognition. *Political Theory* 20 (2): 187–201.
———. 1995/1992. *The Struggle for Recognition: The Moral Grammar of Social Conflicts*. Cambridge, MA: MIT Press.
———. 2002. Grounding Recognition: A Rejoinder to Critical Questions. *Inquiry: An Interdisciplinary Journal of Philosophy* 45 (4): 499–519.
———. 2007/2000. *Disrespect: The Normative Foundations of Critical Theory*. Cambridge, UK: Polity Press.
———. 2008/2005. *Reification: A New Look at an Old Idea*. Oxford: Oxford University Press.
Honneth, Axel, and Hans Joas. 1988/1980. *Social Action and Human Nature*. Cambridge, UK: Cambridge University Press.
Ikäheimo, Heikki. 2009. A Vital Human Need: Recognition as Inclusion in Personhood. *European Journal of Political Theory* 8 (1): 31–45.
———. 2017. Recognition, Identity, and Subjectivity. In *The Palgrave Macmillan Handbook of Critical Theory*, ed. Michael J. Thompson, 567–585. Basingstoke: Palgrave Macmillan.
Jakobsen, Jonas, and Odin Lysaker. 2010. Social Critique Between Anthropology and Reconstruction: An Interview with Axel Honneth. *Norsk Filosofisk Tidsskrift* 45 (3): 162–174.
Linklater, Andrew. 2006. The Harm Principle and Global Ethics. *Global Society* 20 (3): 329–343.
Lysaker, Odin. 2008. Sårbarhet og ukrenkelighet: Filosofisk-antropologisk innholdsbestemte menneskerettigheter. *Nordisk Tidsskrift for Menneskerettigheter* 26 (3): 244–258.
———. 2013. *Menneskeverdets politikk: Anerkjennelse av kroppslig krenkbarhet*. Oslo: Abstrakt forlag.
———. 2015. Democratic Disagreement and Embodied Dignity: The Moral Grammar of Political Conflicts. In *Recognition and Freedom: Axel Honneth's Political Thought*, ed. Odin Lysaker, and Jonas Jakobsen, 147–168. Leiden: Brill.
———. 2017. Institutional Agonism: Axel Honneth's Radical Democracy. *Critical Horizons: A Journal of Philosophy and Social Theory* 18 (1): 1–19.
———. 2020a. Nowhere Home: The Waiting of Vulnerable Child Refugees. In *Vulnerability in Scandinavian Art and Culture*, ed. Margareta Dancus, Maria Karlsson, and Mats Hyvönen, 81–101. Basingstoke: Palgrave Macmillan.
———. 2020b. Dignity in Natality: Hannah Arendt on Human Rights in Dark Times. In *Research and Human Rights*, ed. Jakob Lothe, 141–157. Oslo: Novus Press.
———. 2020c. Ecological Sensibility: Recovering Axel Honneth's Philosophy of Nature in the Age of Climate Crisis. *Critical Horizons: A Journal of Philosophy and Social Theory*.
Petherbridge, Danielle. 2011. Introduction: Axel Honneth's Project of Critical Theory. In *Axel Honneth: Critical Essays With a Reply by Axel Honneth*, ed. Danielle Petherbridge, 1–30. Leiden: Brill.
———. 2013. *The Critical Theory of Axel Honneth*. Plymouth: Lexington Books.
Seglow, Jonathan. 2009. Rights, Contribution, Achievement, and the World: Some Thoughts on Honneth's Recognitive Ideal. *European Journal of Political Theory* 8 (1): 61–75.
Staples, Kelly. 2012. Statelessness and the Politics of Mis-Recognition. *Res Publica* 18 (1): 93–106.
Thompson, Simon. 2013. Recognition Beyond the State. In *Global Justice and the Politics of Recognition*, ed. Tony Burns, and Simon Thompson, 88–107. London: Palgrave Macmillan.

UN. 1954. *Convention Relating to the Status of Stateless Persons*. https://www.unhcr.org/ibelong/wp-content/uploads/1954-Convention-relating-to-the-Status-of-Stateless-Persons_ENG.pdf. Accessed 21 April 2020

———. 2018. *'12 Million' Stateless People Globally, Warns UNHCR Chief in Call to States for Decisive Action*. https://news.un.org/en/story/2018/11/1025561. Accessed 21 April 2020.

Weber, Martin. 2007. The Concept of Solidarity in the Study of World Politics: Towards a Critical Theoretic Understanding. *Review of International Studies* 33 (4): 693–713.

Zurn, Christopher. 2000. Anthropology and Normativity: A Critique of Axel Honneth's 'Formal Conception of Ethical Life'. *Philosophy and Social Criticism* 26 (1): 115–124.

Chapter 6
Claims-Making and Recognition Through Care Work: Narratives of Belonging and Exclusion of Filipinos in New York and London

Rizza Kaye C. Cases

Abstract Based on a larger study on mobility projects and networks of 134 Filipino nurses, domestics, and care workers in New York and London, this chapter examines how Filipino migrant workers "claim" their "rightful" place in the place of destination within the context of underappreciation for the kind of (care) work that they do. Research participants tend to position themselves as valuable by differentiating themselves from the stereotypical image of migrants who are just after the benefits they can get from the "host" country. Their feelings of belongingness and "deservingness to be there" are also validated by the recognition that participants get from their work and perceived greater purpose of their role as care providers.

At the same time, the financial security that they were able to attain allows them not only to become a part of the "host" society but, more importantly, of their "home" country. Being able to afford the "good life" means that they are not (or no longer) located at the periphery in their homeland. However, while care work allows participants to see themselves as deserving to be in the place of destination given their contribution, being employed in what is deemed as low-status job can also make migrants feel that they are living on the margins of the "host" society and that they do not really belong.

Thus, these ambivalences concerning migrants' claims-making and struggles for recognition through care work could be conceptualized as Irene Bloemraad's notion of "structured agency." On the one hand, claiming recognition conforms to certain normative ideals, such as being a "good migrant" or a "good citizen." It is also constrained by immigration regimes and practices of "controlled inclusion." On the other hand, claims-making is also agentic as migrants try to fulfill their migration projects and assert their right to have better futures for themselves and their families.

R. K. C. Cases (✉)
Department of Sociology, University of the Philippines Diliman,
Quezon City, The Philippines
e-mail: rccases@up.edu.ph

G. Schweiger (ed.), *Migration, Recognition and Critical Theory*, Studies in Global Justice 21, https://doi.org/10.1007/978-3-030-72732-1_6

Keywords Filipino migration · Care work · Claims-making · Recognition · Belonging

6.1 Introduction[1]

In the post-colonial world, the movement of peoples from the Global South to the Global North to provide "cheap labor" has not only continued but has intensified in contemporary times. The case of the Philippines illustrates how an export-oriented economic policy has translated not only into the production of goods for overseas demands but also into the reproduction of "exportable" people (Choy 2000). What perhaps changed radically in recent times is the increasing feminization of labor for export. Compared to the earlier migration of predominantly Filipino (single) men to the USA in the early part of the twentieth century, the typical migrant from the Philippines today is a woman who has left her family and her traditional "domestic duties" to take up the same duties – though of another woman from the First World. In the words of Shutes and Anderson (2014, 1):

> In the global North, international migrants have increasingly supplemented the unpaid or low-paid care labour provided by non-migrant women – as domestic workers, nannies, care assistants and nurses – in the private sphere of the home and in publicly and privately funded care services.

This phenomenon has been captured and expounded through the concepts of "global care chain" (Hochschild 2000) and "international division of reproductive labor" (Parreñas 2012).[2] Such "care chain' connects seemingly unrelated care jobs being performed in different areas or countries by various care workers. Parreñas (2000, 561) refers to this as "the three-tier transfer of reproductive labor among women in sending and receiving countries of migration" wherein a "migrant Filipina domestic workers hire poorer women in the Philippines to perform the reproductive labor that they are performing for wealthier women in receiving nation."[3]

---

[1] This chapter is based on the empirical materials and discussions in Cases (2018).

[2] There is a rich literature that discusses and critically examines the concepts of care labor/care work and reproductive labor/social reproduction – their differences, convergence and divergence, as well as the advantages of using one over the other (e.g., *see* Kofman 2012; Kofman and Raghuram 2015; Parreñas 2012). Cognizant of the insights from the works on social reproduction, I primarily discuss "care" and "care work" in this chapter to emphasize how overseas Filipino nurses, domestics, and care workers employ the care work that they do as moral currency to justify their deservingness claims and how they distance and distinguish themselves and assert their status over others.

[3] These concepts have been extended and further developed by other scholars like Nicola Yeates (2011, 2012) who broadened the theoretical scope of the concept (FitzGerald Murphy 2014) and include care provided in institutional settings such as hospitals, men who perform care work, as well racial and ethnic inequalities (Parreñas 2012).

Apart from the contradictory position of the migrant care workers in the care chain, their presence in the places of destination also produces other forms of contradictions and ambivalences. For one, while migrant (care) labor could be rendered as necessary, or even desired, their presence must, at the same time, remain at the fringes. Yen Le Espiritu (2003) and Lisa Lowe (2006) refer to such racialized experiences as differential inclusion. This process of simultaneous (partial) inclusion and exclusion marks migrant labor from the beginning as, first and foremost, colonized subjects. In the post-colonial and globalized world, migrant (care) labor continues to be "differentially included" – needed and unwanted at the same time – in the Global North. For instance, while Filipino nurses and care workers are generally praised for their hard work and caring nature, they can also be discriminated in the workplace by having lower pay, being assigned to unfavorable shifts, or being the first to be blamed or suspected in case something has gone wrong (*see* Choy 2003). As migrant workers, the state can also decide that they are not wanted anymore in the country – i.e., by instituting new immigration policies that do not only make it difficult for new entrants to come but also make it harder for those who have temporary work permits to apply for permanent residency as in the case of changes in the immigration policy of the UK in 2006 (*see* Bach 2010). In this way, Filipino nurses (and other migrant nurses) as well as care workers are desired for a limited time only and seen as disposable when there is no longer a (perceived) need for them or one of the first ones to be targeted when it is perceived that there are "too many migrants" in the country.

Much of the shame associated with care work – especially domestic work and domiciliary elder care – is that it is low-status (feminized) labor. However, Filipinos continue to aspire and decide to migrate given the promise of a better life associated with going abroad – regardless of what kind of work one does. Parreñas (2015, 117) refers to this kind of dislocation that migrants face as *contradictory class mobility* – the "simultaneous experience of upward and downward mobility in migration or, more specifically, their decline in occupational status and increase in financial status." Similarly, the concepts of "transnational frame of reference" (Kelly 2012), "dual frame of reference" (Suárez-Orozco and Suárez-Orozco 1995), and "transnational status paradox of migration" (Nieswand 2011) capture the same condition in different contexts. While these researchers specifically refer to low-status occupations, Filipino nurses who had to "downgrade" to being care workers or nursing aides overseas also experience dissonance and resentment – as they earn more relative to what they are earning in the Philippines but feel degraded at the same time by doing work that does not match their level of education and skill or even their social class in the Philippines. It is therefore important to examine the ways in which Filipinos doing care work overseas resolve such contradictions as part of their overall strategies to negotiate their present position and imagined futures in both sending and receiving contexts.

Given this ambivalent position of migrant care workers in the places of destination, this paper examines narratives of belonging and exclusion that allow migrants to make sense and "carve" what they deemed to be their "rightful" place in the places of destination. Drawing on the literature on deservingness frames (van Oorschot 2000) and moral economy (Näre 2011), I explore migrants' claims-making and struggles for recognition through care work and link such claims to the overarching notion of a "good migrant." Barriers to claims-making, particularity "illegality" and the low-status work that migrants do, will also be discussed to illustrate experiences of exclusion.

Considering that migrants maintain a "dual frame of reference," it is also necessary to explore migrants' feeling of belongingness and struggle for recognition using a transnational lens. As migrants make claims for a place they deserve in the "host" society, they are also justifying their place in their "home" country as they present themselves as the embodiment of the "good life" that can be attained through overseas work. This paper thus extends the extant research and literature on migrants' claims to belong and be recognized by simultaneously examining the ways migrants position themselves in both countries of origin and destination, by employing and referring to either distinct or intertwined deservingness frames.

However, migrants' claims-making is also being structured by macro-level forces that shape the contours and boundaries of what can be possibly attained. For one, recognition entails not only being able to make a claim to be recognized. There is also the question of the extent that such claims are recognized by, for instance, immigration regimes, existing practices of "controlled inclusion," and spaces and groups where they are embedded. Likewise, claiming recognition also push migrants to conform to certain normative ideals, such as being a "good migrant" or a "good citizen." The paper thus concludes by employing Bloemraad's notion of "structured agency" to reflect on both the constraining and agentic dimensions of migrants' narratives of claims-making, recognition, and belongingness.

6.2 Deservingness Frames and Moral Economy

Situated within the debates on social justice and welfare provision (e.g., Reeskens and van Oorschot 2013; van Oorschot 2000), the perceived deservingness of migrants has been explored from the perspective of the receiving society by a number of studies (e.g., Kootstra 2016; Jørgensen and Thomsen 2016; Reeskens and van der Meer 2019; Nielsen et al. 2020). These studies examined to what extent migrants are seen as deserving welfare recipients in destination countries and what particular criteria or cues are being used to assess their deservingness. For instance, van Oorschot (2000) identified five dimensions of deservingness criteria – control, need, identity, attitude, and reciprocity (CARIN) – which serve as the bases for public's opinion on "who should get what and why." These criteria are considered as "shared heuristics" that enable people to classify others of their deservingness, particularly in terms of receiving welfare support (Nielsen et al. 2020, 113). Analyzing nationally representative survey data in the Netherlands, the findings suggest that ethnic minority groups, asylum seekers, and illegal foreigners are considered as less deserving compared to the general Dutch population, particularly along the identity dimension. That is, since they are seen as "not one of us," the public tend to be less willing to provide access to welfare for those who belong in these groups. Extending the earlier study, van Oorschot (2006) also examined deservingness perceptions of the public across European countries using 1999/2000 European Values survey. Among the four target groups (elderly, sick and disabled, unemployed, and immigrants), it has been shown that immigrants ranked lowest in terms of perceived public deservingness among the target groups and such result is consistent across the 23 European countries.

The consistency of public opinion regarding welfare deservingness in cross-national and cross-cultural perspectives has been illustrated using experimental studies that demonstrate how "deservingness heuristic" is being utilized as a basis for welfare support or rejection (Aarøe and Petersen 2014; Petersen et al. 2011). For instance, comparing two different welfare regimes and remarkably contrasting cultural contexts, Aarøe and Petersen (2014) found that the presence of direct deservingness cues (i.e., welfare recipients as "lazy" or "unlucky but hardworking") eliminates the observed differences between the US and Danish samples. This result suggests that while American and Danish respondents predominantly apply different stereotypes to welfare claimants, such stereotypes are overridden by directly providing clear deservingness cues as additional information to favorably (or unfavorably) describe a hypothetical welfare recipient. But what happen if such welfare recipient is a member of an ethnic minority? Will the same deservingness heuristic (i.e., lazy vs. unlucky) outweighs stereotypes attached to migrants and members of ethnic minorities and be therefore seen as deserving welfare recipients? Utilizing vignette experiments to examine how migration status and ethnic background relate to other factors that could affect perceived deservingness among Dutch and British respondents, Kootstra (2016) showed that while ethnic minorities and migrants are perceived as less deserving compared to majority of Dutch and British welfare recipients, additional deservingness cues describing them in more favorable light (e.g., great effort in finding a new job, longer work history, and born in the city) allowed ethnic minorities (but not those born outside of the country) to be also seen by the public as deserving welfare recipients like the British or Dutch claimants (cf. Reeskens and van der Meer 2019[4]). However, when the ethnic minorities are described as exhibiting unfavorable characteristics, they face stricter judgment and harsher penalties compared to their British or Dutch counterpart. Complementing these quantitative studies, Nielsen et al. (2020) qualitatively explored how

[4] Using vignette experiment and between-person design, Reeskens and van der Meer (2019) found that the presence of more positive characteristics does not close the deservingness gap between the foreign-born minorities and native unemployed Dutch majority. The contrasting findings of Kootstra (2016) and Reeskens and van der Meer (2019) could be due to the differences in the study designs (e.g., within-person and between-person designs). Reeskens and van der Meer (2019) suggested that further research could be done to explain and address these inconsistencies.

deservingness criteria toward immigrants are being deployed by the public in four European countries – Slovenia, Denmark, the UK, and Norway. Emphasizing that deservingness criteria are not employed independently of each other, the study has shown how respondents simultaneously utilized deservingness criteria (e.g., control and need) to construct their image of deserving and non-deserving migrants. These studies suggest that other dimensions of deservingness criteria interact with one's identity as an ethnic minority or a migrant that may improve or further worsen how (non-)deserving they are perceived by the public.[5]

However, it must also be noted that everyday encounters are replete with stereotypes attached to migrants and ethnic minorities as these are often reproduced and disseminated from state rhetoric to images perpetuated by various media. As immigration policies become more restricted in destination countries in the Global North, it can be expected that immigrants and ethnic minorities are seen and characterized negatively rather than positively along various dimensions of deservingness criteria. On the other hand, salient events and conditions (e.g., labor shortages) can serve as impetus to push for a more favorable image of immigrants and particular ethnic groups. This opens up a space for negotiating and making claims for one's deservingness and belonging. Thus, it is necessary to also ask if the general public in destination countries employ these deservingness criteria to construct images of (un)deserving migrants that, in turn, justify their normative judgments concerning "who should get what and why" and how will the migrants position themselves vis-à-vis such images and assessment? Against the backdrop of limited and conditional public support for migrants' access to welfare support in destination countries, it is also equally important to understand migrants' notions of their deservingness, which could shape their integration experiences, practices, and feeling of belongingness.

As Osipovič (2015, 733) argued, "literature examining migrants' own views on deservingness in the context of host welfare states is sparse." Extant studies on migrant deservingness point to parallel conditionality of deservingness criteria applied by migrants to themselves, to their groups, and to migrants in general (e.g., Timonen and Doyle 2009; Osipovič 2015; Kremer 2016; Alho and Sippola 2019). Using the same deservingness criteria outlined by van Oorschot (2000), Osipovič (2015) looked into how Polish migrants in London conceptualize their own welfare deservingness. Utilizing desert-based principles, the research participants emphasized their contributions to British society through working, paying taxes, and obeying law. Such contributions serve as their claim for rightful access to welfare. The findings show that such views are not only applied to Polish migrants but are also extended to other migrant groups. By highlighting one's contributions to "host society," migrants do not only make claims concerning their welfare deservingness and

[5] Whether or not deservingness of migrants and ethnic minorities is perceived to be similar to the level of deservingness accorded to natives after they are described more favorably using other dimensions of deservingness criteria, such "improvements" in how they are perceived by the public could still allow migrants and ethnic minorities space to maneuver and make claims that could improve their conditions.

belonging but also highlight their difference from the popular view of undeserving migrants. Similar themes can be found from the interviews conducted by Alho and Sippola (2019) on Estonian migrants in Finland who, apart from embracing and asserting their roles as "good worker" and taxpayer, were also distancing and differentiating themselves from those they considered to be "undeserving" and unworthy migrants who are receiving welfare support without contributing to the system. Thus, utilizing deservingness frames constructs migrants' relationships both with the migration regimes in the destination country and with particular groups in the "host society" (Monforte et al. 2019, 25–26):

> the injunction to deservingness creates a space in which migrants not only perform strategically what they think is expected from them by state representatives, but also invoke and use the deservingness frame more widely, outside of state interactions, through narratives in which lines of distinction between the 'deserving' citizen and the 'undeserving Others'…

Deservingness frames then function as a way for migrants to (legitimately) locate and position themselves vis-à-vis the state, native population, other migrant groups, and co-ethnics in the place of destination (Chauvin and Garcés-Mascareñas 2014; Andrews 2017; Monforte et al. 2019). Such frames are intimately connected to the deservingness criteria that the general public utilize to judge deservingness of welfare recipients and stereotypes that surround the image of undeserving migrants that they try to move away from.

It is therefore unsurprising that examining migrants' views on deservingness is inevitably intertwined with the larger notion of what constitutes a "good migrant" and a "good citizen." In other words, notions of (migrant) deservingness cannot be divorced from normative ideals and moral logics. In order to better understand how notions of deservingness are framed and how a good migrant is being imagined by migrants themselves, it is useful to embed such notions within the concept of moral economy.

The concept of moral economy has been utilized by other scholars in various disciplines to understand diverse research topics. Analyzing food riot in eighteenth-century England beyond what he described as "crass economic reductionism," historian E.P. Thompson (1971, 79) defined moral economy as "a consistent traditional view of social norms and obligations, of the proper economic functions of several parties within the community, which, taken together, can be said to constitute the moral economy of the poor." In his article *Moral Economies Revisited*, Fassin (2009, 1242–1243) further explained Thompson's usage of the concept to highlight not only the system of exchange and distribution of goods and services but also the corresponding norms and reciprocities that guide any social action – economic activities included:

> As a result, if peasants revolt against owners, it is not only because resources are scarce: it is also because norms are not respected and because the implied commitments of rights and obligations are not met. The understanding of their reaction therefore involves not only a political economy in which the market imposes its harsh law, but also a moral economy that reminds us that another form of exchange is possible.
>
> As being constituted by norms and obligations, moral economy is also 'moral' in the sense that it is based on a shared belief and understanding of 'what ought to be' and that

> actions deriving from such consensus is legitimate and supported by the larger community. (Thompson 1971; Fassin 2009)

Applying the notion of moral economy to examine peasant communities and politics in Southeast Asia, anthropologist James Scott (1976) explored the conditions that make peasant uprisings possible. In accounting for such conditions, he shifted the focus of the concept to values and emotions compared to Thompson's emphasis on norms and obligations (Fassin 2009, 1249):

> The focus is no longer purely on norms, obligations, customs, or traditions, but it is also on values and emotions, and especially on the sense of justice. The issue is less about understanding what should and what should not be done (the normative dimension) than about what is and what is not tolerable (the evaluative dimension). The Burmese or Vietnamese peasant is not simply someone who adheres to a tradition he perpetuates. Like the English peasant of the eighteenth century, he is also someone who invokes and claims rights.

Incorporating both norms and values in conceptualizing moral economy, the sociologist and anthropologist Didier Fassin (2005) utilized the concept to analyze immigration policies, particularly in France. He defined moral economy as "the production, distribution, circulation, and use of moral sentiments, emotions and values, and norms and obligations in social space" (Fassin 2009, 1257). Such definition expands the application of moral economy as a heuristic lens to encompass not only pre-modern, pre-market societies but any social systems, groups, and relations.

In this chapter, migrants' claims concerning their deservingness and "rightful" belonging are embedded within the notion of moral economy of care work. Principles of reciprocity and mutuality surround providing care that goes beyond economic exchanges. Care provision, whether paid or unpaid, is also evaluated in terms of empathy and concern given by the carers, and care work is largely seen as "help" instead of purely "work" by those providing the care. As Hess (2008, 148) put it, "[t]he conception of 'help' implied a very important promise as, in line with the moral economy logic, the currency of the reward is not a monetaristic one, but a 'moral' one of appreciation, caring and familial integration." While Hess (2008) was particularly analyzing how au pairs view the kind of work they are doing, the notion of "help" as part of doing care work can be applied to other kinds of care work both in institutional and domiciliary settings.

Thus, following Näre (2011, 401), this paper considers "migrant domestic and care labour as a moral economy by looking at how moral notions of good/bad and just/unjust regard labour practices and relationships." However, instead of focusing on employer-employee relations within the domestic sphere, the chapter examines "care work" in general and reflects on how reproductive labor being done by migrants is perceived in relation to migrants' own positioning and claims-making. What about care work that makes migrants feel that they are necessary part and deserving member of the "host society?" How do they utilize "moral currency"[6]

[6] Applying moral economy to transnationalism, Carling (2008) refers to "the exchange and accumulation of moral 'currency'" (1459) wherein "transnational practices such as sending remittances or facilitating migration can be a source of personal gratification, pride and social prestige" (1461).

from the kind of work they do to assert distinction and difference from those deemed to be "undeserving?" Within the moral economy of care work, notions of what constitutes a "good migrant" are tied with the normative ideal of a "good carer." Migrants could then anchor and derive their deservingness frames through these notions of good/bad, legitimate/illegitimate, and just/unjust providers of paid care labor. Since migrants tend to emphasize their contributions in the "host society" through employment, it is important to closely examine the relationship between the kind of work that migrants do and the deservingness frames that migrants employ to justify various (moral) claims such as welfare benefits, formal citizenship, or belonging. By focusing on migrants working in the care sector (as nurses, caregivers, and domestic workers), this paper explores which idealized and moral aspects of care work and migrant labor are mobilized by migrants to project an image of a "good migrant" and articulate claims for varied forms of inclusion. This is in the context of the little value assigned to care work and negative public perceptions on migrants.

6.3 Claims-Making, Recognition, and Belongingness

Despite the increasing demand for migrant labor in the care sector (Cangiano and Shutes 2010), the stigma of being identified with lowly and poorly recognized job of doing "care work" (Guevarra 2014) points to the general devaluation of care labor (or reproductive labor). The stigma, precarity, and low status accorded to care workers are compounded in the case of migrant care workers, documented and undocumented. It is within this context that migrants' claims for recognition and belonging can be better understood. Following Bloemraad (2018), claims-making as a framework involves paying attention to both the agentic and creative ways of actors as well as to the constraints and barriers that limit actor's agency to make claims and to demand recognition. Bloemraad (2006, 2018) refers to this as "structured agency" to highlight that while migrants make claims for citizenship rights, recognition, or belonging, they do so within the bounds of immigration regimes and practices of "controlled (or differential) inclusion" and by also conforming to the "normative ideals" of a "good migrant" and supposed "good citizen."

> Immigrants and their children can make claims, at times in creative ways, modifying the normative content of citizenship, affecting recognition evaluations, and changing the allocation of status and rights. But they are also constrained by the legal structures that they and their families confront, the institutional practices of a society, and prevailing public perceptions. To be more powerful, claims in the name of citizenship must resonate with normative ideas, and these ideas cannot be invented whole cloth (Bloemraad 2018, 6).

In the same vein, there is also a negative currency of ingratitude tied to the moral responsibility of migrants to continue to care and "give back" to those they left behind. Similar application can be found in the work of Katigbak (2015) on emotional remittances and translocal moral economy in the Philippines.

Thus, connecting to the previous discussion on deservingness and moral economy, employing deservingness frames to effectively make various kinds of claims and establish one's "rightful" place requires resonance and conjunction with existing normative ideals, public sentiments, immigration laws, and market labor demands. As Bloemraad (2018, 5) puts it, "the flip side of claims-making is recognition. Do other actors recognise the claim and claimant as legitimate?" Indeed, claims-making and struggle for recognition are relational in such that actors position themselves vis-à-vis other actors, whether such actors are individual, group, or institutions. This means that the other party can be denied of recognition, disrespected, and denigrated. In the case of migrants, such denial can correspond to desiring to overcome misrecognition or under-recognition of their worth and deservingness in the private and public spheres. For this paper, claims-making and struggle for recognition are located within ambivalences in care work and care labor – where migrants are differentially included, needed but not necessarily wanted. This then creates a state of intersubjective dissonance in which migrant care workers are being recognized within and through certain framework or set of normative ideals (e.g., "good citizen" or "deserving migrant") but their claims are misrecognized within that same framework.

Finally, while previous research specifically explored welfare deservingness both from the perspectives of the general public and migrants, this chapter focuses on how migrant care workers narrate and make sense of their relations to people and institutions, as well as to their degree of rootedness and feeling of belongingness in spaces they inhabit and communities where they participate. Yuval-Davis et al. (2006, 2) defined belonging as "about emotional attachment, about feeling 'at home' and … about feeling 'safe'." They also emphasized the diverse articulations of belonging that go beyond the nation-state and fostered through various affiliations and communities. Belonging can also be constructed (and contested) through boundary-making and exclusionary practices based on social class, gender, religion, ethnicity, and immigration status (among others). Antonsich (2010, 645) referred to these two dimensions of "belonging" as *place-belongingness* ("personal, intimate, feeling of being 'at home' in a place") and *politics of belonging* ("discursive resource which constructs, claims, justifies, or resists forms of socio-spatial inclusion/exclusion").

Given these two dimensions, belonging can be conceptualized in various scales and levels that take into account the coexistence of multiple forms of membership. Apart from legal citizenship, membership can be analyzed in terms of its social (or interpersonal) and cultural dimensions (Bloemraad et al. 2019). Thus, being able to gain legal citizenship and access to welfare system and social protection do not readily translate to favorable attitudes, social ties, interactions, and cultural representations toward migrants and ethnic groups. Likewise, having legal citizenship does not also mean that migrants will feel a sense of being "at home," safe, and secure in the destination country in general and the spaces that they inhabit and relations they have in particular.

Thus, rather than viewing belonging as an outcome (i.e., as a status or feeling), it is also useful to examine and conceptualize it as a process by focusing on "the various ways in which belonging can be performed, displayed, and enacted through individual and collective practice" (Antonsich 2010, 652). However, performativity also takes place within certain boundaries, and practices are enabled and constrained by what is possible at a given moment. As Carrillo Rowe (2005, 36) puts it, belonging cannot be chosen "outside of the bounds of power." This connects to claims-making as "structured agency," which can be seen as an exercise of agency within the limits of structures and normative ideals (Bloemraad 2006, 2018).

In this chapter, I examine place-belongingness in the context of the place of destination (i.e., feeling at home in New York or London) as well as in the continued ties and attachment of the participants to the Philippines (both through imaginations and actual transnational engagements). As migrants inhabit transnational spaces and maintain transnational relations, parallel and interconnected claims-making, recognition, and belonging are also examined in both destination country and place of origin. The chapter concludes by looking into the instances when Filipino migrants encountered or practiced exclusion in their everyday lives as they interact with various actors in different social settings. In this regard, migrants' deservingness claims and struggle for belonging and recognition based on the kind of care work that they do are incorporated within the notion of politics of belonging.

6.4 Filipino Care Workers in New York and London: Care Work and Migration in Global Cities

While global cities are important hubs in a globalized world and profoundly linked, embedded, and integrated within the global economy, they are also in need of cheap labor. As Robinson (2009, 16) puts it, "a global city's 'glamour,' observes Sassen, is often supported by large populations of immigrant workers who perform the blue-collar, industrial, low-wage, menial – in short, the 'dirty work' – of the global economy." Global cities can therefore be typified as having, in the words of Bloemraad (2013, 34), "migrant-attracting labor market structures."

However, Filipinos in London and New York not only are employed in low-skilled occupations but also have considerable presence as professionals and highly skilled workers. This is another reason that makes London and New York suitable as comparative cases in this study. In terms of Filipino professionals, what usually stand out are the health and social care workers, particularly nurses. Both the US and the UK have practiced active recruitment of internationally educated registered nurses due to staff shortages (Matsuno 2009). Apart from nurses, domestic work is also a prominent occupation of Filipinos both in London and New York. In between nurses and domestic workers, there are care workers who could be nurses in the Philippines but are just awaiting to pass the exam or to finish the processing of their certification allowing them to practice as registered nurses in London or New York. There are also former domestic workers who underwent some training and certification or simply gained the experience to become caregivers or nursing aides. Some of them are employed in hospitals as nursing aides or assistants, while others are in nursing or residential homes as care support workers. There also caregivers working and/or living in their employers' residences. Despite the different work settings, they all perform almost the same tasks – primarily personal care (e.g., feeding, bathing, or dressing).

The Philippines, as one of the largest labor exporters in the world (Asis 2006; Guevarra 2010; Rodriguez 2010), responds to staffing shortages in health care and service sectors by deploying Filipino workers overseas. The considerable presence of Filipino migrant workers in the health and care sectors of London and New York made them, for better or for worse, more visible to the public. Whether such visibility contributes to "better" integration and recognition or further exclusion and discrimination remains to be explored.

6.5 Data and Methods

The data utilized for this paper were collected during fieldwork in London from the last week of March to the third week of September 2015; the New York part of the fieldwork was from October 2015 to mid-April 2016. The field sites are expanded to Greater London Urban Area and New York metropolitan area. In the case of London, participants are located (working and/or residing) in Greater London and in a town northwest of London (in east of England).[7] In New York, the areas covered are the five boroughs (Manhattan, Queens, Brooklyn, Bronx, and Staten Island) as well as areas on the other side of the Hudson River, facing Manhattan.[8]

In total, there were 134 completed interviews: 58 completed interviews in London, 20 nurses, 20 domestics, and 18 care workers, and 76 completed interviews in New York, 27 nurses, 26 domestics, and 23 care workers. Most interviews were conducted face-to-face. In New York, however, there were requests for interviews to be conducted through FaceTime or Skype. While scheduling adjustments were made, there were still four interviews that were done via Skype as it was the most convenient for those respondents. Face-to-face interviews took place in different settings – homes or friend's homes, cafes/restaurants, workplaces (hospitals or employer's residence), churches, malls, and own business establishments. The participants were asked if they would agree to be interviewed and for the interview to

[7] Though this town is technically outside the administrative region of Greater London (32 boroughs and the City of London), it is still within the Greater London Urban Area and connected to Central London by the London Underground. One respondent – a nurse – works/resides here, and I decided to include her since this provides an opportunity to also account for conditions that shape the decision of some nurses *not* to move to Central London.

[8] In the case of New York, Filipinos have considerable presence in Hudson and Bergen counties, as well as in other parts of northern portion of New Jersey. Jersey City, for instance, has a thriving Filipino community, reminiscent of Queens. Most of the participants living in these areas are working in New York, while others tried working in New York in the past. For these reasons, I expanded the field site to include these areas.

be recorded after the research was explained. Instead of written informed consent, I asked the participants if I could record their consent instead.[9] Interviews are also simultaneously conducted in both English and Filipino to approximate normal conversation and to make the respondents as comfortable as possible.

While the main focus of the research projects was support networks of Filipino migrants in global cities, interview questions pertaining to life satisfaction and integration experiences were also asked (e.g., satisfaction with current job and income; present concerns; feeling of belongingness; voting participation; Filipino and non-Filipino friendships; participation in Filipino community). Through these questions, themes surrounding care work, deservingness claims, and feeling of belonging were explored.

6.6 Claims-Making and Belonging Through Care Work: The Deserving "Good Migrants"

Consistent with the previous research findings on the conditionality of migrants' own view of their deservingness (Timonen and Doyle 2009; Osipovič 2015; Kremer 2016; Alho and Sippola 2019), the interviewees in this study also utilized desert-based principles (Osipovič 2015) to construct not just their welfare deservingness but their overall worth as part of the "host society." In the participants' narratives, articulations of belonging are intimately tied to what they are able to *give* or *contribute*, most especially through their occupation. Going beyond the general (and concrete) contributions through working, paying taxes, and obeying the law, migrant care workers delve deeper into what makes the work they do important and significant for the "host society" – claiming that they are necessary and indispensable. Nurses feel that they are able to *make a difference* in this part of the world and that they matter because the system cannot properly run without them – the Filipino nurses. Care workers consider themselves as valuable part of the society because of the support and care they give to the elderly and the vulnerable.

> Because we are in health care. Health and social care. Those we are supporting are, let's say, vulnerable individuals. So, we are the ones protecting them. We are there so that they are not abused or get exploited. We are there. So, we are integrated. As in we have a role, we have a role in the community. We are not just maids. We are not just paid to clean [the patients]. It's like, we are essential part, valuable part of this society. That's why they recruited there [in the Philippines], right? Because that's what they lack here, right? Although the view of some is that we are being paid to clean, bathe the patients, and that's it. But behind that, you have a higher purpose, of what you are.
> (Carlo, 40 years old, care worker respondent, London sample)

---

[9] Recording the consent instead of asking the participants to sign a form is a decision based on previous experiences of interviewing Filipinos. There was already a great deal of doubt with regard to being interviewed and being asked about one's personal life (especially for undocumented migrants), and asking them to affix their signature is deemed to be too invasive and will not foster rapport and trust building.

> Because the UK cannot function without us. I think. Because the truth of the matter is, the reason why they are hiring Filipino nurses is not only because it is cheaper to do so but simply because no one [from here] wants to be a nurse. [...] Very few of them. If you would go to our unit, and that of [name of batchmate], our units are just beside each other; 80% are born outside of the UK. Everybody's from another place. And then, from those 80%, 40% are Filipinos. That's how many.
> (Richard, 32 years old, nurse respondent, London sample)

These narratives, however, bring to the fore tensions and ambivalences surrounding interviewees' deservingness claims and struggles to be recognized as valuable members of "host society." First, there is the awareness of the demeaning and low status accorded to doing care work and that it is migrant labor that has to take over these "unwanted" jobs. While the respondents emphasized that their labor is indispensable, such claim is also contextualized given how care work and reproductive labor are generally (under)valued. Second, it should also be noted that deservingness claims can also mean distancing oneself from perceived "undeserving" others (Monforte et al. 2019). Such distancing can be observed given the occupational hierarchy within the care sector. It is within the context of care work that boundaries are most porous and contentious when it comes to distinguishing oneself from others deemed as belonging to lower status. Much of what is "dirty" and "demeaning" about care work centers on dealing with the body – especially its wastes and dirt. Whereas being a staff nurse allows one to lay claim on technical and medical know-how (e.g., inserting I.V.), "care of the elderly often involves dealing with bodily effusions and excrement, and the hands-on intimate care that comes with cleaning elderly bodies" (Huang et al. 2012, 199). Apart from care work being a form of bodywork (Twigg 2000), the proximity of doing care work with domestic work is another dimension that makes caregiving degrading (e.g., cooking, mopping the floor, or washing the dishes). Distinctions can also be found in different sites where care is being provided and, in relation to that, the different job titles given to almost the same kind of tasks that they are expected to perform – only in different settings. Care workers in domiciliary setting are categorized as "low-skilled," while care being provided in institutional spaces is seen as more formalized and professionalized. In the UK, those working in hospitals are called *health-care assistants* (HCAs). In the USA, there are several job titles and corresponding certifications for direct care providers – e.g., certified nursing assistants (CNAs) and home health aides (HHAs). These various forms of creating distinctions in relation to care work point to ambiguities in boundaries and how such boundaries are constantly being challenged and negotiated. As Amrith (2010, 421) puts it, "social and class distinction is most hard won when one's positioning in society is not so different from the positions from whom one is distancing oneself." Thus, by distinguishing oneself and the work one does and to mitigate the lowly status accorded to care work, migrant care workers employ moral currency – their "higher purpose" and their valuable role in the community. In this case, such role is defined as caring and protecting the vulnerable.

What is therefore interesting in these narratives is how the participants carve their "rightful" place in the context of (perceived) underappreciation for the kind of (care) work that they do. It is as if they need to assert and legitimize their sense of belongingness and recognition as migrant workers. They do not just refer to themselves as individuals but as a collectivity – as Filipinos, as migrant care workers – as they make a case for the contribution of the presence of this particular group in the "host" country. In the same way, domestic workers also establish their sense of belonging in New York or London by highlighting their invaluable role – they feel that without them, their employers will also not be able to work properly:

> Because if not for us, our employers will not be able to work properly. So, in that sense, […] we are also helping the economy because they are able to work well. […] And the reason why they are hiring us is because no employer would do the work that we are doing.
> (Rebecca, 50 years old, domestic worker respondent, New York sample)

This feeling of belongingness is also validated by the recognition that participants get from their work and is embedded into the larger narrative underscoring that Filipinos are deemed as "good" workers and are valued for the quality of their work. Nurses feel accepted by their colleagues and do get promoted (albeit, not without issues). Being trusted by their employers and being treated "like a family," domestic workers get a sense of acknowledgment of their presence and contribution (cf. Näre 2011). These constructions of belongingness are more akin to the notion of citizenship, wherein acknowledgment and recognition are deemed as definitive markers of belonging. Such narratives of belonging also include concurrent obligations of paying tax and following rules and law of the country.

> Rizza: Do you see yourself as part of this society?
> Mariel: Yes, of course, because we share the tax and all. Yes, we work. We are not asking benefits from the government, no? We pay our own taxes.
> (Mariel, 44 years old, care worker respondent, London sample)

Given these articulations, belonging can thus be seen as being construed as something that one must earn and deserve. The research participants positioned themselves as valuable and important, thereby also differentiating themselves from the stereotypical image of migrants who are just after the benefits they can get from the "host" country. As Andrews (2017) highlighted, the danger in utilizing and adopting the normative ideals of a "good migrant" for claims-making could reinforce and legitimize exclusion of those who cannot (or choose not to) conform to such normative ideals.

6.7 Belonging and Class Position in Transnational Context

The financial security that they were able to attain as nurses, care workers, or domestics in London and New York allows them not only to become a part of the "host society" but, more importantly, of their home country. Being able to afford the "good life" means that they are not (or no longer) located at the periphery in their homeland – a feeling that is reinforced through visits and symbolic presence. As Aguilar (1999) discussed, overseas employment not only transformed the lives of migrants and their nonmigrant kin. They also acquired "new sense of self" and an elevated status back home. "Regardless of the type of work, the migrant worker attains a prestigious new self in the place of origin, with the status of a generous financial saviour to the kin group and becomes an enviable role model to others" (Aguilar 1999, 105).

Thus, being able to work in London or in America also allowed them to realize their dreams, provide for their families, or even feel free from economic and cultural constraints. Thus, another form of belongingness is articulated through the opportunities and freedom that they were able to enjoy despite their ambivalent position as migrant care workers.

Rizza: Do you feel that you belong here? That you're part of this country?
Erica: Yes, of course.
Rizza: In what way do you feel that you are no longer an outsider?
Erica: In every way because here in New York, you have freedom. You're free to do whatever you want.
(Erica, 41 years old, care worker respondent, New York sample)

What I like here is that I was able to accomplish my promise to my parents [to support them]. But if I was in the Philippines, I wouldn't be able to do that.
(Glenda, 49 years old, domestic worker respondent, New York sample)

Rizza: In what way do you feel that you belong here?
Linda: Of course, the life that you have now here, you're not rich, but as I have told you, what is good here is that … what the poor can do, what the rich can do, you can do that as well. You are able to go to the shows, everything! Like, there is no discrimination. You are able to eat; you are able to do everything. […] You are not being looked down upon, whatever kind of person you are, even though you're poor and regardless of your educational background.
(Linda, 58 years old, domestic worker respondent, New York sample)

I think if I did not come here, I would not go this far. I don't know. [Rizza: And you see yourself as part of this country? Like you feel that you already belong here?] Yeah, I think so. […] Because we are already well-adjusted here. We have stable jobs already. […] There is no hindrance to attain your dream, if ever you have a dream. It's available. It's up to you to grab it.
(Marie, 41 years old, nurse respondent, New York sample)

As the study of Gelatt (2013) on Asian and Latino immigrants in the USA shows, most migrants maintain a "dual frame of reference" – they tend to foster reference groups and assess their social positions in both countries of origin and destination through transnational practices and engagements. While Filipinos experience exploitation and discrimination inside and outside of their workplaces, the "gain" of being able to earn considerably more compared to what they can earn in the Philippines, and to be able to afford and consume goods and leisure that are not available for them (and their families) before, overrides these negative experiences. Therefore, the frame of reference is one's socioeconomic mobility in the Philippines and not one's racialized and discriminated position overseas. This condition has been studied among migrants employed in low-status jobs as captured by Parreñas' (2001, 2015) concept of "contradictory class mobility" to refer to the experience of Filipino domestic workers of being downgraded with the kind of job they do but, at the same time, earning more compared to their previous (professional) work in the

Philippines. In the same vein, the work of Kelly (2012) on Filipino migration in Canada also points out to the importance of "transnational frame of reference" in understanding the experience of "deprofessionalization and deskilling" in destination country as migrants evaluate and re-evaluate their class positions in different contexts. Beyond the experience of Filipinos overseas, the work of Suárez-Orozco and Suárez-Orozco (1995) on Latino adolescents in the USA suggests that the first-generation immigrants use the notion of "dual frame of reference" as they compare their dire condition in the country of origin and the perceived opportunities they have in the USA. In this sense, they are able to bear the hardships they are experiencing in the destination country. Another illustration is Nieswand's (2011) "transnational status paradox of migration." Using his research on Ghanaian migrants, Nieswand (2011, 150) examines forms of status inconsistency as migrants' "status gain in the country of origin relies on a simultaneous loss of status in the receiving country" – echoing the findings of authors previously mentioned.

Thus, despite "contradictory class mobility" (Parreñas 2001, 2015) of being more financially secured while feeling degraded for doing low-status work, Filipino migrants are able to imagine their trajectories and migration projects as successful or at least having the potential to be successful – with "success" being primarily oriented toward one's country of origin. The struggle then to belong and be recognized as "valuable" member of the society could also be a simultaneous struggle in both countries of origin and destination.

6.8 Exclusion and Barriers to Claims-Making: "Illegality" and Care Work

However, despite these various forms and different layers of belonging, there are also research participants who feel that they remain as an outsider in the place of destination – and, to a certain extent, in their home country. The most apparent articulation of being excluded is rooted in one's immigration status. Having no legal papers does not only render undocumented migrants as ineligible for benefits and social protection provided by the host country, but they also feel unsafe and insecure because of their irregular status.

> That's [lack of legal papers] the main essence why I am not happy. [...]
> I feel like you're a second-class citizen and you have nothing to be proud of.
> (Vicky, 53 years old, care worker respondent, New York sample)
> Rizza: Did you ever feel that you don't belong here [even] after 8 years [of being here]?
> Evita: Yeah, because that's [legal papers] really a very big factor because staying here without legal papers means I am always in an uncertain position.
> (Evita, 57 years old, domestic worker respondent, London sample)

Owing to their irregular status, undocumented migrants also tend to voluntary exclude themselves from civic participation and in building meaningful relationships so as to remain invisible and not risk being exposed. As one respondent put it, "you avoid getting involved and being recognized in the society." Their involvement is usually limited within their intimate networks and organizations (such as religious groups and ethnic associations) where they feel safe and secured.

The need to claim for the "right to belong" and "deservingness to be here" becomes more acute for undocumented research participants who highlight that they are not being a burden in this country and are not breaking any law. However, the feeling of being "at home" and safe greatly comes from their participation and membership in organizations such as faith-based groups and hometown associations. For instance, extant research on the role of religion and religiosity in the lives of migrants highlight how churches and ethnic faith-based organizations serve as safe spaces and havens (especially for undocumented migrants) while also providing assistance (such as finding job and accommodation) and serving as places to meet other people (Hagan and Ebaugh 2003; Odem 2004; Lorentzen and Mira 2005; Ley 2008; Nakonz and Shik 2009; Watson 2009; Ahmed 2010). Efren, an undocumented migrant in New York, made similar reference to the importance of the church:

> You know why? Like why you feel that you belong? At church. The church is there. Perhaps, if the church is not there, then you'll feel weak. You would have given up a long time ago.
> (Efren, 52 years old, care worker respondent, New York sample)

As previously discussed, another source of feeling of exclusion is the kind of low-status work that one does. Inasmuch as care work allows participants to see themselves as deserving to be in the place of destination given their contribution, being employed in what is deemed as low-skilled, low-status job can also make migrants feel that they are living on the margins of the "host" society and that they do not really belong. Viola, who was previously employed as a registered nurse in Vienna and currently works as a nanny in New York, shared: "I do not feel that I belong here. [...] My self-confidence has gone down here. [...] Perhaps it's because of my work." In this sense, belonging pertains to being part of one's reference group. Since to say that one is a part of something also means creating distinctions and making boundaries apparent, it is inevitably exclusionary and, thus, political (Anthias 2016). However, since migrants also position and orient themselves in their home country – where most of them eventually intend to return – ensuring "success" in that context remains a top priority.

6.9 Conclusion: Of Ambivalence, Dislocation, and Recognition

This paper seeks to examine how migrants utilized deservingness frames derived from moral dimensions of care work to claim belongingness and recognition in both the place of destination and country of origin. Given that migrant care labor is both needed, undervalued, and, at times, unwanted, migrants are "differentially included" in the "host society" wherein they can face downward mobility and get limited access to resources and state welfare. The particularities of care work provide another layer of ambivalence in the position, lived experiences, and future trajectories of migrant care workers. The shame accompanying doing care work is intertwined with opportunities for mobilities that migrants take to attain good life and better future for themselves and their families.

In order to mitigate "differential inclusion" and negotiate their position as "deserving migrants," interviewees highlighted their contributions as care workers by claiming that what they do is indispensable, valuable, and for the greater good. Beyond the general deservingness frame tied to the notion of a "good citizen" who is employed and law abiding, the moral economy of care work enables migrants to utilize the moral currency of care provision. That is, while care work is paid, the value of such work (e.g., protecting and caring for the vulnerable) should be more than the monetary payment they receive. Therefore, as care workers, migrants are claiming their "rightful" place and worth as part of "host society" through the invaluable contributions that they are able to provide given the kind of work that they do, albeit the nonrecognition (or under-recognition) of such work in the eyes of the general public. Such claims are substantiated by the respect and esteem they gain from their colleagues, clients, and employers.

Likewise, as a salve to both institutionalized and informal exclusionary practices they continue to face as migrants and as care workers, respondents emphasized how they are able to have more freedom and resources to pursue their goals and perform their duty and obligations in their home country. Thus, "dual frame of reference" and transnational engagements allow migrant care workers to also claim belongingness and recognition in their country of origin and mitigate the difficulties and challenges of geographical dislocation and contradictory class mobility. Such ambivalences highlight the tensions and intersections between structural constraints and agentic actions. Thus, locating migrants' deservingness claims and struggle for belonging and recognition within the contexts of differential inclusion, dislocation, and contradictory class mobility demonstrates claims-making as a structured agency (Bloemraad 2018).

In this way, Filipino migrant care workers in New York and London were able to assert what they perceived as their "rightful place" by accumulating moral currency and capitalizing on idealized notions of a "good migrant/carer" amidst the backdrop of devaluation of care work and migrant care workers.

Further research can examine whether in the current time of pandemic, perceptions toward care work can change and migrant care workers could be accorded better treatment by the host society. In contrast to the observed increase support for welfare chauvinism in recent times, it remains to be seen whether the current health crisis would sway public opinion and sentiments to recognize the value of (migrant) care workers. In such context, it is interesting to explore how such value can add to the moral currency that migrant care workers can use to frame their deservingness claims and to struggle for greater recognition.

References

Aarøe, L., and M.B. Petersen. 2014. Crowding Out Culture: Scandinavians and Americans Agree on Social Welfare in the Face of Deservingness Cues. *The Journal of Politics* 76 (3): 684–697. https://doi.org/10.1017/S002238161400019X.

Aguilar, F.V., Jr. 1999. Ritual Passage and the Reconstruction of Selfhood in International Labour Migration. *Sojourn: Journal of Social Issues in Southeast Asia 14* (1): 98–139. http://www.jstor.org/stable/41057014.

Ahmed, A. 2010. 'I Need Work!': The Multiple Roles of the Church, Ranking and Religious Piety Among Domestic Workers in Egypt. *The Asia Pacific Journal of Anthropology* 11 (3–4): 362–377. https://doi.org/10.1080/14442213.2010.514940.

Alho, R., and M. Sippola. 2019. Estonian Migrants' Aspiration for Social Citizenship in Finland: Embracing the Finnish Welfare State and Distancing from the 'Non-deserving'. *Journal of International Migration and Integration* 20 (2): 341–359. https://doi.org/10.1007/s12134-018-0606-9.

Amrith, M. 2010. 'They Think We Are Just Caregivers': The Ambivalence of Care in the Lives of Filipino Medical Workers in Singapore. *The Asia Pacific Journal of Anthropology* 11 (3–4): 410–427. https://doi.org/10.1080/14442213.2010.511631.

Andrews, A. 2017. Moralizing Regulation: The Implications of Policing "Good" Versus "Bad" Immigrants. *Ethnic and Racial Studies* 41 (14): 2485–2503. https://doi.org/10.1080/01419870.2017.1375133.

Anthias, F. 2016. Interconnecting Boundaries of Identity and Belonging and Hierarchy-Making Within Transnational Mobility Studies: Framing Inequalities. *Current Sociology* 64 (2): 172–190. https://doi.org/10.1177/0011392115614780.

Antonsich, M. 2010. Searching for Belonging – An Analytical Framework. *Geography Compass* 4 (6): 644–659. https://doi.org/10.1111/j.1749-8198.2009.00317.x.

Asis, M.M.B. 2006. *The Philippines' Culture of Migration. Migration Information Source*. Washington, DC: Migration Policy Institute. http://www.migrationpolicy.org/article/philippines-culture-migration. Accessed 16 May 2014.

Bach, S. 2010. Managed Migration? Nurse Recruitment and the Consequences of State Policy. *Industrial Relations Journal* 41 (3): 249–266. https://doi.org/10.1111/j.1468-2338.2010.00567.x.

Bloemraad, I. 2006. *Becoming a Citizen: Incorporating Immigrants and Refugees in the United States and Canada*. Berkeley: University of California Press.

———. 2013. The Promise and Pitfalls of Comparative Research Design in the Study of Migration. *Migration Studies* 1 (1): 27–46. https://doi.org/10.1093/migration/mns035.

———. 2018. Theorising the Power of Citizenship as Claims-Making. *Journal of Ethnic and Migration Studies* 44 (1): 4–26. https://doi.org/10.1080/1369183X.2018.1396108.

Bloemraad, I., W. Kymlicka, M. Lamont, and L.S. Son Hing. 2019. Membership Without Social Citizenship? Deservingness & Redistribution as Grounds for Equality. *Daedalus* 148 (3): 73–104. https://doi.org/10.1162/daed_a_01751.

Cangiano, A., and I. Shutes. 2010. Ageing, Demand for Care and the Role of Migrant Care Workers in the UK. *Journal of Population Ageing* 3 (1–2): 39–57. https://doi.org/10.1007/s12062-010-9031-3.

Carling, J. 2008. The Human Dynamics of Migrant Transnationalism. *Ethnic and Racial Studies* 31 (8): 1452–1477. https://doi.org/10.1080/01419870701719097.

Carrillo Rowe, A. 2005. Be Longing: Toward a Feminist Politics of Relation. *NWSA Journal* 17 (2): 15–46.

Cases, R.K.C. 2018. *Changing Ties, Ambivalent Connections: Mobilities and Networks of Filipinos in London and New York Metropolitan Areas*. Doctoral dissertation, University of Trento.

Chauvin, S., and B. Garcés-Mascareñas. 2014. Becoming Less Illegal: Deservingness Frames and Undocumented Migrant Incorporation. *Sociology Compass* 8 (4): 422–432. https://doi.org/10.1111/soc4.12145.

Choy, C.C. 2000. “Exported to Care”: A Transnational History of Filipino Nurse Migration to the United States. In *Immigration Research for a New Century: Multidisciplinary Perspectives*, ed. N. Foner, R.G. Rumbaut, and S.J. Gold, 113–133. New York: Russell Sage Foundation.

———. 2003. *Empire of Care: Nursing and Migration in Filipino American History*. Durham: Duke University Press.

Espiritu, Y.L. 2003. *Home Bound: Filipino American Lives Across Cultures, Communities and Countries*. Berkeley/Los Angeles: University of California Press.

Fassin, D. 2005. Compassion and Repression: The Moral Economy of Immigration Policies in France. *Cultural Anthropology* 20 (3): 362–387. https://doi.org/10.1111/soc4.12145.

———. 2009. Moral Economies Revisited. *Annales. Histoire, Sciences Sociales*, 64 (6), 1237–1266. https://www.cairn-int.info/journal-annales-2009-6-page-1237.htm.

FitzGerald Murphy, M. 2014. Global Care Chains, Commodity Chains and the Valuation of Care: A Theoretical Discussion. *American International Journal of Social Science* 3 (5): 191–199.

Gelatt, J. 2013. Looking Down or Looking Up: Status and Subjective Well-being Among Asian and Latino Immigrants in the United States. *International Migration Review* 47 (1): 39–75. https://doi.org/10.1111/imre.12013.

Guevarra, A.R. 2010. *Marketing Dreams, Manufacturing Heroes: The Transnational Labor Brokering of Filipino Workers*. New Brunswick: Rutgers University Press.

———. 2014. Supermaids: The Racial Branding of Global Filipino Care Labour. In *Migration and Care Labour: Theory, Policy, and Politics*, ed. B. Anderson and I. Shutes, 130–150. Basingstoke: Palgrave Macmillan.

Hagan, J., and H.R. Ebaugh. 2003. Calling Upon the Sacred: Migrants' Use of Religion in the Migration Process. *International Migration Review* 37 (4): 1145–1162. https://doi.org/10.1111/j.1747-7379.2003.tb00173.x.

Hess, S. 2008. The Boundaries of Monetarizing Domestic Work: Au Pairs and the Moral Economy of Caring. In *Migration and Mobility in an Enlarged Europe. A Gender Perspective*, ed. S. Metz-Göckel, M. Morokvasic, and A. Senganata Münst, 141–156. Leverkusen: Verlag Barbara Budrich.

Hochschild, A. 2000. Global Care Chains and Emotional Surplus Value. In *On the Edge: Living with Global Capitalism*, ed. W. Hutton and A. Giddens, 130–146. London: Jonathan Cape.

Huang, S., B.S. Yeoh, and M. Toyota. 2012. Caring for the Elderly: The Embodied Labour of Migrant Care Workers in Singapore. *Global Networks* 12 (2): 195–215. https://doi.org/10.1111/j.1471-0374.2012.00347.x.

Jørgensen, M.B., and T.L. Thomsen. 2016. Deservingness in the Danish Context: Welfare Chauvinism in Times of Crisis. *Critical Social Policy* 36 (3): 330–351. https://doi.org/10.1177/0261018315622012.

Katigbak, E.O. 2015. Moralizing Emotional Remittances: Transnational Familyhood and Translocal Moral Economy in the Philippines' 'Little Italy'. *Global Networks* 15 (4): 519–535. https://doi.org/10.1111/glob.12092.

Kelly, P.F. 2012. Migration, Transnationalism, and the Spaces of Class Identity. *Philippine Studies: Historical and Ethnographic Viewpoints* 60 (2): 153–185. https://doi.org/10.1353/phs.2012.0017.

Kofman, E. 2012. Rethinking Care Through Social Reproduction: Articulating Circuits of Migration. *Social Politics* 19 (1): 142–162. https://doi.org/10.1093/sp/jxr030.

Kofman, E., and P. Raghuram. 2015. *Gendered Migrations and Global Social Reproduction*. New York: Palgrave Macmillan.

Kootstra, A. 2016. Deserving and Undeserving Welfare Claimants in Britain and the Netherlands: Examining the Role of Ethnicity and Migration Status Using a Vignette Experiment. *European Sociological Review* 32 (3): 325–338. https://doi.org/10.1093/esr/jcw010.

Kremer, M. 2016. Earned Citizenship: Labour Migrants' Views on the Welfare State. *Journal of Social Policy* 45 (3): 395–415. https://doi.org/10.1017/S0047279416000088.

Ley, D. 2008. The Immigrant Church as an Urban Service Hub. *Urban Studies* 45 (10): 2057–2074. https://doi.org/10.1177/0042098008094873.

Lorentzen, L.A., and R. Mira. 2005. *El milagro está en casa*: Gender and Private/Public Empowerment in a Migrant Pentecostal Church. *Latin American Perspectives* 32 (1): 57–71. https://doi.org/10.1177/0094582X04271852.
Lowe, L. 2006. Foreword. In *Positively No Filipinos Allowed: Building Communities and Discourse*, ed. A.T. Tiongson Jr., E.V. Gutierrez, and R.V. Gutierrez, vii–ix. Philadelphia: Temple University Press.
Matsuno, A. 2009. Nurse Migration: The Asian Perspective. In *ILO/EU Asian Programme on the Governance of Labour Migration Technical Note 1*. Bangkok: International Labour Organization (ILO) Regional Office for Asia and the Pacific. http://www.ilo.org/wcmsp5/groups/public/%2D%2D-asia/%2D%2D-ro-bangkok/documents/publication/wcms_160629.pdf. Accessed 2 June 2020.
Monforte, P., L. Bassel, and K. Khan. 2019. Deserving Citizenship? Exploring Migrants' Experiences of the 'Citizenship Test' Process in the United Kingdom. *The British Journal of Sociology* 70 (1): 24–43. https://doi.org/10.1111/1468-4446.12351.
Nakonz, J., and A.W.Y. Shik. 2009. And all Your Problems Are Gone: Religious Coping Strategies Among Philippine Migrant Workers in Hong Kong. *Mental Health, Religion and Culture* 12 (1): 25–38. https://doi.org/10.1080/13674670802105252.
Näre, L. 2011. The Moral Economy of Domestic and Care Labour: Migrant Workers in Naples, Italy. *Sociology* 45 (3): 396–412. https://doi.org/10.1177/0038038511399626.
Nielsen, M.H., M. Frederiksen, and C.A. Larsen. 2020. Deservingness Put into Practice: Constructing the (Un) Deservingness of Migrants in Four European Countries. *The British Journal of Sociology* 71 (1): 112–126. https://doi.org/10.1111/1468-4446.12721.
Nieswand, B. 2011. *Theorising Transnational Migration: The Status Paradox of Migration*. New York/London: Routledge.
Odem, M.E. 2004. Our Lady of Guadalupe in the New South: Latino Immigrants and the Politics of Integration in the Catholic Church. *Journal of American Ethnic History* 24 (1): 26–57.
Osipovič, D. 2015. Conceptualisations of Welfare Deservingness by Polish Migrants in the UK. *Journal of Social Policy* 44 (4): 729–746. https://doi.org/10.1017/S0047279415000215.
Parreñas, R.S. 2000. Migrant Filipina Domestic Workers and the International Division of Reproductive Labor. *Gender and Society* 14 (4): 560–580. https://doi.org/10.1177/089124300014004005.
———. 2001. *Servants of Globalization: Women, Migration, and Domestic Work*. Stanford: Stanford University Press.
———. 2012. The Reproductive Labour of Migrant Workers. *Global Networks* 12 (2): 269–275. https://doi.org/10.1111/j.1471-0374.2012.00351.x.
———. 2015. *Servants of Globalization: Migration and Domestic Work*. 2nd ed. Stanford: Stanford University Press.
Petersen, M.B., R. Slothuus, R. Stubager, and L. Togeby. 2011. Deservingness Versus Values in Public Opinion on Welfare: The Automaticity of the Deservingness Heuristic. *European Journal of Political Research* 50 (1): 24–52. https://doi.org/10.1111/j.1475-6765.2010.01923.x.
Reeskens, T., and T. van der Meer. 2019. The Inevitable Deservingness Gap: A Study into the Insurmountable Immigrant Penalty in Perceived Welfare Deservingness. *Journal of European Social Policy* 29 (2): 166–181. https://doi.org/10.1177/0958928718768335.
Reeskens, T., and W. Van Oorschot. 2013. Equity, Equality, or Need? A Study of Popular Preferences for Welfare Redistribution Principles Across 24 European Countries. *Journal of European Public Policy* 20 (8): 1174–1195. https://doi.org/10.1080/13501763.2012.752064.
Robinson, W.I. 2009. Saskia Sassen and the Sociology of Globalization: A Critical Appraisal. *Sociological Analysis 3* (1): 5–29. Retrieved from http://escholarship.org/uc/item/44j854qc. Accessed 29 May 2018.
Rodriguez, R.M. 2010. *Migrants for Export: How the Philippine State Brokers Labor to the World*. Minneapolis: University of Minnesota Press.
Scott, J.C. 1976. *The Moral Economy of the Peasant: Rebellion and Subsistence in Southeast Asia*. New Haven: Yale University Press.

Shutes, I., and B. Anderson. 2014. Introduction. In *Migration and Care Labour: Theory, Policy, and Politics*, ed. B. Anderson and I. Shutes, 1–7. Basingstoke: Palgrave Macmillan.
Suarez-Orozco, C., and M.M. Suárez-Orozco. 1995. *Transformations: Immigration, Family Life, and Achievement Motivation Among Latino Adolescents*. Stanford: Stanford University Press.
Thompson, E. 1971. The Moral Economy of the English Crowd in the Eighteenth Century. *Past & Present* 50 (1): 76–136.
Timonen, V., and M. Doyle. 2009. In Search of Security: Migrant Workers' Understandings, Experiences and Expectations Regarding 'Social Protection' in Ireland. *Journal of Social Policy* 38 (1): 157–175. https://doi.org/10.1017/S0047279408002602.
Twigg, J. 2000. Carework as a Form of Bodywork. *Ageing and Society* 20 (4): 389–411. https://doi.org/10.1017/S0144686X99007801.
van Oorschot, W. 2000. Who Should Get What, and Why? On Deservingness Criteria and the Conditionality of Solidarity Among the Public. *Policy and Politics* 28 (1): 33–48. https://doi.org/10.1332/0305573002500811.
———. 2006. Making the Difference in Social Europe: Deservingness Perceptions Among Citizens of European Welfare States. *Journal of European Social Policy* 16 (1): 23–42. https://doi.org/10.1177/0958928706059829.
Watson, S. 2009. Performing Religion: Migrants, the Church and Belonging in Marrickville, Sydney. *Culture and Religion* 10 (3): 317–338. https://doi.org/10.1080/14755610903287716.
Yeates, N. 2011. Going Global: The Transnationalization of Care. *Development and Change* 42 (4): 1109–1130. https://doi.org/10.1111/j.1467-7660.2011.01718.x.
———. 2012. Global Care Chains: A State-of-the-Art Review and Future Directions in Care Transnationalization Research. *Global Networks* 12 (2): 135–154. https://doi.org/10.1111/j.1471-0374.2012.00344.x.
Yuval-Davis, N., K. Kannabiran, and U.M. Vieten. 2006. Introduction: Situating Contemporary Politics of Belonging. In *The Situated Politics of Belonging*, ed. N. Yuval-Davis, K. Kannabiran, and U.M. Vieten, 1–14. London: Sage.

Part II
Recognition, Migration Policies and the State

Chapter 7
Work to Be Naturalized? On the Relevance of Hegel's Theories of Recognition, Freedom, and Social Integration for Contemporary Immigration Debates

Simon Laumann Joergensen

Abstract In democratic welfare states, a growing number of immigrants form a permanent part of the citizenry but are not recognized as full members of the demos as they fail to meet the demands set for full political membership. For instance, states may demand work and economic self-support as proof of integration. This paper aims to show that the theory of recognition developed by G.W.F. Hegel could further the contemporary demos-debates concerning legitimate justifications of in- and exclusion substantially. Hegel articulated theories of recognition, freedom, and social recognition of great relevance for evaluating contemporary policies of integration such as demands of work and self-support as a prerequisite for naturalization. Normative political theoretical debates about policies of in- and exclusion of immigrants seldom confront whether work could be understood as a relevant driver and marker of integration. Much can be learned from Hegel's attempt to describe what he took to be a realistic utopia of integrated citizens reproducing freedom through recognition. This analytical model asks whether policies of integration help linking individual and communal identities and types of freedom in ways that promote rather than undermine the freedom of the individuals and the community. The central motor of integration is here taken to be recognition once this takes institutional forms that promote both individual and communal freedom, while transforming both in manners that allow them to link.

Keywords Recognition · Social integration · Freedom · Hegel · Critical theory

S. L. Joergensen (✉)
Department of Politics and Society, Aalborg University, Aalborg, Denmark
e-mail: simonl@dps.aau.dk

G. Schweiger (ed.), *Migration, Recognition and Critical Theory*, Studies in Global Justice 21, https://doi.org/10.1007/978-3-030-72732-1_7

7.1 Introduction

Naturalization policies[1] are contested and complex expressions of states' strong concern with social integration of immigrants (Bosniak 2006; Offe 2011; Owen 2013; Vink and Bauböck 2013). In democratic welfare states, a growing number of immigrants form a permanent part of the citizenry but are not recognized as full members of the demos (Bauböck 2018, 257). Their democratic exclusion is often grounded in their failing to meet the often very high demands to qualify for enfranchisement. Citizenship is normally linked to certain general duties (Mason 2012), but for immigrants, full citizenship is often preconditioned on a set of very specific demands which other citizens do not need to meet. Immigrants need to flag their signs of social integration. Since full citizenship includes the right to vote, it may appear justified to test immigrants for some formalized familiarity with the political system, its norms, and the language of the country (D. Miller 2016). More controversially, some countries demand that aspiring immigrants become economically self-supporting as a requirement for enfranchisement. For instance, Danish policies demand that immigrants "as soon as possible become self-supporting and productive though employment." Though the Danish language requirements are even harder to meet (Jensen et al. 2019), the demand for self-support seems less related to contemporary understandings of *democratic* citizenship than (even strict) language requirements. Is there any potentially justified connection between economical self-support and full citizenship (Kymlicka and Norman 1994, 289)?

Classical republican and liberal thinkers did link economic status to free, democratic participation. Aristotle worried that the rich and the poor might be preoccupied solely with their own narrow interests rather than obeying "the rule of reason" (Politics IV.11.1295b4–6, cf. F. Miller 2017), while John Stuart Mill claimed that persons should "acquire the commonest and most essential requisites for taking care of themselves, for pursuing intelligently their own interests and those of the persons nearly allied to them" (Dahl 1989, 125). Nevertheless, though political and economic status have historically been connected (Marshall 1992), how might contemporary societies justify making economic status a precondition for political status? After all, "'Democratic citizenship' in the 21st century tends to evoke an embracing sort of membership, no longer premised on such factors as socio-economic class, wealth, ethnicity or sex" (Fine 2011, 623). The UN Declaration on Human Rights states that "Everyone has the right to take part in the government of his country." There is no mentioning of self-support or productivity as a precondition.

There is, however, a tension between the states' right to sovereignty and the individuals' right to citizenship, as also Hannah Arendt pointed out (J. Cohen 1996, 172). According to the International Covenant on Civil and Political Rights, "peoples have a right of self-determination" which many take to include the right to determine who should be members of the people (D. Miller 2015, 178). As Jean

[1] Policies regarding the preconditions for acquisition of citizenship on the basis of consent rather than principles of territory (jus soli) or origin (jus sanguinis) (Benhabib 2002, 166f).

Cohen puts it, "the membership component of the citizenship principle is thus particularistic, exclusionary *and* exclusive" (J. L. Cohen 1999, 250). A more optimistic approach to this so-called boundary problem in contemporary political theory is to see the determination of the ideal of citizenship as a "self-correcting learning process" (Habermas 2001, 774; cf. Goodin 2007, 44; compare Honneth 1995; Marshall 1992). Still, as long as the right of immigrants to vote is contested (Beckman 2009; E. F. Cohen 2009), there is a question of justification of power over the process of inclusion as well as of the concrete naturalization policies.

To Sarah Song, peoples have a legitimate pro tanto right to determine who should be part of the people due to the fact that the people is obliged to take care of the reproduction of social institutions necessary for democratic self-determination on which legitimate political authority and political freedom depend (Song 2018). Song adds that once residents have participated in a subset of such institutions over time, they should be granted the right to full participation (ibid.). This approach raises some interesting questions. What level and form of participation are needed and what mechanisms are in play here?

This paper claims that valuable resources for such an analysis can be found in the theories of recognition, freedom, and social integration developed by G.W.F. Hegel.[2] To Hegel, humans are in a constitutive sense vulnerable to the ideas that others have of them (1821/22, §47f). This means that humans have a potential for freedom that is dependent on numerous types of protection and recognition (1821/22, §43, 71, 120, 147, 200, 207, 230, 235f, 253). The basic claim here is that in certain forms, recognition is a precondition for individual self-development toward freedom and a precondition for social integration and that justice refers to the reproduction of such relations of recognition. Thus, justice demands the recognition of individuals in specific ways necessary for freedom, but to realize this, citizens need to take upon themselves the burdens of recognition. In relation to this, societies have a complex task in institutionalizing the rights and duties of recognition. Since the ideal is freedom, the duties cannot be merely enforced from above. The individuals need to come to see the duties as preconditions for their own freedom. This sharpens the focus on reproducing the right kinds of transformative institutions. Basically, individuals are recognized in different ways that allow them to constitute themselves intersubjectively as free. Social integration is the process of transforming individual aspirations and identities to a state of freedom in which their freedom does not contradict the freedom of others, but rather promotes and furthers it.

Focusing on Hegel's theories of recognition, freedom, and social integration, the route taken here is a bit unusual. Normative analyses of immigrants' right to naturalization often center on *justice* and its related principles (Abizadeh 2008; Bauböck 2015; Erman 2013; Goodin 2007; Owen 2011; Song 2016; Walzer 1983; Seglow 2009). Hegel's theory of recognition is seldom related to contemporary debates about immigration and naturalization policies. This is surprising since being

[2] Abbreviations are used in references to Hegel's work. PR = (Hegel 2003); 1821/22 = (Hegel 2005); 1822/23 = (Hegel 1973, iii); 1824/5 = (Hegel 1973, iv).

recognized as a full citizen clearly marks a fundamental normative status (Arendt 2017) which is vital for self-respect (Honneth 1995; Marshall 1992; Rawls 1971; Taylor 1994). One reason might be that Charles Taylor, who set the stage for discussing immigration and recognition, argued that defenders of "the politics of equal dignity" such as Rousseau, Kant, and Hegel left no room for difference and cultural identity (Taylor 1994, 41–52). On this basis, their writings appear superfluous to discussions of the *recognition* of immigrants, which are thought to concern the recognition of cultural identities and difference (Carens 2013, 83; Fraser 1995; D. Miller 2000, 62ff) or questions about how mainstream cultures should react to cultural difference (Galeotti 2002; Kymlicka 1995; Laborde 2008; Parekh 2006).

Whereas the term *recognition* has been applied in migration and citizenship studies of *civil* rights of minority citizens (Modood 2013; Tully 2004) or politics of *difference and identity* (Markell 2003; Patten 2014), the term is rarely applied in studies of the *political* rights of immigrants, not to mention the contemporary policies of naturalization of contemporary democratic welfare states. This is surprising since ascribing others the status of "an equal normative authority" (Forst 2019, 308f) or as mutual participant in the processes of justifying the basic normative principles of society (Rawls 2001, 202) could be seen as a precondition for self-respect and autonomy (Anderson and Honneth 2005), whereas its denial "can inflict damage on those who are denied it" (Taylor 1994, 36). Furthermore, many will claim that questions of malleable institutions and tendencies of moral psychology should be part of a full normative analysis (Gilabert and Lawford-Smith 2012).

As an exception to the general neglect of recognition in relation to political rights, Mark E. Warren recently remarked that the so-called demos-paradox (whether the demos should determine who belongs to the demos) should be analyzed in terms of *recognition* (Warren 2017, 47). Another exception is Rainer Bauböck, whose conception of *stakeholdership* leads to claims for inclusion into a democratic policy (Bauböck 2018, 3) on the basis of human's "strong stakes in being recognized as members of a particular political community" since "membership in a polity is a necessary condition for human autonomy and well-being" (Bauböck 2018, 40; compare Bauböck 2015, 825). Additionally, many researchers would probably also agree that at least concepts related to recognition are important to both empirical and normative studies of such policies. For instance, the empirical studies of access to citizenship focus on what applicants need to fulfill to achieve full status and rights. Likewise, the political theoretical literature focuses on *who* should have (i.e., be recognized) a say in the process of establishing who should belong to the demos (i.e., be recognized), and on *what* normative models and ideals to apply in determining legitimate forms of in- and exclusion (D. Miller 2009).

This paper aims to show that the theories of recognition, freedom, and social integration developed by Hegel could further the contemporary demos-debates substantially. In particular, the paper claims that the normative problems related to naturalization and non-naturalization of immigrants can fruitfully be reviewed and analyzed through Hegel's theories of recognition, freedom, and social integration.

Once it is realized that the objects of recognition are themselves constituted as subjects of recognition through recognition, the question of just recognition relates to questions concerning the practices of social integration and intersubjective individualization (Habermas 1994, 113).

On this basis, it becomes possible to analyze questions of justice and feasibility within a comprehensive, dynamic model. Furthermore, since Hegel links citizenship to social integration as related to social and economic reproduction, going back to Hegel helps rethinking citizenship as linked to not only civil, political, social, and cultural rights but also the integrating practices of work (compare Somers 1994). The fact that Hegel was strongly concerned with social integration to the extent that, for instance, day laborers were left disenfranchised (Beiser 2005, 258) actually speaks in favor of interpreting his political writings in search of potential justifications of making naturalization preconditioned on social integration.[3] What most other contemporary normative approaches to integration miss is the economic integration of immigrants that preoccupies many receiving countries. It is worth turning to Hegel for a nuanced approach to social integration through work.

Though the paper is not about *grounding* principles and rights (compare Benhabib 2013), this approach is not without critical bite, and following critical theorists, I will claim that it is possible to apply the developed theory of recognition as an approach of immanent and integrative critique of contemporary practices (Benhabib 1986) since the different ideals of recognition are already inherent to liberal, democratic welfare states.

In the following sections, I interpret Hegel's theories of recognition, freedom, and social integration. The aim is to sketch a valuable analytical model for analysis of naturalization policies such as those in Denmark. Though short of a full analysis, which would involve not only all relevant legal regulations and their implementation through concrete practices, focusing on the recent concern that immigrants should be economically self-supporting, the paper illustrates how this theoretical approach can reveal problems inherent to making naturalization depend on self-support. Importantly, if the policies miss the dynamic and transformative modes and logics of recognition, they may undermine the very aims of naturalization.

7.2 Hegel on Recognition, Freedom, and Social Integration

Hegel believed that the central individual, social, and political end is the realization of freedom and that it is this end that provides political rule with legitimacy. Hegel attempted to integrate individual aims with the community's need to reproduce itself normatively as a free society. The central point here is that societies can realize

[3] Notice that Hegel criticized the idea that the right to political participation should depend merely on possessions, as was the case in France and England, but defended instead that representative politicians should in a trusted and recognized manner possess virtues on the basis of practical experience with public affairs (1824/5 §309f). The people should afterward provide check on the virtues and qualities of such representatives through the publicity of their political affairs and arguments (ibid., §315).

a set of ambitious normative aims if and only if they institutionalize a background culture of recognition needed for the free flourishing of individuals that will also motivate individual support for the freedom-enabling institutions and practices of recognition. At the level of the normative ideal, the claim is that individual and collective wills can unite without the one giving in to the other. How might this be possible? The short answer is that bringing up citizens well demands a good state (PR §153; 1822/23 §153). Any society that claims that it cannot integrate citizens thus has to confront the challenge that *it* is failing in some central sense.

Confronted with the complex texts of Hegel, understanding Hegel's thoughts on recognition, freedom, and social integration is very much an interpretative challenge. For instance, Hegel speaks of *true* freedom and of *being with oneself* in contrast to *abstract* freedom or being outside of oneself. He talks of needs and aspirations that have undergone a *transforming mediation* (*Vermittlung*) and of being able to combine *universal* and *particular* concerns (1821/22, §11, §186). He also states that the closer rights and duties can be bridged in freedom-realizing modern states, the *stronger* the state (PR §155; 1824/25 §260f). This ideal suggests that we try to organize societies systematically in ways that stimulate bridging particular and universal desires and needs. The following is an attempt to lay out central aspects of this story focusing on Hegel's theories of recognition, freedom, and social integration.

Whereas the idea that individual and social ends could go hand in hand has its roots in the works of Plato and Aristotle, it was Rousseau and Hegel who defended the view that such an ambitious standard of justice could go hand in hand with the individual good including a fully fleshed theory of individual freedom (Neuhouser 2000). Following Rousseau, freedom is a way to live among others with human neediness (Rousseau 1979). Jürgen Habermas writes that "Hegel was the first to argue that we misperceive the basic moral phenomenon if we isolate" the just from the good (1990b, 201). Habermas also seemed to have captured a Hegelian insight when he wrote that the basic human phenomenon is vulnerability which is why we should protect "the web of individual relations of mutual recognition by which these individuals survive as member of a community" (Habermas 1990b, 200). Individual freedom is thus dependent on recognition. Similarly along a Hegelian line, in "Justice and Solidarity," Habermas claimed that

> moral provisions for the protection of individual identity cannot safeguard the integrity of individual persons without at the same time safeguarding the vitally necessary web of relationships of mutual recognition in which individuals can stabilize their fragile identities only mutually and simultaneously with the identity of their group (1990a, 46).

The institutional promotion of individual flourishing and well-being should thus take shape within the frame of protecting and supporting freedom through webs of recognition (Neuhouser 2000, 237). Apart from this preoccupation with freedom, vulnerability, and recognition, importantly, as we will see in the following, Hegel's approach accommodates many of the concerns of central political and ethical theories such as respect for individual freedom and non-domination, the development of autonomy, the realization of welfare as a whole, and the development of virtues as well as identities of membership.

7.3 Three Necessary Images of Recognition

Based on respect for individual freedom and belief in the malleable nature of human needs, Hegel rejected utilitarian and statist attempts to satisfy human desires from above. Instead, he pointed to the constitutive dimensions of recognition, the intersubjective dimensions of freedom, and the individualizing dimensions of socialization and social integration.[4] On this basis, he developed a potentially integrated, dynamic model of freedom that involved both private and civic dimensions. In a central paragraph of *The Philosophy of Right*, Hegel points out how individuals in the modern world are recognized as being persons, subjects, family-members, burghers (i.e., citizens "in the sense of bourgeois"), and human beings and that this recognition matches different institutional forms (PR §190). In combination, these »images« illustrate the complexity of human vulnerability and potentials for freedom. Humans do not only have a body with certain vulnerabilities that needs protection and certain needs to be satisfied, but also need to take possession of their body and mind which includes using their body and producing things, all of which entail different degrees of vulnerability (PR §43, 47f, §52, §57; 1821/22 §43f, §57). To Hegel, humans can realize justice understood as complex freedom if they can combine individual and communal images of recognition such as those he calls »persons«, »subjects«, and »members«.

Recognizing a human as a *person* means that one recognizes that the individual is both "body and spirit" with a potential for and a right to rationality, morality, ethical life, and religion (PR §66). Likewise, it entails the negative recognition that one will not violate the person's body and spirit through arbitrary power and authority. The recognition of the potential for personhood does not come without some (more or less outspoken) obligations, duties, and expectations. As Hegel puts it, the imperative for persons is "be a person and respect others as persons" (PR §36). Empirically, humans are unequal, but as persons, they should all be recognized as equal (*Berliner Enzyklopädie*, §531). What does *respecting a person* mean? To put it negatively, you can fail to respect a person's body or spirit. For Hegel, bodily misrecognition concerns physical movements, control over body, time, and belongings (PR §66). Misrecognition of the spiritual side of the person means "the alienation of intelligent rationality, of Morality, Ethical Life, and religion." This form of alienation is found in "superstition, "when power and authority are granted to others to determine and prescribe" [...] "what actions I should perform" [and] "how I should interpret the dictates of conscience, religious truth, etc." (PR §66). Humans are thus seen as vulnerable to the will and recognition of others as well as their own (PR §48, 51,

[4] As Habermas puts it, humans are "individuated through socialization" (Habermas 1990b, 200).

57, 71, 112). The political system should be established such that the inalienable right of personality is protected (PR §66). However, the vulnerability to alienation cannot be protected by negative means alone nor by the duties understood as the persons' own which leads to concern about the preconditions for developing subjective freedom.

Recognizing somebody as a *subject* involves recognizing the reflective side of the person as well as the side of interests and concerns about leading a good life. It involves asking for reasons and supporting the individual's attempts to lead a good life. In a sense, the recognition of personhood is the recognition of a »negative« freedom of individuality, whereas the recognition of *subjectivity* is the »positive« recognition of reflexive individuality. Thus, in the recognition of another or myself as a person, I *refrain from* taking over the other person or from leaving my own judgment to some random authority. In the recognition of a subject, I ask for reasons, underlying principles, and reflective interest or make sure that public institutions ask for reasons and interests. Hegel calls it the right of knowing, willing, thinking, and judging agents (PR §120, §137f). The right of the subjective will means that "the will can *recognize* something or *be* something only in so far as that thing is *its own*, and in so far as the will is present to itself in it as subjectivity" (PR §107, §132). It is thus the right to find oneself reflected in the outside world, including the right to find satisfaction and realizing the good as a whole (PR §235). Humans are thus recognized as vulnerable to not finding support for their attempts to discover, revise, and fulfill their life plans. This right includes the right to find one's subjective interests (PR §121f), including one's desire for recognition (PR §124), as well as one's reflexive capacities (PR §120) reflected and recognized in and through one's occupation,[5] and consists generally in the idea that citizens have "rights in so far as he has duties" (PR §155). This right to subjectivity relates to duties "to promote *welfare*, one's own welfare and welfare in its universal determination, the welfare of others" (PR §134). This concern with the good as a whole also marks the limits of being recognized as a subject and points toward the recognition of membership.

The recognition of *membership* for Hegel involves the recognition that the person has a social identity (PR §158; 1821/22 §158; 1822/23 §209). The individual needs to take part in mediating institutions that can provide them with communal characters through which they may realize that their personal needs cannot be met on their own (PR §161, 182). If the person participates in the mutual satisfaction of individual needs (rather than merely receiving benefits from others), the person can find self-esteem in belonging to society as a recognized member. Here, Hegel seeks to make care for others and protection of their vulnerability integrate with self-interests (PR §170) as well as virtue and habit (PR §150f) while acknowledging that citizens may sometimes need love and care without being able to reciprocate (PR §242). Care for the protection of the institutions of recognition spring from knowing

[5] As long as this occupation "is valid in and for itself" (PR §124) and in so far as the individual is free which for Hegel means that it does not contradict the basis on which it rests (PR §126).

that membership in the society is a precondition for one's existence, dignity, and possibility for realizing one's particular ends (PR §152). As Hegel puts it, "The universal is a system of all desires and needs" ["Das Algemeine is ein System aller Triebe"] (1822/23, §17). For Hegel, this is a *substantial* rather than an *atomistic* approach to citizenship policies (1822/23 §156). The ideal is thus a *self-conscious* realization of freedom (1822/23 §142). It involves at the individual and collective level integrating membership with personhood and subjectivity in institutionalized relations of freedom and welfare of all over time.

Remember that Hegel's membership image took different forms and that he mentioned family-members, members of political and civil society, and human beings. We can thus be members of different spheres of reproduction, production, and oversight as well as the human society. Still, Hegel does not say that being recognized full political membership was necessary for the freedom of all adults. Now, rather than setting Hegel's theory aside as a normative theory of political integration, I suggest we acknowledge that he is not fundamentally at odds with strong contemporary defenders of democracy such as Habermas and that though we may disagree with some of the things he wrote 200 years ago, there is still a lot to learn from his thoughts on freedom, recognition, and social integration.

Clearly, Habermas much more directly than Hegel states that "political rights must guarantee participation in all deliberative and decisional processes relevant to legislation and must do so in a way that provides each person with equal chances to exercise the communicative freedom to take a position on criticizable validity claims" (Habermas 1996, 127). Still, what is striking is how strong allies the two theorists are. Though acknowledging that Hegel primarily connected the right to elect political representatives to membership in a cooperation (Neuhouser 2000, 205), his views on voting rights were complex and progressive for his time. For instance, Hegel argued that if the principle of suffrage would be the possession of a free will, then women should be included as well "since they are individuals, that is, human beings with a free will" (1824/5, §308).

As many commentators have pointed out, though Hegel may be one of the strongest critics of liberalism in some of its forms (Pippin 2008, 31), he was not an anti-liberal or merely a defender of the institutions of his time (Taylor 1975, 452). He defended constitutional principles, the rule of law, private property, the division of power and public engagement, and freedom of speech along with social protection (Schnädelbach 1997, 258ff). His constitutional monarchy which entailed democracy (1824/25, §272) may not be that different from the present-day polities of the United Kingdom, Spain, the Netherlands, Belgium, Norway, Denmark, and Sweden.

Hegel argued that politics should be based on reasons and constitutional principles rather than brute power or tradition, leaving plenty of room for public exchange of reasons (Hegel 1824/5 §316). Hegel called public opinion the noblest and holiest since it entails insight into justice and human well-being as well as entails subjective freedom which gives humans dignity (1824/5 §316f), and he understood the importance of representation and voice (compare Hegel 1824/5 §301f, 308). Furthermore, Hegel saw the importance of bringing in citizens' practical insight and oversight as well as their expectation to meet criticism from those affected by laws

(PR §301; compare Habermas 1996, 426; compare Young 2000). As he points out, "just as one need not be a shoemaker to know whether one's shoes fit, so is there no need to belong to a specific profession in order to know about matters of universal interest" (PR §215).

Hegel would clearly agree with Habermas that the right to voice also has instrumental grounds. The public sphere partly works as a warning system and must be supported as such. Here, public problems should not only be detected and identified but also thematized in ways that bring them forth in the democratic system (Habermas 1996, 359). Thereby, the political system should recognize the experiences of members of society concerning the failures of public institutions and let them express their grievances through the public sphere (Habermas 1996, 365). To Habermas, the normative criteria for the public sphere should be the inclusion of relevant perspectives and voices (Habermas 1996, 183). These inputs form the background for struggles over recognition that form the basis for the "concrete relations of recognition mirrored in the mutual attribution of rights" (Habermas 1996, 426). Such struggles over recognition take form as contests over the interpretation of needs (Habermas 1996, 426). The public sphere is thus a sphere of both struggles over needs, struggles over recognition, and struggles over the interpretation of needs (Habermas 1996, 314) through which it plays a central role in integrating highly complex societies (Habermas 1996, 352). Hegel might have considered the right to voice more important than the right to vote given that voice is an omnipresent and full-time right that potentially allows you great influence based on the high degree of public openness, Hegel defended, whereas voting takes place rather seldomly and your particular vote is most likely to have no practical influence at all.

Thus, the Hegelian model has close affinities with the political and social thought of Habermas, though the overall relation between their theories is complex (Benhabib 1986; Honneth 2014, 304ff; Pippin 2008, 240, 265, 276f; Williams 2000, 13ff). Importantly, the Hegelian theory is not fundamentally at odds with Habermas' proposal to link individual freedom to a conception of citizenship that "developed out of Rousseau's concept of self-determination" (Habermas 1996, 495; Neuhouser 1993). This is "the idea of autonomy according to which human beings act as free subjects only insofar as they obey just those laws they give themselves in accordance with insights they have acquired intersubjectively" (Habermas 1996, 445f). As Habermas explains, "popular sovereignty signified [...] the transformation of authority into *self-legislation* [...] Such an association is structured by relations of mutual recognition in which each person can expect to be respected by all as free and equal" (Habermas 1996, 496). Hegel and Habermas both agree that individuals can be cognitively overburdened by reflecting on how the overall system fits together and might be repaired. On the other hand, both stress the importance of insight as well as the fact that what goes on at the social and political level is felt by individuals, all of whom could gain from being integrated and not feeling alienated from the common project of promoting individual and collective freedom.

Though Hegel like Habermas favors deliberative democracy and reason in contrast to habits, custom, religion, or authority, his approach is a more epistocratic one than Habermas' (PR §270, 309ff, 316). Still, I believe Hegel would agree with

Habermas that the legitimacy of law depends on the collective and reflexive use of private and civil autonomy (Habermas 1996, 83f) which involves perspective-taking (Habermas 1996, 92), reflections on one's own life (Habermas 1996, 96), and reflection on the norms of interaction (Habermas 1996, 97). He would probably also agree that legitimate laws are based on the reflexive consideration of "the interests of each person" (Habermas 1996, 109), and that these interests should not be taken at mere face value. They should be reviewed as "inputs that, open to the exchange of arguments, can be discursively changed" (Habermas 1996, 181).

Concerned about the reproduction of the freedom-realizing system, Hegel was worried that the people should rule in the form of a mere mass or aggregation of atomistic individuals (1822/23 §303; 1824/5, §308; *Berliner Enzyclopädie*, §544). For this reason, political institutions should be arranged to make it likely that reasoning and insight into the people's needs (rather than immediate preferences) would become central to politics. It is his insights into the formative practices and institutions that make the state a precondition for freedom rather than oppression that are in focus here.

7.4 The Transformation of the Free Will

Hegel developed a powerful theory of social integration in combination with individual transformations toward freedom through mutual practices of recognition. He followed a tradition from Rousseau according to which "a 'remarkable change in man' is necessary before true citizenship can be possible" (*The Social Contract*, I, 7 cf. Pippin 2008, 252). To Hegel, the transformation is a transformation of the will through which being something determined with others in a manner that supports the realization of overall freedom becomes a being with oneself as one wants to be as a particular being (1822/23, §4). In this regard, he talks of a rebirth (PR §151).

How might such an organized translation of the will toward freedom take place in a manner that integrates the three different images of recognition as dimensions of freedom? Hegel describes freedom as entailing the continuous possibility of a full withdrawal from and abstraction from everything one wants and takes part in as well as the will making up its mind. The trick is to entail both in one's actions (PR §5–7). For this to happen, the individual needs to interact with the world and others in a manner that allows it to translate its existence as free sensing and reflective will both into the world and within itself (PR §8ff, 21). The process, however, is not about expressing an already present authentic »I« or about merely learning to perform what others take to be objectively right (PR §26), but about working on and reflecting on one's already present desires and aspirations and the conflicts they entail (PR §11–17, 20, 28, 34f, 39).

However, this can also remain a mere action and a determination reflecting a will without freedom. This brings us back to the point about living in a good state. According to Hegel, you are not on your own in this process if you belong to a good state (PR §33, 258) in which the rights of subjectivity are objectively recognized

(PR §40), and in which mediating institutions of production, reproduction, and transformation are organized and protected (PR §129). In fact, as member of a good society, you may achieve a form of freedom in which freedom is your "object" of reflection, sensation, desires, and strivings wherever you turn your reflection (inward or outward) (PR §258).

The rights of citizens should thus be institutionalized in ways that stimulate transformations of the individual wills toward freedom. For instance, Hegel argues that the deeper reason for recognizing persons' rights of ownership relates to the fact that taking something in possession is a precondition for upholding a free will as well as means for making oneself an external object of reflection (PR §45). Within such freedom-promoting arrangements, citizens may gain a material form of existence and well-being in which they can satisfy their desires without depending on something alien to themselves (PR §24).

Given the right institutions and transformations, the institutions may be seen as reflecting one's own freedom and interests. As Robert Pippin puts it, "In the proper institutional context, acknowledging the claims of the other is a form of self-acknowledgement and self-realization, and taking the other into account as I reflect what to do is no more a compromise with something 'alien' than taking the particularities of my own interests into account" (Pippin 2008, 140). He adds that "various forms of what seem human dependencies are now understood to determine such a will only in so far as such considerations are understood not as qualifications or limitations on the subject, but as aspects of its actualization" (Pippin 2008, 141). Freedom is then both a reflexive act of making up one's mind and acting upon it ("I want this kind of education") and a reflexive act of looking back on one's actions and activities "and take it up or not" (Pippin 2008, 146; compare Kierkegaard 1971). Often individuals find themselves in conflict with their own desires as well as public norms and need to make up their own mind (1822/23 §151). This stimulates reflection. My relation to myself is, however, mediated not only by such activities but also my relation to others (Pippin 2008, 149) including relations of justification (ibid., 152).

The model sketched above formulates a tension between the freedom of individuals and members in terms of different types of recognition and confronts the question of linking individual and communal ends as well as normative reproduction through institutions and modes of recognition that stimulate transformations of the will. From the perspective of society, the aim is both to reproduce itself materially "and the formation of conscious agents of social reproduction who are free as persons and moral subjects" (Neuhouser 2000, 132). It is important that citizens generally come to identify with this project if the system is to reproduce itself. The idea is then reflectively to install and maintain institutions that recognize citizens as individuals while these institutions promote solidarity and respect more or less behind the backs of the participants (Neuhouser 2000, 150). A central role is ascribed to the identity transformation of citizens within institutions that recognize them and help them shape themselves as free individuals with particular interests and concerns about their welfare (PR §261; Pinkard 1998, 324f).

The different forms of recognition (of personhood, subjectivity, and membership) provide the participants with different grounds for supporting the reproduction and institutionalization of formal and material recognition. *Persons* depend on material and formal recognition to protect citizens against domination from others, whereas *subjects* depend on material and formal recognition to stimulate their free reflections on how they want to lead their lives. *Members* depend on material and formal recognition to protect their sense of belonging to a community that cares for their well-being and freedom (Neuhouser 2000, 74ff). They also need institutional practices and settings in which they can express themselves as members. As members of families, work, civil society, and the state, citizens are able to give concrete expression to their conceptions of themselves as individuals with particular and reflexive ends (Neuhouser 2000, 98f, 115f).

If institutions provide citizens with the *preconditions* for their own freedom, they have a justified reason to protect and work to uphold these institutions (Neuhouser 2000, 80). For Hegel, however, it is not enough to *claim* that such institutions are objectively justified. Though "rational social institutions" are "both precondition and embodiment" of freedom, "the subjective affirmation of the principles that govern social life is a necessary component" (Neuhouser 2000, 81). The right to insight into the rationality of society means that the "rationality and goodness of the rational social order should be apparent to its members" (Neuhouser 2000, 256), but the members of society also have a right to demand that the validity and rationality of public institutions can be shown to be truly rational and thoroughly justified (Neuhouser 2000, 243). To Hegel, "individuals are capable of embracing the ends of the state as their own only if they are able to experience their role as citizens as source of their own selfhood" (Neuhouser 2000, 137).

The problem of reproducing a free society is furthermore that while it depends on individuals generally forming public or general wills, having a general will or expressing public freedom cannot be demanded by law (Neuhouser 2000, 141). In respect to citizens' personal and subjective freedom, there are limits to what the state may demand of its citizens. Hegel criticizes the Platonic state in which citizens are distributed to different types of work by administrators (PR §206). In addition, he criticizes attempts to legislate concerning the inner qualities and thoughts of citizens, including their loyalty to the state (PR §213). Thus, the strategy is to shape institutions to make the transformative modes of recognition likely. The central, integrating institutions are expected to transform the self-image of children as they grow up and mature up to a level where they see themselves and others as both individuals and members. This transformation can take place if the shared, rationalized lifeworld of the individual with its central public institutions integrates the social members in practices that express recognition of individuals as persons, subjects, and members. In this manner, the participants will come to see themselves and others as persons, subjects, and members. Such institutions will send two messages to the participants. On the one hand, they are special, unique, and irreplaceable with their own particular activities, plans, and ways of expressing themselves. On the other hand, they are one among many and take part in a system in which it is in their overall interest to support the preconditions for the freedom of all (Neuhouser 2000,

164f) and where they ought to engage in open interpretations over needs (Habermas 1990a, 49). The challenge is that in complex, modern societies, individuals have little involvement with the workings of the state (1822/23 §253). Still, as Habermas puts it, at different levels they may take part in struggles in the public sphere about "the appropriate interpretation of needs" which should be understood as struggles over recognition (Habermas 1994, 115).

To Habermas, "Legally guaranteed relations of recognition cannot reproduce themselves of their own accord. Rather, they require the cooperative effort of a civic practice that no one can be compelled to enter into by legal norms" (1996, 499). Here, Hegel and Habermas are partners of a specific approach to social integration (Jean L. Cohen and Arato 1992, 100), where social integration has "a place second to none on the political agenda" (Habermas 1996, 352). As Habermas puts it, "Law [...] is nourished by the 'democratic *Sittlichkeit*' of enfranchised citizens and a liberal political culture that meets it halfway" (Habermas 1996, 461, compare Habermas 1996, 58 and 499). As Habermas explains, "the legally constituted status of citizen depends on the *supportive spirit* of a consonant background of legally noncoercible motives and attitudes of a citizenry oriented toward the common good" (Habermas 1996, 499).

Hegel took care to describe the different membership types and associated institutions since he believed that the modern state made it possible to be vulnerable and needing human beings developing all their capabilities and realizing their subjective, substantial, and concrete welfare in free relations to others. In this perspective, work – even as something one is strongly encouraged to do – can, paradoxically, be a precondition for independence of mind or for achieving "the relation between such dependence and independence" (Pippin 2008, 161, see also 187). I am what I am as independent being in my dependence on others (both in the form of material and organizational practice as well as their recognition) (ibid., 200f, 208, 214ff). Participating in practices with others may also stimulate my thinking and practical reasoning (ibid., 242; compare Honneth 1998). Participating in organized freedom, however, also allows me to achieve a kind of objective freedom in which I take part in a self-sufficient organization of life (ibid., 242; Neuhouser). This self-sufficient organism is, moreover, dependent on the self-awareness of the participants (or citizens) for its freedom (ibid., 250). Its universal or justified existence is dependent on the free and freedom-promoting activities of the participants as well as their mutual recognition and self-awareness (ibid., 250f; Neuhouser 2000). Let us look further into how work might support recognition, freedom, and social integration.

7.5 Work as a Transformative Institution

Why would Hegel link recognition, freedom, and social integration to work? As Hegel rightly saw, humans have a potential for freedom that they cannot achieve on their own (ibid., §153, 186). As the above sketch has shown, to realize and reproduce freedom, the societal members should come to "possess the basic subjective capacities that are needed" for persons and subjects and "must value [...] their freedom as persons and moral subjects" (Neuhouser 2000, 148). To Hegel, the formation of such capacities is likely to take place also through practices that are not fully voluntary. For instance, one does not choose one's family, and living in a family involves some degree of disciplinary action. Similarly, Hegel points out that humans are forced to work out of necessity and basic needs (PR §170; Neuhouser 2000, 149f). This may not contradict freedom if the satisfaction of the individual involves the right to choose occupation and not to be involved in alienating types of work in which the individual is reduced to stupefying machinelike activities (ibid., §198). Hegel is fascinated by the ways in which following one's personal and subjective self-interest as in choosing an occupation because it satisfies a set of more or less immediate desires may strengthen the individuals' sense of personhood and subjectivity as well as lead to a membership identity. As he puts it, the first principle of civil society is the satisfaction of the individual, but this takes place through work and consumption in a system where working and consuming involve the work and satisfaction of everyone else (ibid., 188, 199). If this is the case, then working may support the person, subject, and membership dimensions of freedom.

Hegel believes that it is good for the individual to live a practical life, including relating to things that give the individual's life a certain character and shape, though this also involves missing other options (1822/23 §197). Something important happens when the satisfaction of individual needs and interests is translated through relations to others. It becomes a relation of mutual recognition (1822/23 §192). The individual is then no longer merely dependent on itself and is no longer a mere private person or isolated individual, and a concern for well-being as a whole (both individually and collectively over time) emerges (1821/22, §203). To illustrate, Hegel speaks of the family wage. The wage received from working can satisfy a number of individual needs, but if the person receiving a wage belongs to a family, the wage may stimulate a sense of being a member rather than an isolated subject. Thereby, they may achieve a social identity and a sense of only being themselves through others and that others only are what they are through others (1822/23 §184). Caring for others through one's possessions also changes the nature of one's possessions (1822/23 §170).

Since working can be motivated by oneself on the basis of our needs for survival, it is not fundamentally alien to us. Since institutionalized and organized (or corporated) types of work may allow us to take care of our freedom and well-being over a lifetime, allow us to express a membership identity, and make us realize the degree to which our well-being and freedom depend on the cooperation with others, as well as help us train our ability to reach our personal goals, work can be seen to contribute to personal and public autonomy (1822/23 §303). Not only is it not necessarily alien to us, but may help us realize our potential for the higher ideal of reflexive and social freedom. In that case, it is a form of living in which we are *with ourselves* within organized and institutionalized relations of mutual recognition that are offered as historical achievements (Pippin 2008, 54). Such a state of freedom is "a form of individual and collective mindedness, and institutionally embodied recognitive relations" that is also "an active state, a state of doing" (Pippin 2008, 39).

For Hegel, work is thus a central mediating practice (PR §196f, 199, 206). To live up to its functions, however, the institutions of work are in need of support. Hegel thought of corporatist unions of families serving the overall functions of society often thought of today as belonging to the private and public sector (PR §250, 263f). These could, as incorporated within the state together with the families, serve the purpose of protecting and reproducing the preconditions for the overall welfare and interests of the individuals and families associated with them (PR §302). As member of such a corporation, one would thus express that one is self-sufficient without being self-interested in a purely selfish manner.

The central functions which Hegel ascribes to the corporations indicate that Hegel wanted to bridge the modern principle of free labor (including the right to *human* in contrast to machine-like work, freedom of choice, education, health, honor, income, and non-alienation) with a form of normative functionalism according to which the individual and social good is linked up with the performance of social functions necessary to protect and realize the welfare and freedom of all.

To Hegel, the state, the political system, and public servants should facilitate this overall system of transformation of the ways in which needs are satisfied and identities shaped (1821/22, 236, 254; 1822/23 §205; *Berliner Enzyclopädie*, §537). The system should maintain itself also by protecting subjective freedom (1822/23 §273), by supporting lower level institutions and mechanisms that stimulate the transformative mediation between individual and communal concerns (1822/23 §262, §283), and by protecting against unemployment and poverty (PR §230, 239ff; 1821/22, 241; 1822/23 §252), the right to work and the avoidance of domination (ibid., §244), and thereby the status and recognition of citizens (ibid., §245, 253, 258).

Thus, Hegel introduced the functions of the state for purposes of oversight, protection, and coordination of the continuous reproduction of the preconditions for freedom and welfare, as well as self-conscious identification and reflection (PR §268ff, 289f, 301). According to this ideal, the state contains and protects the other unions, thereby creating an onion-shaped union of unions (corporations) of unions (families) just as the unions and circles should shape the state politically (PR §303). Having membership here is fundamental as it marks an identification with protection of all the unions rather than merely what serves one's own self-interests (PR §258).

These unions are worth protecting if they are rational in the sense of combining individual, human interests, and if they reflectively prepare the expression of the developed capacities necessary for realizing freedom (PR §258, 265). If we think of the state as containing these unions, we can think of it as entailing extremes of individuality as well as mediating, transformative institutions that stimulate that the members come to have their self-conscious identity in the union of unions as well as shape political power in direction of reproducing freedom and welfare (PR §302). In this way, citizens may both express themselves as self-sufficient individuals on their own and in their smaller unions (such as families) and realize that true self-sufficiency is something larger which they take part in as members of a good state (PR §260). The state itself contains divisions of power that should also not see themselves as self-sufficient on their own (PR §272, 278). Hegel speaks of the unity of duty and rights in the state based on the singular principle of freedom as being of highest importance. He stresses, however, that the different parts of society may have different duties and rights (workers may pay different levels of taxes as a response to the right of welfare). Though they follow the same principle of freedom, they may be different in content (PR §261).

Furthermore, the state, should also support that the reflexive concern for the universal end of realizing freedom is kept alive by the citizens such that the individuals have this realization of freedom as their individual end (ibid., §260). One of the complex aspects of this is the human desire for recognition and the complex concerns we have for both being equal and different to others (1822/23 §193). Hegel believed that settling the question of whether the individual was a recognized valuable contributor to the realization of freedom over time was fundamental for the individual's well-being and that the freedom-realizing society could be in danger if too many people lacked this sense of recognition (1822/23 §207). Given the right setup, however, subjective interests and plans can work as "the sole animating principle" (PR §206). One way of integrating individuals and supporting their self-esteem is by providing them with work that is linked to the reproduction of the "self-sufficient, rational whole that realizes the ideal of self-determination more completely than any human individual" (Neuhouser 2000, 197).

Without fully developing a model of a universal welfare state, Hegel was worried that the unemployed would lack "the feeling of self-sufficiency and honour" if their livelihood was to be provided "without the mediation of work" (PR §245). He was also concerned that the membership identity in the form of solidarity should find expression and recognition. Hegel believes that mediating institutions had a role to play for these purposes. Again, the idea is that individuals could be brought through their particular interests and human needs to work and consume in ways that would also satisfy the needs of others. For Adam Smith, this was also the solution to the problem of recognition. To this, Hegel added mediating institutions believing that the recognition one could achieve as individual, self-sufficient consumer and worker would lack stability and lack the membership character which would express solidarity with others rather than selfishness (PR §253).

7.6 How to Apply Hegel's Theory to Naturalization Policies

How might we apply Hegel's central ideas to debates about naturalization policies? Though Hegel linked voting rights to work, his story is complex. This also raises some questions, including what to make of the particular, mediating institutions, Hegel defended. We may share Hegel's worries about ruling by merely aggregating people's immediate preferences (Brooks 2013, 131), and his concern about the role of public institutions in forming and integrating citizens toward a cooperative form of freedom, and still believe that contemporary public institutions could have a stronger integrative effect than one could realistically imagine in his days (Jørgensen

2017). If the institutions have changed, where does this leave Hegel's theory? Importantly, from early on, Hegel was not in favor of a state that organized everything in detail centrally in a top-down manner, leaving little to voluntary institutions of society (Avineri 1972, 48f). His story is very much about the aims and fundamental logics of society. Furthermore, contemporary welfare states remain concerned with these ends and functions. Many states have in this process made themselves stronger by taking upon themselves the tasks of education and care that Hegel ascribed to the family and civil society. This, however, only strengthens the responsibility for the state to *justify* its demands on citizens. By making itself stronger and weakening corporations, however, the state has made itself more dependent on the market and has lost sight of the mediating functions of institutions in between. The still valid claim is that in justifying its demands on citizens, it should take seriously its own part in shaping public institutions in light of the mechanisms of recognition.

Hegel is telling us that all the modern images of recognition (persons, subjects, members of reproduction, members of production, members of oversight, and members of humanity) are transformative. He is telling us that some of us need to take these tasks upon ourselves and that we all need to integrate membership identities with freedom to become truly free. It is practically open whether everyone will support the reproduction of the freedom supporting society given that it cannot be compelled without contradicting individual freedom (Habermas 1996, 130). In the case of children, they are taken *generally* to mature and achieve political preparation through the expressive institutions of recognition. In analogy, immigrants could be taken generally to come to understand and care about the reflexive and communal logics of recognition (included in the recognition of individuality) through taking part in such practices. On the other hand, even if societies provided immigrants with meaningful jobs that allowed for choice, protected against domination and alienation, stimulated reflection, and provided a membership identity in civil society and a sense of self-esteem as member of society, there might still be more to be done to promote freedom and social integration. In short, work may not in itself do away with the problem of motivating support for the freedom-promoting institutions. Also, the reflexive struggles over needs can be blocked by "dogmatic worldviews and rigid patters of socialization" (Habermas 1996, 325). Public institutions should serve to challenge dogmatic worldviews and rigid patterns of socialization and be accommodating in ways that make them attractive alternatives for immigrants who may suffer from such rigid patterns of socialization in the families.[6]

Nevertheless, we should not lose sight of what Hegel's overall approach to recognition has to offer. Hegel asks us to evaluate all policies in light of whether they promote or counter freedom. Perhaps, it makes sense to see nonideological and emancipatory types of work as a central precondition for social integration (Durkheim 2013; Honneth 2012). As Hegel pointed out, contributing to shared ends

[6] And thereby perhaps meet the challenge of accommodating questions of difference and identity (Taylor 1994, 41–52).

is clearly a sign of social integration (Honneth 2012, 65). On the other hand, it is not clear that individuals cannot develop the necessary civic attitudes through other institutionalized practices. Participating in democratic institutions of education or civil society institutions may on their own support the achievement of the capacities required.

In contrast to contemporary state's focus on immigrants, this approach shifts the burden of proof from individuals to public institutions. As Habermas put it, individual and public freedom depend on a system of right as well as the background culture of a rationalized lifeworld. Habermas added that "[s]ocialized individuals could not maintain themselves as subjects at all if they did not find support in the relationships of reciprocal recognition articulated in cultural traditions and stabilized in legitimate orders – and vice versa" (Habermas 1996, 80). Later he argues that "political autonomy is an end-in-itself that can be realized not by the single individual privately pursuing his own interests but only by all together in an intersubjectively shared practice. The citizens' legal status is constituted by a network of egalitarian relations of mutual recognition" (Habermas 1996, 498).

Of particular concern here is the fact that for Hegel and Habermas, the normal processes of social integration begin with a recognition of individuals only with the hope that the receiver will come to understand the expectation of reciprocated respect and potential transformation entailed by the gift. In contrast, naturalization policies often follow a logic of performance and reward. This strategy might miss the transformative potentials of the recognition model.

It is in this sense I suggest we understand Hegel's considerations concerning the integrative forces of work. Clearly, even if we took our outset in Hegel's specific ideas on work as a precondition for enfranchisement, we would not have to give in to every detail of his story. As Hegel pointed out, there are groups with personal autonomy which this model leaves out (women and day workers in Hegel's time). Not allowing people outside a specific part of the workforce, the vote may turn out to be counterproductive if the aim is to integrate these groups socially, just as their valid and important insights into the workings of the overall social system may be lost from public debate without their fully recognized political status. Likewise, we could think it unfair to withhold full citizenship from adult members of the citizenry. What if someone is in the process of integration into the workforce (undergoing education); if the person is doing valuable but unpaid tasks of reproduction, care work, etc.; or if the person is unable to get a job, or only has a part-time job? Summing up, if a society should consider work a central vehicle of social integration, it would be expedient to evaluate its practices on the basis of Hegel's theories of recognition, freedom and social recognition and to broadly support that immigrants could come to take part in the central mediating, transformative institutions of the particular society. It should observe also that transformative, integrating mechanisms may be found in many other types of institutions and practices involving recognition and freedom and that the modern relations of work in its more precarious forms may be the least likely place to find them.

Furthermore, it should be noticed that for Hegel as well as Habermas, the ascription of rights to individuals is a precondition for the state meeting its own ends.

Thus, for Hegel, having political rights reflect one's objective status of being part of the system that reproduces freedom, and though Hegel tried to downplay the value of having voting rights for the individual, this recognition is constitutive in itself. Neither author would argue that the mere formal recognition of a certain status transforms citizens in the manners needed. A thorough analysis of the practices of integration would thus need to investigate the overall mechanisms of recognition practiced and institutionalized in a given society. The focus should thus be both on *just* recognition and on society's transformative institutions. The important purpose is to reproduce the freedom-promoting, reflexive institutions including its political ones. Here, recognition works as a transformer of individual ends through its expressive nature. Withholding recognition thus threatens both the normative ideal and its reproduction.

Neither Hegel nor Habermas would subscribe to enforced loyalty, but would defend the human need for and right to recognition. As Hegel puts it, individuals need to know, want, and believe in the purposes of the state (Neuhouser 2000, 202). Being recognized has transformative and social integrative force, but it does allow individuals to act as mere individuals. They also defended the need for transformative institutions that would stimulate the bridge between private and public autonomy. Participating in a democratic culture of dialogical interpretations of needs and working to satisfy one's private ends and there through realizing one's dependence on others are mechanisms that generally need to be in place for the state to reproduce itself normatively. As Sarah Song pointed out, if you participate in institutions of normative reproduction for some time, you should be ascribed the right to full participation in the structuring of such institutions. In this light, institutions can be evaluated in light of whether they promote and protect the integration of personal, subjective, membership, and public autonomy, trusting that it is participation in such institutions that makes it likely that one achieves the qualities and self-understandings necessary for normative reproduction.

Noticeably, this model could be said to bring out the underlying values of the democratic welfare states. Even naturalization policies will tend to affirm the rights of individuality as well as concerns about communal identities. The strategy can thus be formulated as an immanent critique of current practices. For instance, Danish laws and policies indirectly affirm the individual freedom of citizens since newcomers have to respect the individual freedom of others (i.e., UIM Udlændinge- og Integrationsministeriet 2019). For a further analysis, a central concern will be whether the practices and institutions of naturalization are in line with the models sketched above. Importantly, for Hegel and Habermas, the rights of individuality should not be compromised. As Hegel puts it, individuals should shape their particular way of contributing to the common project on "their own accord […] knowingly and willingly" (PR §260).

This model indicates that state policies should be reviewed in light of an overarching purpose of realizing freedom and that this depends both on *protecting* individuals and certain institutions, but also on institutionalized forms of recognition that help to *transform* individuals to a state of freedom. As Habermas puts it, "[b]oth sides of autonomy are essentially incomplete elements that refer to their respective complement. This nexus of reciprocal references provides an intuitive standard by which one can judge whether a regulation promotes or reduces autonomy" (Habermas 1996, 417). The model stresses the importance of integrating citizens into practices in which they will come to cherish individual and civic freedom – practices through which the democratic welfare state reproduces itself normatively. Institutional reforms and central policies should aim at making sure that institutional practices fully respect the integrity of individuals as persons, subjects, and members and shape and protect their relevant individual capacities (Neuhouser 2000, 198) while reproducing society institutionally and materially in ways that allow individuals "to maintain a free will while contributing in their own particular way to reproducing their society" (Neuhouser 2000, 207).

Clearly, these demands are not always met at modern relations of work. Municipalities are often more concerned about making immigrants enter any kind of work (think night shift cleaning or factory work with no communication with any co-workers). Economic self-sufficiency is thus far from Hegel's ideal of integrating institutions of work. In short, mere self-sufficiency is the wrong marker of social integration.

Contemporary states may build on the overall principles of the recognition model, but then turn to demand self-support as a prerequisite for full citizenship for immigrants. This may appear to support an image of individual freedom as well as solidarity. Nevertheless, this strategy may in fact contradict the ideal of membership if it promotes an image of individuals not relying on the community for their obtainment of self-determination and self-sufficiency (Fraser and Gordon 1994). In that case, it would contradict Hegel's view (Pippin 2008, 222f; Markell 2003, 11f, 16, 21). The point being that the policies of making full citizenship depend on economic self-support promote an image of self-reliance that contradicts the transformative strategies of the recognition model. This worry relates to how individuals should relate to their own dependence and freedom. Rather than defending economic self-reliance in individual term, Hegel argues that participating in the reproduction of the institutions that secure and support freedom is both a collective and individual good of a type that supports the transformation of individual needs and aspirations toward private and public autonomy (compare Patten 1999, 195; 1821/22 §178). For Hegel, in relation to individual persons, self-sufficiency is something that society can allow for – but only if society makes sure that individuals who are allowed to think of themselves as self-sufficient are »brought« back identity-wise to having a sense of purpose and self-esteem in the common project of reproducing freedom-realizing institutions and their preconditions together (PR §260). Otherwise, the individual ends up with what he calls a *Pöbel*-identity which is a threat to the social cohesion of society (something he sees both at the lower and top strata of society).

Though Hegel speaks of membership in civil society as "an association of members as self-sufficient individuals" (PR §157, 182), and of marriage as initially involving a contract between two self-sufficient individuals, who create a new self-sufficient unit (PR §172, 181), he believed that only society as a whole could in an important sense be self-sufficient (PR §258) and that society cannot exist if these identities of individualized self-sufficiency are not overcome (PR §163). On the other hand, he criticizes despotism in which citizens are not treated as self-sufficient adults (PR §174, 185, 206) or merely made to work for the public good (PR §236) and argues that families should raise children "to self-sufficiency and freedom of personality" (PR §175). The point is thus that only within a community can you be a self-sufficient individual recognized as a person, subject, and member (Neuhouser 2000, 118ff). Families and labor corporations become institutions of recognition through which self-sufficient individuals come to see that they depend on the cooperation and mutual recognition of others to achieve their own ends (PR §182f, 192) and come to recognize the fragility of individuals if they more or less arbitrarily relate to such institutions (PR §185). The state should be absolutely concerned with institutions that may help individuals in the "hard work" of transforming particular subjectivity from the shape of private personality to a reflexive kind of civic personality and subjective civility (PR §187). Institutions such as family and work can base this transformation on the individuals' immediate private and personal desires (e.g. for sex, love, consumption, and status) in order to free these desires of their immediacy, vanity, and arbitrariness and move them to self-conscious participation in institutions that over time will protect the full realization of individual freedom (ibid.). Through practices of recognition, our attempts to satisfy ourselves gain a social character (PR §192). As he puts it, you can only be what you are as recognized.

7.7 Conclusion

To conclude on how we ought to analyze the strict demands on immigrants for naturalization in contemporary European welfare states, this paper stresses the analytical potential of applying theories of recognition to the analysis of integration policies including naturalization policies. Recognition is a transformer of individual and collective interpretations of needs through the institutional expressive acknowledgment of shared vulnerability at the level of individual and communal lives. Though citizens will need to take part in transformative practices, this approach shifts the burden of proof from aspiring immigrants to the people reproducing and protecting their institutions of recognition. The approach can be used to analyze the dynamic mechanisms of recognition which central institutions need to set in place if societies are to reproduce themselves as free. Having taken part in central social institutions for a while, the institutions should be accommodating and expressive in ways that stimulated the transformation of aspirations and the development of identities and capacities that would make it safe for a democracy to include immigrants fully in the reflexive normative reproduction of society. If they are not, the major burdens of upholding and shaping transformative institutions are on the demos. As mentioned, private autonomy develops out of conflicts between individuals' desires and public norms. This experience of conflict may be particularly present among people who move from one culture to another. Such conflicts may stimulate self-conscious reflection about both private and public ends. In conclusion, it is the burden of the demos to ensure that the transformative institutions support the development of private and public autonomy. For this purpose, much can be learned from Hegel's theories of recognition, freedom, and social integration.

References

Abizadeh, Arash. 2008. Democratic Theory and Border Coercion. *Political Theory* 36 (1): 37–65. https://doi.org/10.1177/0090591707310090.
Anderson, Joel, and Axel Honneth. 2005. Autonomy, Vulnerability, Recognition, and Justice. In *Autonomy and the Challenges to Liberalism: New Essays*, ed. John Christman and Joel Anderson, 127–149. New York: Cambridge University Press. https://doi.org/10.1017/CBO9780511610325.008.
Arendt, Hannah. 2017. *The Origins of Totalitarianism. Modern Classics*. London: Penguin Books.
Avineri, Shlomo. 1972. *Hegel's Theory of the Modern State. Cambridge Studies in the History and Theory of Politics*. London: Cambridge University Press.
Bauböck, Rainer. 2015. Morphing the Demos into the Right Shape. Normative Principles for Enfranchising Resident Aliens and Expatriate Citizens. *Democratization* 22 (5): 1–20. https://doi.org/10.1080/13510347.2014.988146.
———, ed. 2018. *Democratic Inclusion: Rainer Bauböck in Dialogue*. Manchester: Manchester University Press.
Beckman, Ludvig. 2009. *The Frontiers of Democracy: The Right to Vote and Its Limits*. London: Palgrave Macmillan.
Beiser, Frederick C. 2005. Hegel. In *Routledge Philosophers*, ed. Frederick C. Beiser, 1st ed. London: Routledge.
Benhabib, Seyla. 1986. *Critique, Norm, and Utopia: A Study of the Normative Foundations of Critical Theory*. New York: Columbia University Press.
———. 2002. *The Claims of Culture: Equality and Diversity in the Global Era*. Princeton: Princeton University Press.
———. 2013. Reason-Giving and Rights-Bearing: Constructing the Subject of Rights. *Constellations* 20 (1): 38–50. https://doi.org/10.1111/cons.12027.
Bosniak, Linda. 2006. *The Citizen and the Alien: Dilemmas of Contemporary Membership*. Princeton: Princeton University Press.
Brooks, Thom. 2013. *Hegel's Philosophy of Right*. 2nd ed. Edinburgh: Edinburgh University Press.
Carens, Joseph H., ed. 2013. *The Ethics of Immigration*. Oxford: Oxford University Press.
Cohen, Jean. 1996. Rights, Citizenship and the Modern Form of the Social: Dilemmas of Arendtian Republicanism. *Constellations* 3 (2): 164–185.
Cohen, J.L. 1999. Changing Paradigms of Citizenship and the Exclusiveness of the Demos. *International Sociology* 14 (3): 245–268. https://doi.org/10.1177/0268580999014003002.
Cohen, Elizabeth F. 2009. *Semi-Citizenship in Democratic Politics*. Cambridge: Cambridge University Press.
Cohen, Jean L., and Andrew Arato. 1992. *Civil Society and Political Theory. Studies in Contemporary German Social Thought*. Cambridge, MA: The MIT Press.
Dahl, Robert. 1989. *Democracy and Its Critics*. New Haven: Yale University Press. https://doi.org/10.1017/S002238160004799X.
Durkheim, Emile. 2013. *The division of labour in society*. 2nd ed. Houndmills: Palgrave Macmillan.
Erman, Eva. 2013. Political Equality and Legitimacy in a Global Context. In *Political Equality in Transnational Democracy*, 61–87. New York: Palgrave Macmillan US. https://doi.org/10.1057/9781137372246_4.
Fine, Sarah. 2011. Democracy, Citizenship and the Bits in Between. *Critical Review of International Social and Political Philosophy* 14 (5): 623–640. https://doi.org/10.1080/13698230.2011.617122.

Forst, Rainer. 2019. Two Bad Halves Don't Make a Whole: On the Crisis of Democracy. *Constellations* 26 (3): 378–383. https://doi.org/10.1111/1467-8675.12430.
Fraser, Nancy. 1995. From Redistribution to Recognition? Dilemmas of Justice in a 'Post-Socialist' Age. *New Left Review* 212: 68.
Fraser, Nancy, and Linda Gordon. 1994. A Genealogy of Dependency: Tracing a Keyword of the U.S. Welfare State. *Signs* 19: 309.
Galeotti, Anna E., ed. 2002. *Toleration as Recognition*. Cambridge: Cambridge University Press.
Gilabert, Pablo, and Holly Lawford-Smith. 2012. Political Feasibility: A Conceptual Exploration. *Political Studies* 60 (4): 809–825. https://doi.org/10.1111/j.1467-9248.2011.00936.x.
Goodin, Robert E. 2007. Enfranchising All Affected Interests, and Its Alternatives. *Philosophy & Public Affairs* 35 (1): 40–68. https://doi.org/10.1111/j.1088-4963.2007.00098.x.
Habermas, Jürgen. 1990a. Justice and Solidarity: On the Discussion Concerning 'Stage 6'. In *Hermeneutics and Critical Theory in Ethics and Politics*, ed. Michael Kelly, 32–52. Cambridge, MA: MIT Press.
———. 1990b. *Moral Consciousness and Communicative Action*. Cambridge, MA: MIT Press. https://doi.org/10.1007/s13398-014-0173-7.2.
———. 1994. Struggles for Recognition in the Democratic Constitutional State. In Taylor, Charles, K. Anthony Appiah, Jürgen Habermas, Steven C. Rockefeller, Michael Walzer, and Susan Wolf. *Multiculturalism: Examining the Politics of Recognition*, ed. Amy Gutmann, 107-148. Princeton: Princeton University Press.
———. 1996. *Between Facts and Norms: Contribution to a Discourse Theory of Law and Democracy*. Cambridge: Polity Press. https://doi.org/10.7551/mitpress/1564.001.0001.
———. 2001. Constitutional Democracy: A Paradoxical Union of Contradictory Principles? *Political Theory* 29 (6): 766–781.
Hegel, Georg Wilhelm Friedrich. 1973. *Vorlesungen Über Rechtsphilosophie: 1818–1831*, ed. Karl-Heinz. Ilting. Stuttgart: Firedrich Frommann.
———. 2003. *Elements of the Philosophy of Right*, ed. Allen W. Wood. Cambridge: Cambridge University Press.
———. 2005. *Die Philosophie Des Rechts. Vorlesungen von 1821/22*. Frankfurt a.M.: Suhrkamp.
Honneth, Axel. 1995. *The Struggle for Recognition: The Moral Grammar of Social Conflicts*. Cambridge, MA: The MIT Press.
———. 1998. Democracy as Reflexive Cooperation: John Dewey and the Theory of Democracy Today. *Political Theory* 26 (6): 763–783. http://www.jstor.org/stable/191992.
———, ed. 2012. *The I in We: Studies in the Theory of Recognition*. Cambridge: Polity Press.
———. 2014. *Freedom's Right: The Social Foundations of Democratic Life*. New York: Columbia University Press.
Jensen, Kristian Kriegbaum, Per Mouritsen, Emily Cochran Bech, and Tore Vincents Olsen. 2019. Roadblocks to Citizenship: Selection Effects of Restrictive Naturalisation Rules. *Journal of Ethnic and Migration Studies*: 1–19. https://doi.org/10.1080/1369183X.2019.1667757.
Jørgensen, Simon Laumann. 2017. Between Integration and Freedom: School Segregation in Critical Perspective. *Scandinavian Political Studies* 40 (3): 265–288. https://doi.org/10.1111/1467-9477.12088.
Kierkegaard, Søren. 1971. *Either/Or. New Edition*. Princeton: Princeton University Press.
Kymlicka, Will. 1995. *Multicultural Citizenship: A Liberal Theory of Minority Rights. Oxford Political Theory*. Oxford: Clarendon Press.
Kymlicka, Will, and Wayne Norman. 1994. Return of the Citizen: A Survey of Recent Work on Citizenship Theory. *Ethics* 104 (2): 352–381. https://doi.org/10.1086/293605.
Laborde, Cécile. 2008. *Critical Republicanism: The Hijab Controversy and Political Philosophy*. Oxford: Oxford University Press.
Markell, Patchen. 2003. *Bound by Recognition*. Princeton: Princeton University Press.
Marshall, T.H. 1992. *Citizenship and Social Class*. London: Pluto Press Ltd.
Mason, Andrew. 2012. *Living Together as Equals the Demands of Citizenship*. Oxford: Oxford University Press.

Miller, David. 2000. *Citizenship and National Identity*. Cambridge, MA: Polity Press.
———. 2009. Democracy's Domain. *Philosophy & Public Affairs* 37 (3): 201–228. https://doi.org/10.1111/j.1088-4963.2009.01158.x.
———. 2015. Is There a Human Right to Democracy? In *Transformations of Democracy: Crisis, Protest and Legitimation*, ed. Robin Celikates, Regina Kreide, and Tilo Wesche, 177–191. Lanham: Rowman & Littlefield Publishers.
———. 2016. *Strangers in Our Midst: The Political Philosophy of Immigration*. Cambridge, MA: Harvard University Press. https://www.statsbiblioteket.dk/au/#/search?query=recordID%3A%22sb_6286628%22.
Miller, Fred. 2017. Aristotle's Political Theory. In *The Stanford Encyclopedia of Philosophy*, Winter, ed. Edward N. Zalta. https://plato.stanford.edu/archives/win2017/entries/aristotle-politics/.
Modood, Tariq. 2013. *Multiculturalism*. 2nd ed. Cambridge: Polity Press.
Neuhouser, Frederick. 1993. Freedom, Dependence, and the General Will. *The Philosophical Review* 102 (3): 363. https://doi.org/10.2307/2185902.
———. 2000. *Foundations of Hegel's Social Theory: Actualizing Freedom*. Cambridge, MA: Harvard University Press.
Offe, Claus. 2011. From Migration in Geographic Space to Migration in Biographic Time: Views from Europe. *Journal of Political Philosophy* 19 (3): 333–373. https://doi.org/10.1111/j.1467-9760.2011.00394.x.
Owen, David. 2011. Transnational Citizenship and the Democratic State: Modes of Membership and Voting Rights. *Critical Review of International Social and Political Philosophy* 14 (5): 641–663. https://doi.org/10.1080/13698230.2011.617123.
———. 2013. Citizenship and the Marginalities of Migrants. *Critical Review of International Social and Political Philosophy: The Margins of Citizenship* 16 (3): 326–343. https://doi.org/10.1080/13698230.2013.795702.
Parekh, Bhikhu. 2006. *Rethinking Multiculturalism: Culture Diversity and Political Theory*. Vol. 2. Cambridge, MA: Harvard University Press.
Patten, Alan. 1999. *Hegel's Idea of Freedom*. Oxford: Oxford University Press.
———, ed. 2014. *Equal Recognition: The Moral Foundations of Minority Rights*. Princeton: Princeton University Press.
Pinkard, Terry. 1998. *Hegel's Phenomenology: The Sociality of Reason*. Cambridge: Cambridge University Press.
Pippin, Robert B., ed. 2008. *Hegel's Practical Philosophy: Rational Agency as Ethical Life*. Cambridge: Cambridge University Press. https://doi.org/10.1017/CBO9780511808005.
Rawls, John. 1971. *A Theory of Justice*. Cambridge, MA: The Belknap Press of Harvard University Press.
———. 2001. *Justice as Fairness: A Restatement*. 2nd ed. Cambridge, MA: Harvard University Press.
Rousseau, Jean-Jacques. 1979. *Emile, or On Education*. Vol. 48. New York: Basic Books.
Schnädelbach, Herbert. 1997. Die Verfassung Der Freiheit (§§ 272–340). In *Grundlinien Der Philosophie Des Rechts*, ed. Ludwig Siep, 243–266. Berlin: Akademie Verlag.
Seglow, Jonathan. 2009. Arguments for Naturalisation. *Political Studies* 57 (4): 788–804. https://doi.org/10.1111/j.1467-9248.2008.00768.x.
Somers, Margaret R. 1994. Rights, Relationality, and Membership: Rethinking the Making and Meaning of Citizenship. *Law & Social Inquiry* 19 (01): 63–112. https://doi.org/10.1111/j.1747-4469.1994.tb00390.x.
Song, Sarah. 2016. The Significance of Territorial Presence and the Rights of Immigrants. In *Migration in Political Theory*, ed. Sarah Fine and Lea Ypi, 225–248. Oxford: Oxford University Press. https://doi.org/10.1093/acprof:oso/9780199676606.003.0011.
———. 2018. *Immigration and Democracy. Oxford Political Theory Series*. https://doi.org/10.1093/oso/9780190909222.001.0001.
Taylor, Charles. 1975. *Hegel*. Cambridge: Cambridge University Press. https://doi.org/10.1017/CBO9781139171465.

———. 1994. The Politics of Recognition. In Taylor, Charles, K. Anthony Appiah, Jürgen Habermas, Steven C. Rockefeller, Michael Walzer, and Susan Wolf. *Multiculturalism: Examining the Politics of Recognition*, ed. Amy Gutmann, 25-73. Princeton: Princeton University Press.

Tully, James. 2004. The Challenge of Reimagining Citizenship and Belonging in Multicultural and Multinational Societies. In *Demands of Citizenship*, ed. Catriona McKinnon and Iain Hampsher-Monk, 212–234. London: Continuum.

UIM Udlændinge- og Integrationsministeriet. 2019. *Residence and Self-Sufficiency Declaration*. 2019. https://uim.dk/filer/integration/opholds-og-selvforsorgelseserklaering/opholds-og-selvforsoergelseserklaering-engelsk.pdf.

Vink, Maarten Peter, and Rainer Bauböck. 2013. Citizenship Configurations: Analysing the Multiple Purposes of Citizenship Regimes in Europe. *Comparative European Politics* 11 (5): 621–648. https://doi.org/10.1057/cep.2013.14.

Walzer, Michael. 1983. *Spheres of Justice: A Defence of Pluralism and Equality*. Oxford: Basil Blackwell.

Warren, Mark E. 2017. "A Problem-Based Approach to Democratic Theory." American Political Science Review 111 (1): 39–53. doi: https://doi.org/10.1017/S0003055416000605.

Williams, Robert R. 2000. *Hegel's Ethics of Recognition*. Berkeley: University of California Press.

Young, Iris Marion. 2000. *Inclusion and Democracy*. Oxford: Oxford University Press.

Chapter 8
German and US Borderlands: Recognition Theory and the Copenhagen School in the Era of Hybrid Identities

Sabine Hirschauer

Abstract This chapter explores securitization and recognition theories as lenses into conflict, emerging from specific migration practices. In current critical scholarship, these practices are, for the main part, interpreted as disrupting and breaking down Axel Honneth's institutionalized recognition order. They are fundamentally undermining migrants' self-trust, legal rights, and abilities for self-determination and autonomy. By approaching the intersection of recognition and migration—and global mobility generally—from an interdisciplinary, critical, post-structural security studies perspective, this chapter makes two contributions to existing scholarship: First, it highlights the linkages between *mis*recognition and the logic of subjective intentionality of security, also understood as the securitization of subjectivity through negative securitization logic; and second, it draws attention to a normative opening—an inclusive, positive securitization, which moves misrecognition injuries as emancipation toward a more advanced, more complete, social and moral progress. By utilizing two similarly constructed "crises" environments—the migration discourse in Germany and the USA—this chapter takes its epistemological point of departure from both countries' containment and deterrence migration policies and practices. Specifically, it interprets the US Remain-in-Mexico and Germany's ANKER Center policies as liberal dispositifs of security. The struggle for recognition is a struggle of becoming "a self." If the processes and lived experiences of becoming a self are disrupted, we speak of misrecognition. Instead of facilitating trust, respect, and esteem through undifferentiated rights, agency, and positive security for migrants, German and US migration practices violate the two countries' own liberal and moral aspirations of universal rights, justice, social and moral progress. A way out of the negative security and misrecognition logic can be achieved through emancipation. The proposed interdisciplinary recognition and securitization lens, including the concept of thick recognition, provides an opening toward alternative approaches. It *is* an approach toward positive securitization as the

S. Hirschauer (✉)
Department of Government, New Mexico State University, Las Cruces, NM, USA
e-mail: shirscha@nmsu.edu

G. Schweiger (ed.), *Migration, Recognition and Critical Theory*, Studies in Global Justice 21, https://doi.org/10.1007/978-3-030-72732-1_8

maintenance of just, core values lived and realized through diversity, self-determined, and multi- and transcultural agency. This chapter evaluates these claims of a new normative opening increasingly asserted by a twenty-first-century, progressive era of transcultural, hybrid identities.

Keywords Securitization · Germany · US · Borderlands · Misrecognition · Emancipation

8.1 Introduction

Securitization and recognition theories provide intriguing lenses into conflict as social and moral development, emerging from specific migration practices. In current critical scholarship, these practices are, for the main part, interpreted as disrupting and breaking down Axel Honneth's "institutionalized recognition order" (Cox 2012), for one, by fundamentally undermining migrants' self-trust, legal rights, and abilities for self-determination and autonomy. By approaching the intersection of recognition and migration—and global mobility generally—from an interdisciplinary, critical, post-structural security studies perspective, this chapter makes two contributions to existing scholarship: First, it highlights the linkages between *mis*-recognition and the logic of subjective intentionality of security (Burgess 2019, 103), also understood as the securitization of subjectivity (Kinnvall 2004) through a negative securitization logic; and second, it draws attention to a normative opening—an inclusive, positive security logic, which moves misrecognition injuries as emancipation toward a more inclusive, advanced, more complete, social and moral progress. It *is* an approach toward positive securitization as the maintenance of just, core values (Roe 2008, 793) lived and realized through diversity, progressive multi- and transcultural agency, and forms of social creativity (Pieterse 1992, 13). A positive security logic and positive securitization highlights how migration practices, policies, and their outcomes represent features that not only deconstruct the migrant as a threat but also enable "conditions for human wellbeing" (Nyman 2016, p. 822). This chapter then evaluates these claims of a new normative opening, increasingly asserted by a twenty-first-century, progressive era of transcultural, hybrid identities.

By utilizing two similarly constructed "crises" environments—the migration discourse in Germany and the USA—this chapter takes its epistemological point of departure from both countries' containment and deterrence migration policies and practices. Specifically, it interprets the US Remain-in-Mexico and Germany's ANKER Center policies (ANKER stands for Arrival, Decision, and Return) as *liberal dispositifs of security* (Bigo 2008, 110). The Foucauldian concept of *dispositifs* broadly defines "a complex assembly of self-governing measures" (Dillon and Neal 2008, 14), an apparatus of practices and technologies of power. In the context of migration, they revolve around "life's circulations" and human mobility. The term "liberal" as a qualifier refers to the tension between the freedom of movement versus the limits imposed on mobility by the states through these dispositifs as technologies of power. These dispositifs misrecognize human beings by organizing and pressing them into dehumanized and immobile abstracts. Through Remain-in-Mexico as well as the ANKER Center policies, for example, migrants are internally bordered and othered—marked segregated and kept delineated from the outside selves such as citizens and legal residents—in indefinite, static, inhumane, if not life-threatening and dangerous borderlands of containment "at the boundaries of democracies […] with the 'freedom' to leave to where one does not want to go" (Bigo 2008, 110). This chapter defines borderlands not as fixed, territorial delineations that describe the perimeters and edges of states, but as normative concepts that shape interaction. Borderlands are "processes, practices, discourses, symbols, institutions or networks through which power works" (Johnson et al. 2011, 62). This chapter also uses migrants and migration as an inclusive, overarching term. It will not differentiate between irregular, forced, or regular migrants or migrants and refugees, or immigrants. It does so because these definitions as generalized categories are, at best, fluid, negotiated and renegotiated, shaped and reshaped (Crawley and Skleparis 2018), hence, in constant flux depending on political circumstances—political alliances and national interests, including the "politics of human rights"—such as seen in the US and Germany.

Misrecognition is understood as "an injury to the self" (Epstein 2018, 808), which transpires once recognition of one's self-understanding fails to occur. These liberal dispositifs of security then in a Foucauldian-fashion—similar to centrifugal devices from an internal axis outward—sort human beings time and again into "dangerous and non-dangerous, risky and non-risky" (Aradau 2016, 565) intangible categories as globally mobile abstracts. Through their solely subjective intentionality, the dispositifs turn migrants into negatively securitized human beings, into a threat. The negative logic of securitization enables these dispositifs and facilitates normalized, ongoing, reiterated misrecognition processes. In Germany and the USA, for example, the Remain-in-Mexico and the ANKER Center policies facilitate a daily display of the "spectacle of sovereign protection and impotence at the same time" (Aradau 2016, 569). The paradoxical power spectacle of concurrent protection and actual paralysis in the context of misrecognition and negative securitization—with the global and local intimately implicated—has thus far received only scant attention in current scholarship.

This chapter then evaluates the misrecognition and negative securitization dilemma and theorizes about a normative opening toward emancipation and positive securitization. This "way out" strategy could provide a more complete realization of social and moral progress in a twenty-first-century world of increasingly transcultural, hybrid identities. Transcultural refers to the shift of people's identities between multicultural settings. Increasingly, people do not identify with one monolithic, cultural background, but adapt to multiple ones and can shift between these cultural settings. Why does a more complete realization of social and moral progress matter? As "the desire for recognition is on par with security" (Wendt 2003, 514), unequal recognition, or non- or misrecognition will continue to challenge and undermine human, global relations. As Georg Wilhelm Friedrich Hegel has famously reasoned, only reciprocal, intersubjective, equal recognition received and given provides recognition of value. One's full self-realization only occurs within the "state of societal solidarity" (Honneth 1995, 128). Reciprocal recognition of individuals, groups, and also states, therefore, is imperative for an interdependent, peaceful global society since its social interdependence will re-constitute collective solidarity (Wendt 2003, 512): Two Selves will become one, a We (Williams 1997, 293). By mitigating and transforming migrants' misrecognition through emancipation and positive securitization, it changes global, human interaction and relationships, in a positive way, whereby collective trust is established within societies and populations and among each other. Emancipation becomes foundational in sustaining cooperative human global relations.

This chapter proceeds in three parts: It will firstly define misrecognition and the Copenhagen School's securitization theory in the context of migration. It will then secondly explore how misrecognition and negative securitization can be identified in the two case studies—Remain-in-Mexico and ANKER Centers. Thirdly, it will evaluate its positive securitization as emancipation claim as a possible "way out" of the normative and normalized negative security-misrecognition dilemma.

8.2 Recognition, Misrecognition, Security, and Securitization

Securitization scholars have from early on questioned security as an *objective* reality. The reality of security remains ultimately unknowable. Security and insecurity are what a subject—actors, agents, or security dispositifs—makes of them. Securitization—and its twin-concept de-securitization, for example—is, therefore, broadly viewed as the *making* and *unmaking* of a security issue, the construction and de-construction of an existential threat through the intersubjective engagement between securitization actors and audience. Successful securitization, for example, generally prescribes specific steps: a) viewing an issue/referent object as threatened *or as an actual existential threat*, b) convincing an audience through a speech act of an existential threat, and c) responding to the threat by deploying extraordinary measures, which suspend normal rules. The *securitized* migrant as the threatening, invading "other" in the US and German context, for example, is constructed through a distinct intersubjective interplay between the state—through its normative institutions (policies, legal rights, etc.)—and an audience such as citizens, politicians, civil society, or the media. The audience ought to accept, and hence legitimize, the existential threat. Political grammars of mobility (Aradau 2016)—specific speech acts—support the securitization of the intersubjective interplay. They rhetorically reduce migration flows to imaginaries: uncontrollable infestations, invasions, caravans, mobs, and boatloads, for example. Similarly, ill-designed, ad hoc asylum policies as liberal security dispositifs also become speech acts. They imagine migration as "asylum tourism" (Die Welt 2018) or the Malthesian overpopulation. Securitization is, therefore, often essentialized as a negative outcome. It is a "failure to deal with issues of normal politics" (Buzan et al. 1998, 29), while in comparison securitization's twin-concept de-securitization epitomizes "what security should – or not be" (Taureck 2006, 55). De-securitization then becomes the *unmaking* of the existential threat. It is the process of returning an issue to the political discourse, a reopening of "the political game" (Balzacq 2015, 85), a space where the bargaining and negotiations of ideas take place, where issues are resolved through policy.

Yet, critical scholarship has increasingly questioned these deterministic, binary properties that fail explaining how security and securitization are actually "understood and practiced" (Nunes 2015, 143) in their everyday realities. This kind of ontological suspicion toward the singularity of the *always already negatively securitized* mobile body and space needs more careful attention. Migrants and migration are made synonymous with threat and danger to states and citizens' ontological identity selves. Migrants, borderlands, and other abstractions of mobility are immediately termed as "securitized" and, therefore, deemed negative without more thoroughly interrogating its assumed negative logic. Specifically, migration scholarship, for example, has adopted this linearity without more fully parsing out the ontological make-up of security and acts of securitization.

Human mobility is understood as "an absolute right of man" (Cresswell 2006), yet the liberal dispositifs of security as migration practices time and again curtail these rights. The institutionalized orders of recognition—historically differentiated in modern capitalist societies (Cox 2012, 193) as self-trust, legal rights, self-determination, and autonomy—are repeatedly violated. They become injuries to the self. They violate by preventing migrants to circulate through the institutionalized order of recognition—through self-trust, legal rights, self-determination, and autonomy. Broadly seen, these injuries are deeply embedded, for example, in the logic of capital as a distributional form of misrecognition. Migrants are misrecognized in their relationship to materiality, for one through the Karl Marxian "army of labor" (Marx 1976), readily exploitable at will. The logic of capital as a distributional mode of misrecognition is also seen in the current US criminalization and mass incarceration of migrants. Detention facilities and prisons function as vessels of labor control "to store reserve labor when it is undesirable" (Martinez and Slack 2015, 544). Migrants are also misrecognized through states' integration laws where the political call for forced assimilation or "assimilation without compromise" (Chin 2017), for example, presses migrants into cultural and societal sameness and invisibility. By flattening migrants' multi- or transcultural idiosyncrasies through "integration at all cost," receiving states, for example, are cast as a "can of soup," "with newcomers being expected to blend in and dissolve as fast as possible" (Rytter 2019, 681). Integration is increasingly seen as part of the problem and not its solution. It epitomizes and reiterates a distinct, asymmetrical relationship of privilege and superiority between majority and minority populations. For migrants, integration remains unattainable as majority populations constantly seem to shift and re-erect new "invisible fences." Integration becomes a perpetual double-bind (Rytter 2019, 688).

Likewise, policies as dispositifs of security have only perpetuated the injuries and dangers of misrecognition in Germany and the USA. They created, reiterated, and maintained an image of one-self—the migrant as an individual or migrants as a group—so differently from a migrant's own identity, one's own preferred social status, and one's value in society. Misrecognition, however, is not only performative, but productive. Both, Honneth and Hegel, view the struggle for recognition—and its failure in misrecognition, for example—as central to social conflict (Epstein 2018, 808). As this kind of conflict displays forms or mechanisms of progress or development (Wendt 2003, 493) rather than fixedness, it is though fundamentally of an emancipatory quality (Cox 2012, 193). Honneth sees such potential of moral and social progress specifically instantiated through "individualization" and social inclusion (Honneth 2003, 185), both increasing people's potential for recognition. Similarly, others interpret the Hegelian struggle for recognition as a clear effort to temper egoistic, self-interested identities and to encourage and broaden collective solidarity (Wendt 2003, 493). The productivity emerging from misrecognition as social conflict and the potential for transformation then toward emancipation can be more fully understood through a careful interrogation of the "subject/object divide [...] the Cartesian subject-object structure" (Burgess 2019, 96 & 104) at play.

The imagined threat through the rhetorical "othering"—surrounding migration in the USA and Germany—normalized a *securitization of subjectivity* (Kinnvall 2004, 749). Policies as liberal dispositifs of security, for example, only imagine and contain migrants' bodies, experiences, and agency as "an object-of-threat *affirmed by a subject*" (Burgess 2019, 105). The constant visibility of migrants in real or perceived "borderlands of containment"—in Germany's ANKER Centers or US detention facilities, shelters, or makeshift tent camps—creates "quasi-immobilisation in borderzones" (Aradau 2016, 571). They reify different forms of subjectivities—border control, state agencies, non-governmental organizations, advocacy groups, migrants, and citizens—each entailing "the simultaneous attribution of subjectivity" (Aradau 2016, 570). As such, the securitization of subjectivity, the dominant position of the subject, actually renders the object *objectless*. The solely imagined threat and its negative securitization remain uniquely intersubjective—between subjects—as the actual presence of the threatening object is not required. "Security may be subjective, but the subject is secure" (Burgess 2019, 96).

Kinnvall defined *securitization of subjectivity* further as an "intensified search for one, stable identity" (Kinnvall 2004). To securitize subjectivity then, for example, supports a certain (subjective) threat creation—regardless of its real presence. Similar to the struggle of recognition, a stable identity means that one is confident and certain of one's self-value as it is also a value recognized by others. Often, however, the securitization of subjectivity through the quest for stability and assuredness encompasses the condition "where an individual or group might appeal and become attached to an oversimplified Self-narrative entailing a juxtaposition to an Other" (Kinnvall 2017, 4). This oversimplified "self-other" narrative then can endanger physical and/or conceptual identity security. To feel ontologically secure, for example, means that the self has "answers to fundamental existential questions which all human life in some way addresses" (Giddens 1991, 36). It stands in contrast to chaos, danger, and threat and exemplifies the freedom from deeply embedded existential anxieties. An oversimplified self-narrative, for example, of a citizen, then becomes one that constitutes a person's "biography" (Giddens 1991, 54). It is a narrative of one's "ways of life" (14). In perhaps a more extreme manifestation, the securitization of subjectivity can also be seen in current far-right, racist, populist movements (such as in the USA, Europe, and Latin America) where people seek out the attachment to a larger, collective to affirm one's self-identity. Ontological security and insecurity can, however, also apply to the state. An ontologically secure state-self then can gain a stable state identity through multiple autobiographical state narratives. Different national biographical narratives—often in competition with each other (Delehanty and Steele 2009)—are constructed and then endure, for example, through a collective memory (Mälksoo 2015) or, in the context of migration, through "willful exclusion and untruthful enforcements" (Mälksoo 2015, 5). The "other" is excluded to protect "the self" narrative. Misrecognition emerges, for one, from such forced exclusions and arbitrary enforcements.

The negative securitization of subjectivity and the *objectlessness* (Burgess 2019) it produces (the predicated logic of the subject's intentionality, the threat is *always already* intended or anticipated, *always already* "pre-consciously conscious" (Burgess 2019, 106)) shape the injuries of misrecognition. They construct the emotions of disrespect experienced by migrants. George Mead in 1967 differentiated between the "I" and the "Me" as two ontologically different interpretations of "the self." While the "I" referred to an inner, subjective potential that operated outside of action, the "Me" in comparison relates to how one sees oneself through the eyes of others—as part of a group or community, for example (Mead 1967). The tension between the "I" and the "Me" then drives the inherent insecurity of one's ontological self-identity. One seeks ontological security through a stable identity. Such stability—positive security or positive securitization—is though achieved only through the intersubjective pursuit of recognition. In turn, according to Hegel, for example, such stable self-certainty can only be fulfilled through one's relationship with another self-consciousness. This social interdependency underscores the desire and need for recognition. "The struggle for recognition is born" (Murray 2019, 34). Misrecognition as a danger or injury to the self occurs when such social interdependence remains either unacknowledged or resisted. According to Hegel, the social interdependency is nothing more than an asymmetrical recognition between the independent subject and dependent object. Misrecognition then is not how the *Object* suddenly counts or how the *Self* suddenly accepts certain normative constrains about how the *Object* ought to be treated (Williams 1997, 511). Misrecognition happens when the individual encounters difficulty to enact one's desired social status. It unfolds when the self is represented in a way, which does not concur with how one knows one's own identity. Misrecognition then tends to reject the required intersubjectivity of mutual recognition.

Recognition and securitization are both fundamentally relational and firmly located in and dependent on such an intersubjective world. Humans and all human-filled abstracts or vessels, including states, wish for self-certainty. Therefore, they desire recognition of one's identity from others (e.g., groups or states). Honneth defines one's stable identity as a practical relation to self (Honneth 1995, 79) or as a certainty and assuredness of the value of one's social identity (Murray 2019, 11). However, social constructivists have offered new logics. Alexander Wendt, for one, showed how intersubjective human relations—the recognition game people and states play (Ringmar 2002)—can produce more nuanced forms of recognitions (Wendt 2003). Wendt delimits these new types as thin and thick recognition. These more refined degrees are particularly instructive in how migrants' desires for recognition shift through misrecognition (the injuries to the self) to mitigate social conflict. These variations transform negative, frustrated, and rejected recognition desires, for example, into positive ones—into a positive, inclusive sense of self through confidence, self-assuredness, and certainty. They can transform these frustrated desires into positive security and positive acts of securitization.

Thin recognition, for example, views self-identity of an individual through a sheer sameness framed, for example, by law. One is a human being—a sovereign individual, a person, and a subject and not just an object—simply because of one's humanness. One is a human being just because one is human. "Everyone who has this status is the same, a 'universal person'" (Wendt 2003, 511). Similarly, a state (subject), for example, is only recognized as a state by other states (subjects) because of one's sheer "stateness." Stateness and humanness then mean one is recognized as a state or a human being simply because of the traits and qualities that are assumed of making one a state or a human being. In comparison, the struggle for thick recognition goes beyond these narrow, reductionist assertions. Thick recognition is defined by how an individual gains and commands respect, individualization, and social inclusion through "what makes a person special or unique" (Wendt 2003, 511). In the broader context of migration, recognition, and securitization, thick recognition would then mean the humanness of a migrant is recognized beyond its juridical parameters. A migrant strives for and receives recognition based on one's unique and richly diverse, ever-evolving ontological identity self. The struggle for thick recognition, similar to thin recognition, is centered on subjectivity and the intersubjective exchange it demands. Yet, it is of a much more progressive and relentlessly generative quality. While thin recognition is clearly delineated, for example, by legal norms—such as by defining universal humanness of a person through human rights—thick recognition, in comparison, is "open-ended and never-ending in a way struggles for thin recognition are not" (Wendt 2003, 512). While one is of a distinct, fixed universal feature, the other one is of a fluid particularity. The struggles of thin and thick recognition are also interrelated. While thin recognition would provide migrants with the fundamental, juridical locus of human needs, rights, and degrees of agency, only the interwovenness with thick recognition allows for the necessary and sufficient condition required for emancipation as a way out of misrecognition. Thin recognition alone would not be sufficient. It requires its prolific, thick counterpart.

Since the struggles of recognition are about mutually *receiving* rather than *giving* recognition, the distinctions between thin and thick recognition are then especially useful in deconstructing the negative security logic of global mobility. As the desire for recognition is based on embracing one's confidence in one's distinctiveness—one's diversity and ontological uniqueness—it has the capacity to transform social conflict (emerging from misrecognition and negative securitization) as an emancipatory force in support of collective solidarity. If one self is valued through one's particular ontological richness and particular, diverse qualities (Murray 2019, 13), in the context of migration, this infers that migrants are *not always already* the imagined threat or the perennially threatening Other, but "two Selves in effect become one" (Wendt 2003, 512). Specifically, thick recognition transforms human relations toward Honneth's individualization and social inclusion. A migrant then is not only "thinly" recognized as a human being because of one's juridically defined "humanness," but one is recognized for one's specific, unique role one could play due to one's diversity. The globally mobile human being then is recognized as a member of the community because of one's uniquely diverse qualities—and not universal sameness. Mutual recognition transforms the "other" (migrants) into a confident, certain "self" because of the unique role one can assert as a fully valued member of a community or society. Mutual, interdependent, "thick" recognition of diversity and plurality establishes an assuredness of one's value in society. It builds societal, collective solidarity. The reciprocal, symmetrical thick recognition emerges organically from myriad streams of political and social actors or agencies such as the government, civil society, and non-governmental organizations. Yet, similar to Hegel's master-slave analogy, it needs to be grounded in "worth and dignity" (Wendt 2003, 513) and cannot emerge from a logic of coercion.

Thick, mutual recognition of human beings is then the emerging foundation of positive human relations. Such positive human relations across communities, groups, and states mitigates, if not resolve misrecognition, become a form of emancipation. The struggle for recognition and against misrecognition's injuries becomes the normative opening toward emancipation and positive securitization. Positive security and acts of positive securitization compare so distinctly to security's negative logic because they produce the "collective trust populations have to society and to each other" (Gjorv 2012, 843). Positive securitization draws in part on human security literature. It recognizes the individual human being as the central referent. It understands trust and social bonds as relational, interdependently established by daily routines. Positive security and acts of positive securitization focus on intersubjective relationships and do not exclusively involve the self only (McSweeney 1999, 99). Thick recognition counters the destabilizing effect of the "ever-present possibility of misrecognition" (Murray 2019, 33) through emancipation.

8.3 Misrecognition and Negative Securitization Through Liberal Security Dispositifs

The following section outlines how the US Remain-in-Mexico and Germany's ANKER Center policies as *liberal dispositifs of security* have negatively securitized and misrecognized migrants. As previously mentioned, misrecognition—as "an injury to the self" and as the failure of recognizing one's self-understanding—distinctly expresses itself through dispositifs as technologies of power. For example, the dispositifs sort and organize migrants into immobile and dehumanized entities by routinely denying them access to universal human rights. As such, they violate the fundamental, juridical definition of humanness. They undermined the desire for thin recognition, the fundamental locus of needs, rights, and agency that juridically frame what it means to be human. The intentional lack and non-granting of one's humanness—its subjective intentionality—repeatedly frames migrants as a security threat. Through both state policies, migrants are kept segregated in indefinite borderlands of containment, exposed to refoulement, danger, and injuries that destabilize one's image of oneself. The dispositifs press migrants and their life experiences into a reiterated, paradoxical spectacle where the rejection of states' legal responsibilities to protect human life and human dignity unfolds. Time and again, the spectacle is replayed, the threat normalized and institutionalized—and states' subjectivity negatively securitized. As Peter Burgess articulated earlier, security may be subjective (and negative), "but the subject is secure" (96). Through the rejection of thin recognition, these technologies of power have institutionalized the weakening of a migrant's quest for individualization and social inclusion. As such they fundamentally undermine thin (and thick) recognition and obstruct an opening to emancipation, to a more advanced, inclusive moral and social progress.

8.4 Borderlands of Containment: Remain-in-Mexico

International refugee policies such as the 1951 UN Convention and Protocol to the Status of Refugees have established peoples' universal rights to claim asylum. Explicitly, the freedom from persecution is protected by international law. The US Remain-in-Mexico policy as a "safe third country" arrangement acknowledges its legal commitment not to expose migrants to refoulement, for example. The non-refoulement principle as a legal prerogative refers to not returning asylum seekers to harm and danger. Non-refoulement also includes not exposing migrants to persecution. However, international human rights organizations have consistently argued that the US Remain-in-Mexico policy, for example, gravely violates not only the UN Convention and its 1967 protocol, but also US law that incorporated the non-return, non-refoulement principle in 1980 (Human Rights Watch 2020a, b). The US policy rational, similar to Europe, aims for migrants to seek—apply and potentially receive—asylum protection in explicitly deemed safe third country such as Mexico or Guatemala instead of in the USA. The rational is generally based on administrative efficacy:

1. It reduces stress on existing overwhelmed US asylum systems.
2. It makes the handling of asylum applications (eventually) faster.
3. It is easier returning migrants to third countries than deporting them to their home countries (Fratzke 2019).

However, the cumbersome, non-transparent, and exhaustively prolonged US asylum process—all now outsourced abroad to indefinite, violent makeshift settings in

Mexico—clearly aims to accomplish one primary, proactive goal: to deter other migrants from leaving their home country in the first place (Fratzke 2019).

Third-country arrangements in the USA, Europe, or Australia, for example, have for the most part failed in all of their most basic functions, including in simplifying refugee processes. Since 2015, for example, the European Union (EU) negotiated such arrangements with Turkey and Libya. However, between 2016 and 2019, only 1.7% of migrants were returned from Greece to Turkey (Fratzke 2019). Despite the dismal EU precedent, the USA emulated similar arrangements, primarily in the wake of growing, global criticism of its notorious 2018 family separation migration policy. By early 2019, the Trump administration began increasingly to pressure transit countries such as Guatemala and Mexico to agree to "safe" third-country stipulations in order to halt migrations flows at the USA-Mexico border.

Migration policies, principally designed to assert deterrence, are not unique, but common, global practices. In the USA, long-standing, contemporary migration policies such as Operation Streamline have consistently injured, dehumanized, and denied migrants' thin recognition and migrants' universal legal rights and abilities for basic, human agency. Since 2005, Operation Streamline—a zero-tolerance George W. Bush policy—has criminalized illegal border crossings and turned asylum court hearings into "assembly-line justice." Immigration judges are sentencing migrants in large groups instead of individually. Migrants are prohibited to testify and state their own cases, to introduce their own evidence, and as such judges—people of juridical privilege and authority—are fundamentally discrediting migrants' human claim for agency. Such assemble-line justice, currently the main driver of the criminalization and mass incarceration of migrants in the USA, is deeply steeped in the country's racialized legacies of slavery and Jim Crow when African-Americans were barred to testify on their own behalf.

However, different from Operation Streamline, the Remain-in-Mexico policy stands out in how it has only further decimated US juridical commitments to international law and human rights conventions. The USA's official 2019 announcements of its Remain-in-Mexico policy (US Department of Homeland Security 2019) point to the country's continuing concern about the protection of migrants, including to uphold the non-refoulement principle. The everyday, empirical reality of Remain-in-Mexico, however, fundamentally deprives migrants of their universal rights to "freely dispose over their own body" (Cox 2012, 205). Migrants' humanness and agency as humans—their human authenticity as "genuine bearers of any presumptive (purportedly universal) 'human right'" (Koca 2019, 187) such as the right to claim asylum—are rejected. Migrants' daily lives and experiences, filled with agency, self-confidence, and self-trust are discredited. They are "dishonored and defamed" (Solik 2016, 77). Orthodox, legal asylum principles such as "protected classes or categories" or long-established processes such as "credible fear screening" are arbitrarily applied. In the USA, misrecognition subsumed under "the politics of human rights" has violently proliferated.

The US Department of Homeland Security (DHS) administratively framed the Remain-in-Mexico policy through the 2019 Migrant Protection Protocols (MPP). The MPP, lauded by conservatives as a migration policy "game changer," define specific steps to keep migrants outside of the USA during the entire duration of their court proceedings. As a result, the MPP has changed criteria and raised benchmarks, for example, of who can apply and receive asylum. By end of the fiscal year 2019, for one, the number of denied asylum increased from 9716 in fiscal year 2014 to 46,735 in 2019 (Syracuse University 2019).

By November 2019, more than 566,000 asylum seekers, including 16,000 children (about 500 of them under the age of 12), were returned to Mexico from the US border (Human Rights Watch 2020a, b). Remain-in-Mexico has exposed thousands of migrants to unsanitary conditions in shelters and makeshift tent camps in border cities such as Ciudad Juarez across from El Paso, Texas. These shelters are not operated by the Mexican government, but often informally organized by non-governmental organization, including religious groups. As such they are known to function outside of state or local oversight. Besides non-hygienic, sub-standard, and often violent conditions inside these shelters and ad hoc camp settings, MPP's tightly scripted rules also deny migrants' access to legal counsel. They systematically perpetuate the non-transparency of US immigration procedures through ever-shifting rules and regulations. The MPP, for example, is constantly narrowing the "eligibility grounds for asylum" (Human Rights Watch 2020a, b), including changing requirements for work permits and fee structures.

The MPP require migrants to report frequently to their initial US port of entry. From the port of entry, migrants then are transported or "paroled" (US Department of Homeland Security 2019) to their scheduled court hearings in the USA. During these logistically burdensome and convoluted procedures, migrants (including parents, or single mothers and fathers with their children, or unaccompanied children and young adults) remain in the custody of ICE (US Immigration and Customs Enforcement). Migrants then are constantly forced to drift between the crammed detention centers in the USA and ad hoc shelter settings in Mexico in order to comply with US regulations. Failure to abide by the MPP, including to appear to court hearings, would automatically result in removal proceedings and deportations.

The Remain-in-Mexico policy has also only exacerbated grave insecurity and crisis environments in underfunded Mexican border municipalities. MPP and Remain-in-Mexico have subjected thousands of migrants, for example, to gang and criminal violence, including sexual violence, and human trafficking in Juarez. The injuries of misrecognition then also emerge from absent or failing administrative infrastructures in Mexico and Guatemala as so-called safe, third countries, for one, as both states cannot assure the protection of migrants. Both countries continue to struggle with inefficient, underfunded local governments, informal economies, and corruption. Both wrestle with high crime rates and remain structurally ill-equipped due to the lack of institutional resources to sustain an effective asylum administration. For example, by mid-2019 Mexico's asylum agency had processed nearly 60,000 asylum applications (Fratzke 2019). This was twice the number compared to 2018, all with a mere $1.3 million agency budget. Juarez' already fragile governmental structures are only breaking down even further with the task to accommodate the current influx of migrants. Remain-in-Mexico, however, has not deterred migration flows into the USA. To the contrary, it has created obscure, unintended consequences. It has only incentivized large smuggling and human traffic rings and increased lucrative kidnapping shams in Juarez and other border towns, for example. The suffering of migrants as a perverse form of deterrence has not stopped population mobility. It has only made the agents and mechanisms, operating within these spaces of mobility, more violent and more profitable.

By spring 2020, the coronavirus pandemic has further increased the negative securitization of migrants at the USA-Mexico border. The COVID-19 outbreak has, for example, indefinitely postponed all MPP hearings first through March 2020 and then indefinitely into summer 2020. Yet, the US government continues to force migrants to travel from their detention facilities or shelters in Mexico to their US port of entry. According to a DHS statement in April 2020, the USS required migrants to "present themselves at their designated port of entry on their previously scheduled date to receive a tear sheet and hearing notice containing their new hearing dates" (US Department of Homeland Security 2020). The COVID-19 outbreak, however, has even further made migrants' vulnerability in Mexico's overcrowded shelter and tent settings—lacking basic sanitary resources such as running water, for example—even more precarious (Human Rights Watch 2020a, b). In May 2020, the US government reported the first coronavirus death of a migrant in ICE custody.

8.5 Borderlands of Containment: Anker Centers

In 2015, at the peak of the so-called European migration "crisis," the continent took in approximately one million migrants. A large portion of these migrants applied for asylum in Germany, where a specific quota system organized their resettlements based on a state's population and tax income. Bavaria, one of Germany's wealthiest and most conservative states, accounted in 2016 for the second largest number of incoming migrants. While overtime in Germany the number of asylum applications consistently fell below 2015 levels, far-right xenophobia continues to influence public and political discourse (LSE 2019). Due to these political pressures, in summer 2018, the German interior ministry issued a controversial migration "masterplan." It proposed so-called ANKER Centers (ANKER stands for Arrival, Decision, and Return), oversized mass deportation facilities for newly arrived migrants, designed to facilitate rapid asylum procedures and decision-making processes (Seehofer 2018). As of 2020, most of these mass deportation centers (seven out of nine) are located in Bavaria, with each housing about 1000 migrants. Only two additional ANKER Centers each are placed in the German states of Saarland and Sachsen.

Still in 2020, thousands of migrants continue to live in container-villages dispersed throughout Germany. Differently from these more residential container settings, however, the ANKER Centers stand out in their unique design and ability to press migrants into indefinite immobility that degrades their "practical relationship" with their self-value and respect. Similar to the US Remain-in-Mexico policy, these centers are an ultimate deterrence tool to shut down legal asylum avenues and to expel migrants quickly. They also fundamentally reject migrant's desire for thin recognition, the juridical locus of needs, rights, and agency that frame migrants' basic claim of humanness. Due legal process is compromised in lieu of deportation expediency. As many migrants lack documents, including passports or birth certificates, for example, it has stalled asylum procedures. It has turned many of these centers into indefinite holding infrastructures, completely segregated from surrounding communities. Migrants continued not to be allowed to work; are prohibited to attend integration courses, including language classes; and are unable to afford legal assistance. Similar to the US Remain-in-Mexico policy, these ANKER Centers as liberal security dispositifs segregate, dehumanize, and isolate migrants. They uniquely disrupt the struggle for basic, thin recognition—for the distinct, fixed universal locus of human needs, rights, and agency as juridical inscribed in humanitarian law—and dislocate the institutionalized recognition order. Since 2018, German non-governmental organizations have called out these ANKER Centers as rightless places of violence. They asked the Bavarian government for the centers' immediate closure (Münchner Flüchlingsrat 2019).

In particular, international, regional, and local NGO health and migration advocacy groups such as REFUGIO Munich and Ärzte der Welt (Doctors of the World) criticize these centers' segregated characters; crammed, unsanitary living conditions; and the ad hoc, streamlined deportation proceedings (Munich Field Notes Folder II 2018). Ärzte der Welt argued that many of these conditions represent blatant human rights violations. Facility shower areas and restrooms are not lockable. Migrants are not allowed to prepare their own meals, and children were regularly forced, due to the lack of privacy, to witness physical altercations. Such trauma profoundly undermines the safeguarding of children as it "renders the parent impotent and leaves the child without protection" (Mares et al. 2002). Ärzte der Welt has also criticized these centers' lack of basic mental health resources. The centers' segregated, isolated character has only further aggravated migrants' often grim, hopeless outlook toward the future and only worsened their underlying migration trauma experiences (Deutsches Ärzteblatt 2020b). Within 10 months, mental health staff of Ärzte der Welt was only able to attend to 41 patients. Many migrants have received their first mental health assessments only after staying for more than a year in these Anker Centers. In October 2019, Ärzte der Welt discontinued its mental health support due to the centers' substandard conditions, including the challenging working environments for its mental health staff. Other organizations such as the European Council for Refugees and Exiles (ECRE) have also attested for similarly grim conditions (Deutsches Ärzteblatt 2020a).

8.6 The Era of Transcultural, Hybrid Identities: Emancipation—A Way Out?

This final section will evaluate the proposed claim of emancipation toward positive securitization as a strategy of normative commitment. Such commitment would provide an opening that tackles negative securitization and misrecognition—states' refusal to acknowledge people on the move as "genuine bearers of human rights"—in a unique way. Building on Wendt's concept of "thick recognition," emancipation and positive securitization as a transformative force moves misrecognition's social conflict and negative securitization toward a more inclusive, advanced, more complete moral and social progress. As aforementioned, these normative commitments are foundational to sustainable, cooperative human global relations. The struggle for and the pursuit of recognition, including its injuries, can transform twenty-first-century human relations in a positive, emancipatory way. Emancipation then is the post-structural, post-modern process through which one can exercise freedom from traditional structures and relationships toward human agency and self-determination. It rearticulates "people's ability to determine the course of their lives" (Nunes 2015, 144). The normative commitment toward positive securitization—creating and facilitating opportunities of well-being for all humans—is not the universal call for open borders. It is, however, a call to restore and reinject (thin and thick) recognition into the policies and everyday processes, and practices of migration. Positive securitization is then the positive, social co-existence and the realization of how people's lives "fit together with others" (Taylor 2004, 23) and how expectations of the self and one's social world are met.

Critical theory has long interpreted emancipation—in comparison to the "recognition or security games" states play—of a much more sophisticated, ambitious, and expansive quality (Booth 2005, 10). Emancipation frees people from structural constraints, including oppressive, exclusionary, and violent state policies (Booth 1991) and promotes human liberation and fulfillment. Emancipation can be granted by norms (domestic or global institutions) or fought for by individuals or groups (as self-liberation) (Balzacq 2015). In relation to security, emancipation is logically incompatible with negative securitization, for example, and is as such much more independent and self-sustaining. The negative logic of securitization through states' security dispositifs turns "politics into a matter of command… [and national identity into] a narcissistic paranoia" (Dillon 1996, 130). It exacerbates states' fantasies of homogeneity (Bigo 2008, 109).

Migrants direct their struggle for recognition to states since states are "the primary addressees of the claims that follow from the existence of human rights" (Honneth 2007, 209). Yet, misrecognition occurs as migrants are time and again exposed to states' arbitrary application of human rights, asylum norms, rules, and regulations, for example. Misrecognition reduces a migrant's claim for humanness to the *politics* of human rights. Migrant-refusing countries such as the U.S. or Fortress Europe then consequently lose their own sense of self—their ontological security—through their Hegelian "parasitic existence" (Lindemann 2018, 926). As social interdependence suggests, freedom and positive security, for example, only embody worth—and are only worth having—if they are of value and valued by all selves. Reciprocity, for example, through mutual recognition becomes a precondition for freedom (Wendt 2003, 514). It is also essential for human agency. It allows for the shift from "the means of living *to* the actual opportunities a person has" (Sen 2001, 253). Therefore, state oppression "needs to be struggled against by [...] emancipation" (Aradau 2004, 401)—a struggle for human well-being for all humans. Emancipation emerges because desires of recognition will eventually undermine discriminatory systems "that do not satisfy them" (Wendt 2003, 514). The de-stabilizing effect of misrecognition, the inescapable, intolerable uncertainty of migrants—similar to Hegel' frustrated self-consciousness—becomes "the driving force of social development" (Solik 2016, 76). Therefore, the recognition of humanness, including through human rights, cannot be left to the politics of states.

Forms of emancipation—forms of positive securitization and thick recognition, for example—have already taken place to various degrees in Germany and the USA. They include the US 2012 Deferred Action for Childhood Arrivals (DACA) program that allowed young, undocumented immigrants a path to legal residency. Degrees of emancipation also include Germany's 2000 and 2005 citizenship reforms or the German 2016 and 2019 integration laws, for example. These policy changes have emerged from migrants own political agency such as seen through Germany's multi-generations of former guest worker families or the millions of undocumented migrants in the USA. These forms of emancipation have already to some extent increased the potential for individualization and social inclusion of migrants and encouraged mutual recognition toward social interdependence (Murray 2019, 15).

However, most interestingly current forms of emancipation in the US and Germany increasingly include the involvement of non-state actors such as advocacy groups and non-governmental organizations (NGOs). From early on, in Germany watchdog groups such as the Munich Refugee Council (*Flüchtlingsrat*) called for the immediate closure of all ANKER Centers (Münchner Flüchlingsrat 2019). During the coronavirus 2020 pandemic, similar to the USA, these mass deportation centers have rapidly become dangerous infection hot zones (Münchner Flüchtlingsrat 2020b). In April 2020, the Council pushed the Bavarian government for comprehensive transparency, regarding the number of coronavirus infections in all shelters, including the ANKER Centers, and demanded a detailed quarantine action plan. It asked for the separation and quarantining of infected and not-infected migrants and the transfer of migrants to hotel rooms and on April 2, 2020 filed a lawsuit against the Bavarian government (Münchner Flüchtlingsrat 2020a), again demanding the immediate closure of all ANKER Centers (Münchner Flüchtlingsrat 2020a).

NGOs have played not only instrumental roles in bringing visibility to the misrecognition of migrants. They also increasingly facilitate intersubjective, mutual recognition toward thick recognition. In the USA, protests for comprehensive immigration reform—organized nationally by the American Civil Liberties Union, Human Rights Watch, or locally El Paso's Border Network for Human Rights or Hope Border Institute, for example—have for year pushed to expand migration rights. Most recently, NGOs' activism has grown profoundly effective in drawing attention to the tragic struggles for recognition as seen in the draconian US 2018 family separation policy. Resistance against these policies, for example, has forced the Trump administration to reunite parents with their children. Likewise, NGOs due to the 2020 pandemic have demanded to stop mandatory detentions. It allowed some migrants to stay with family members in the USA to await their court hearings. End of April 2020, non-state government watchdog groups such as Washington--D.C.-based Project on Government Oversight (POGO) drew public attention to the dismal health and medical care conditions in ICE detention facilities in Louisiana and New Jersey. In spring 2020, both facilities had outbreaks of COVID-19 cases among staff and detainees and are now closely monitored by several advocacy groups (Project on Government Oversight 2020).

The emancipation as positive securitization argument, however, also raises a number of issues. For example, scholarship has highlighted the explanatory weaknesses of the Honneth's recognition theory about the role of NGOs as factors or mediators of intersubjective, mutual recognition (Heins 2008). Furthermore, the emancipation and positive securitization argument also does not suggest the "death of the independent state" as propagated by anti-statism scholarship (Linklater 1989). It does not interpret open borders, borderlessness, or generally a "deterritorialized world" (Newman 2006) as the absolute goal of autonomy and self-determination. Emancipation as transformation also does not imply that emancipation is equal to resistance (Habermas 1987). Resistance often infers the maintenance of a certain status quo, while emancipation is of a much more progressive and future-oriented quality (Nunes 2015, 139).

Yet, the emancipation and positive securitization argument is effective as a "way-out" of the negative security and misrecognition dilemma precisely because it provides an escape from the asymmetric subject-object determinism. It displays a unique pragmatism toward mutual, intersubjective, and symmetrical recognition while at the same time understanding the "corporeal, material existence and experiences of individual human beings" (Jones 2005). As illustrated through the Remain-in-Mexico and ANKER Center policies, emancipation understands very well how insecurities stem from real and actual misrecognition injuries and experiences lived, felt, and processed by human beings. Emancipation understands very well the everyday violence embedded in liberal security dispositifs. While security and insecurity ultimately always remain unknowable—there is not a universal truth to both—this does not mean, however, that outcomes and effects of positive securitization and the struggle for recognition cannot account for real, tangible efforts toward a better and improved, material reality. As shown, the experiences of emancipation, insecurity, security, recognition, and misrecognition are real, material, and authentic versions of the truth.

8.7 Conclusion

This chapter made two contributions to existing scholarship: First, it highlighted the linkages between *mis*recognition and the logic of subjective intentionality of security (Burgess 2019), also understood as the securitization of subjectivity (Kinnvall 2004) through negative securitization logic. Second, it drew attention to a normative opening: an inclusive, positive security logic, which moves the struggle for recognition and its misrecognition injuries beyond social conflict toward emancipation—toward an advanced, inclusive, and more complete social and moral progress, increasingly valued, if not demanded in today's progressive era of transcultural, hybrid identities.

The struggle for recognition is a struggle of becoming "a self." If the processes and lived experiences of becoming a self are disrupted, we speak of misrecognition. Instead of facilitating and supporting trust, respect, and esteem through undifferentiated rights, agency, inclusive and self-determined, positive security for migrants, German and US migration practices time and again violate the two countries' own liberal and moral aspirations of universal rights, justice, and social and moral progress. A way out of the negative security and misrecognition logic can be achieved through emancipation. The proposed interdisciplinary misrecognition and securitization lens, including the concept of thick recognition, provides an opening toward alternative approaches. It moves misrecognition injuries beyond social conflict toward emancipation, toward progressive agency "to open up space for people to make decisions and to act on matters pertaining to their own lives" (Nunes 2015, 151). Emancipation is creating an inclusive, transformative framework that supports the security of selves in "progressive and integrative fashions" (Nunes 2015, 140) as thick recognition. Without emancipation, people and states are no longer able to feel secure and safe but both, people and states, will become profoundly vulnerable. Both will co-constitutively cause and only further aggravate each other's vulnerabilities. Therefore, Germany and the USA, for example, ought to pursue emancipatory strategies of comprehensive mutuality—full, mutual recognition and mutual esteem—as inclusive, positive securitization practices, for example, to reduce the essentialization of the "other," to accommodate the transcultural, hybrid state and citizen identity selves of the future.

References

Aradau, C. 2004. Security and the Democratic Scene. *Journal of International Relations and Development* 7: 388–413.

———. 2016. Political Grammars of Mobility, Security and Subjectivity. *Mobilities* 11 (4): 564–574.

Balzacq, T. 2015. *Contesting Security: Strategies and Logics*. New York: Routledge.

Bigo, D. 2008. Security: A Field Left Fallow. In *Foucault on Politics, Security and War*, ed. M. Dillon and A. Neal, 93–114. Basingstoke: Palgrave Macmillan.

Booth, K. 1991. Strategy and Emancipation. *Review of International Studies* 17: 313–326.
———. 2005. Beyond Critical Security Studies. In *Critical Security Studies and World Politics*, ed. K. Booth, 259–278. Boulder: Lynne Rienner.
Burgess, P. 2019. The Insecurity of Critique. *Security Dialogue* 50 (1): 95–111.
Buzan, B., O. Waever, and J. de Wilde. 1998. *Security: A New Framework for Analysis*. Boulder: Lynne Rienner.
Chin, R. 2017. *The Crisis of Multiculturalism in Europe*. Princeton: Princeton University Press.
Cox, R. 2012. Recognition and Immigration. In *Recognition Theory as Social Research*, ed. S. O'Neill and N.H. Smith, 192–212. London: Palgrave Macmillan.
Crawley, H., and D. Skleparis. 2018. Refugees, migrants, neither, both: Categorical fethishism and the Politics of Bounding in Europe's migration crisis. *Journal of Ethnic and Migration Studies*: 48–64.
Cresswell, T. 2006. *On the Move. Mobility in Modern Western World*. London: Routledge.
Delehanty, W., and B.J. Steele. 2009. Engaging the narrative in ontological (in)security theory: Insight from femnist IR. *Cambridge Review of International Affairs* 22 (3): 523–540.
Deutsches Ärzteblatt. 2020a. Geflüchtete Menschen in Deutschland: Ankerzentren machen krank. *DeutschesÄrzteblatt* 117(3):A-70/B-64/C-62.RetrievedApril19,2020,fromhttps://www.aerzteblatt.de/archiv/211911/Gefluechtete-Menschen-in-Deutschland-Ankerzentren-machen-krank.
———. 2020b. *Traumafolgestörungen und Asylrecht*. Retrieved April 19, 2020, from https://www.aerzteblatt.de/archiv/213604/Traumafolgestoerungen-und-Asylrecht-Eine-besondere-Herausforderung.
Die Welt. 2018. *Wir müssen endlich den Asyl-Tourismus beenden*. Retrieved from Die Welt: https://www.welt.de/politik/deutschland/article177596828/Markus-Soeder-Wir-muessen-endlich-den-Asyl-Tourismus-beenden.htm.
Dillon, M. 1996. *The Politics of Security: Towards a Political Philosophy of Continental Thought*. London: Routledge.
Dillon, M., and A. Neal. 2008. Introduction. In *Foucault on Politics, Security and War*, ed. M. Dillon and A. Neal, 93–114. Basingstoke: Palgrave Macmillan.
Epstein, C. 2018. The Productive Force of the negative and the desire for Recognition: Lessons from Hegel and Lacan. *Review of International Studies* 44: 805–828.
Fratzke, S. 2019. *Migration Policy Institute*. Retrieved January 25, 2020, from Migration Policy Institute: https://www.migrationpolicy.org/news/safe-third-country-agreement-would-not-solve-us-mexico-border-crisis.
Giddens, A. 1991. *Modernity and Self-Identity: Self and Society in the Late Modern Age*. Standford: Stanford University Press.
Gjorv, Gunhild Hoogensen. 2012. Security by Any Other Name: Negative Security, Positive Security, and a Multi-Actor Security Approach. *Review of International Studies* 38: 835–859.
Habermas, J. 1987. *Theory of Communicative Action*. Vol. 2. Boston: Beacon Press.
Heins, V. 2008. Realizing Honneth: Redistribution, Recognition, and Global Justice. *Journal of Global Ethics* 4 (2): 141–154.
Honneth, A. 1995. *The Struggle for Recognition: The Moral Grammar of Social Conflicts*. Cambridge, MA: Polity Press.
———. 2003. Redistribution as Recognition: A respond to Nancy Fraser. In *Redistribution and Recognition? A Political Philosophical Exchange*, eds. N. Fraser, H. Axel, N. Fraser, and A. Honneth, Trans. J. I. J. Golg, 110–197. London: Verso.
———. 2007. *Tis Universalism a Moral Trap?* In *Disrespect*, ed. A. Honneth, 197–217. Cambridge, MA: Cambridge University Press.
Human Rights Watch. 2020a, January 29. *Trump Administration's Remain in Mexico Program*. New York, USA. Retrieved April 19, 2020, from https://www.hrw.org/news/2020/01/29/qa-trump-administrations-remain-mexico-program.
———. 2020b, April 2. US Covid 19 Policies Risk Asylum Seekers Lives. *US Covid 19 Policies Risk Asylum Seekers Lives*. Retrieved April 19, 2020, from https://www.hrw.org/news/2020/04/02/us-covid-19-policies-risk-asylum-seekers-lives#.

Johnson, C., et al. 2011. Interventions on Rethinking 'the Border in Border Studies'. *Political Geography* 30 (2): 61–69.
Jones, R.W. 2005. On Emancipation: Necessity, Capacity, and Concrete Utopias. In *Critical Security Studies and World Politics*, ed. K. Booth. Boulder: Lynne Rienner Publishers.
Koca, B. 2019. Bordering Practices across Europe: The Rise of "Walls" and "Fences". *Migration Letters* 16 (2): 183–1944.
Kinnvall, C. 2004. Globalization and Religious Nationalism: Self, Identity, and the Search for Ontological Security. *Political Psychology* 25 (5): 741–767.
———. 2017. Ontological Security and Foreign Policy. In *Oxford Research Encyclopedia of Politics*, ed. J. Mitzen and K. Larson, 1–26. Oxford: Oxford University Press.
Lindemann, T. 2018. Agency (Mis)Recognition in International Violence: The Case of French Jihadism. *Review of International Studies* 44: 922–943.
Linklater, A. 1989. *Beyond Realism and Marxism*. New York: St. Martin's Press.
Mälksoo, M. 2015. Memory must be Defended: Beyond the Politics of Mneumonical Security. *Security Dialogue* 46 (3): 221–237.
Mares, S., et al. 2002. Seeking Refuge, Losing Hope: Parents and Children in Immigration Detention. *Australasian Psychiatry* 10 (2): 91–96.
Martinez, D., and J. Slack. 2015. What Part of 'Illegal' Don't you Understand? The Social Consequences of Criminalizing Unauthorized Mexican Migrants in the United States. *Social & Legal Studies* 22 (4): 535–551.
Marx, Karl. 1976. *Capital: Volume I. In Marx and Engels Collected Works*. Vol. 35. London: Lawrence and Wishart.
McSweeney, Bill. 1999. *Security, identity and interests. A Sociology of International Relations*. Cambridge: Cambridge University Press.
Mead, G.H. 1967. *Mind, self and society: From the standpoint of a social behaviorist*. Chicago: University of Chicago Press.
Münchner Flüchlingsrat. 2019. *Ankerzentren Bilanz nach einem Jahr*. Munich: Bavaria. Retrieved April 19, 2020, from https://www.fluechtlingsrat-bayern.de/bilanz-nach-einem-jahr-ankerzentren.html.
Münchner Flüchtlingsrat. 2020a. *Corona-in-fluechtlingsunterkuenften*. Munich, Bavaria, Germany. Retrieved April 19, 2020, from https://www.fluechtlingsrat-bayern.de/beitrag/items/corona-in-fluechtlingsunterkuenften.html.
———. 2020b. *Durchseuchung-in-bayerischen-fluechtlingsunterkuenften*. Retrieved April 19, 2020, from https://www.fluechtlingsrat-bayern.de/beitrag/items/durchseuchung-in-bayerischen-fluechtlingsunterkuenften.html.
Munich Field Notes Folder II. 2018. *Munich Field Notes Folder II, 2018*. Munich: S.H.
Murray, M. 2019. *The Struggle for Recognition in International Relations: Status, Revisionism and Rising Powers*. Oxford: Oxford University Press.
Newman, D. 2006. The Lines that continue to separate us: Borders in our Borderless World. *Progress in Human Geography* 30 (2): 143–161.
Nunes, J. 2015. Emancipation and the Reality of Security. In *Contesting Security*, ed. T. Balzacq, 141–153. New York: Palgrave.
Nyman, J. 2016. What is the value of security? Contextualizing the negative/positive debate. *Review of International Studies* 42 (5): 821–839.
Pieterse, J. 1992. *Emancipation, Modern and Postmodern*. London: Sage.
Project on Government Oversight. 2020, April 28. *ICE Death Reviews for Roger Rayson*. Retrieved April 29, 2020, from Project on Government Oversight: https://www.pogo.org/document/2020/04/ice-death-reviews-for-roger-rayson/.
Ringmar, E. 2002. The Recognition Game. *Cooperation and Conflict*: 115–136.
Roe, P. 2008. The Value of Positive Security. *Review of International Studies* 34: 777–794.
Rytter, M. 2019. Writing Against Integration: Danish Imaginaries of Culture, Race, and Belonging. *Ethnos. Journal of Anthropology* 84 (4): 678–697.

Seehofer, C.-H. 2018. *Migration Policy Masterplan Maßnahmen zur Ordnung, Steuerung und Begrenzung der Zuwanderung*. CSU. Munich: CSU – Horst Seehofer. Retrieved July 28, 2018, from https://www.csu.de/common/download/Masterplan.pdf.

Sen, A. 2001. *Development as Freedom*. Oxford: Oxford Paperbacks.

Solik, M. 2016. Axel Honneth's Notions of Social Recognition and Normative Theory of Recognition. *European Journal of Science and Theology*: 73–84.

Syracuse University. 2019. *Newshouse School of Public Communations and Whitman School of Management*. TRAC Immigration Data Research Center. https://trac.syr.edu/immigration/reports/588/

Taureck, R. 2006. Securitization Theory and Securitization Studies. *Journal of International Relations and Development* 9: 53–61.

Taylor, C. 2004. *Modern Social Imaginaries*. Durham: Duke University Press.

U. S. Department of Homeland Security. 2020, April 1. Joint DHS/EOIR Statement on MPP Rescheduling. Retrieved April 19, 2020, from https://www.dhs.gov/news/2020/04/01/joint-dhseoir-statement-mpp-rescheduling.

U.S. Department of Homeland Security. 2019, January. Migrant Protection Protocols. *Memo to Field Directors*. Washington, DC: U.S. Department of Homeland Security. Retrieved February 1, 2020, from https://www.ice.gov/sites/default/files/documents/Fact%20sheet/2019/ERO-MPP-Implementation-Memo.pdf.

Wendt, A. 2003. Why a World State is Inveitable? *European Journal of International Relations* 9: 491–542.

Williams, R. 1997. *Hegel's Ethics of Recognition*. Berkley: University of California Press.

Chapter 9
Recognition and Civic Selection

Onni Hirvonen

Abstract Large-scale immigration and the refugee crisis have caused many states to adapt ever stricter civic selection processes. This paper discusses the challenges arising from civic selection from the perspective of recognition theories. The argument is that recognition theories provide good conceptual tools with which to critically analyze civic selection and immigration. However, the paper also aims to highlight that many current institutional practices are problematic from the perspective of recognition. In the context of civic selection, it is helpful to understand recognition as something that comes in two analytically distinct modes: horizontal (or interpersonal) and vertical (or institutional). Many rights depend on institutionally given statuses (skilled worker, refugee, permanent resident, etc.). For a person to have a relevant social standing, she needs to be recognized by a relevant governmental institution. However, in vertical relationships, immigrants are faced with a lack of reciprocity. They need to one-sidedly recognize the institutions, which, in turn, have full power to withhold recognition. Migrants also face challenges in the interpersonal horizontal spheres of recognition. Institutional status being granted does not guarantee interpersonal solidarity or care. As recognition is tied to a particular institutional setting and a particular lifeworld, large-scale immigration sets two challenges. The first is the challenge of multiculturalism and recognition of diverging cultural practices of esteem. The second is the challenge of integration and obtaining recognition from the pre-existing cultural context. It is argued here that from the perspective of esteem-recognition, this is very much a question of working rights and providing opportunities for contributing in the new context. From the perspective of care-recognition, in turn, rights to healthcare and family unifications are central. Thus, achieving meaningful personal relationships is not guaranteed by giving rights, but it is nevertheless dependent on institutional recognition.

Keywords Recognition · Immigration · Civic selection · Citizenship · Liberalism

O. Hirvonen (✉)
Department of Social Sciences and Philosophy, University of Jyväskylä, Jyväskylä, Finland
e-mail: onni.hirvonen@jyu.fi

G. Schweiger (ed.), *Migration, Recognition and Critical Theory*, Studies in Global Justice 21, https://doi.org/10.1007/978-3-030-72732-1_9

9.1 Introduction

Large-scale immigration and the so-called refugee crisis have caused many states to adopt ever stricter immigration policies. Migrants face a selection process to determine who can enter a country and with what rights. These issues can be described using the general terms of *civic selection* and *civic stratification*. Civic selection is here taken in the broadest sense: when a person migrates to a new social setting, civic selection is the process through which she either becomes or does not become incorporated/accepted into the new social setting. The paradigmatic and the most important cases are migrant workers and refugees (from climate, conflict, or both) emigrating to a different nation state. Related to civic selection is the concept of civic stratification – the practice of including or excluding certain individuals or groups with respect to legal rights. Although most nations have committed to uphold equal human rights, civic stratification is something that is widely practiced, with a whole range of different statuses and packages of rights available to migrants. In short, civic stratification is an issue that is constricted to state-given rights in the legal sphere, whereas civic selection is a broader umbrella-term which includes both formal (legal) and informal practices of inclusion and exclusion.

This chapter does not aim to show the empirical existence or extent of the problems in civic selection and civic stratification. The breadth of the issues has been well analyzed in migration literature (see, e.g., Morris 2003). Rather, the approach here is mainly theoretical: the aim is to analyze and discuss the challenges arising from civic selection from the perspective of recognition theories. This is done partly in order to determine whether the language of recognition is helpful in construing what exactly is taking place in civic selection and civic stratification, what might be harmful in practices of civic selection and civic stratification, and what kind of challenges different agents are facing. Secondly, the aim is to evaluate whether recognition theories could provide normative grounds for evaluating current practices of civic selection.

Legal recognition has a natural link to civic selection as migration-related rights depend on institutionally given statuses. Skilled worker, refugee, permanent resident, and – ultimately – citizenship are all legal statuses, given through an institutional process that differs from one country to another. For a person to have a relevant legal standing, she needs to be recognized by a relevant governmental institution. However, the issues of immigration are not limited to the legal sphere. There are also challenges that relate to the private sphere as well as cultural recognition and non-legal social standing of migrants.

This contribution starts with a short discussion of what recognition means (Sect. 9.2.) and how it is tied to migration (Sect. 9.3.). This discussion presents problems of multiculturalism and integration, which are then analyzed in greater detail (Sect. 9.4.). The article finishes with a short discussion on the normative grounds of recognition in the context of civic selection (Sect. 9.5.) and a conclusion (Sect. 9.6.).

9.2 Recognition: Interpersonal and Institutional

Recognition is here understood in its Hegelian sense as the reciprocal actions and attitudes that constitute and respond to personhood. Recognition denotes those relationships that support and construct our identities.[1] In the contemporary literature, it is often understood as a human need (Taylor 1994, 26) or a quasi-transcendental human feature (Honneth 2003, 174). It can also be conceptualized as a resource that can be given out equally (as in the case of equal respect) or more distinctively (as with esteem), based on achievements and comparisons (McBride 2013). This is a conception that includes the relationships and attitudes that make up our legal statuses, but it also extends beyond them to private life and wider social relationships and statuses. Recognition theories vary in their details, and here I want to commit only to some of its more generally accepted features.[2]

(a) *Recognition is responsive.* As Heikki Ikäheimo (2007, 227–228; see also Laitinen 2002, 2006) outlines, recognition is a response to central features that make a person. This does not mean that we all have a shared understanding of these features and a clear-cut definition of what they include. However, most recognition theorists seem to accept the general conditions of personhood like reason-responsiveness, intentionality, communication skills, second-order attitudes, and the like. It is also possible to add the embodied aspect of human life and needs into this category.

If recognition is a response, what sort of response is it? On the one hand, we might theorise some quality in a person that *causally* requires another to recognize her. On the other hand, recognition can be understood as a *normative* response to the relevant features of the other. If recognition were a causal response, normative demands for recognition would become mere epistemic issues of noticing if others have the relevant features, which would then engender recognition. Perhaps this is the reason why most contemporary recognition theorists see recognition as a normative rather than causal response. As Arto Laitinen (2002, 468) formulates it, recognition is a response to normatively relevant features of the other.

(b) *Recognition is constitutive.* Another commonly accepted central feature in recognition theories is that recognition constitutes personhood. Following Ikäheimo (2019) again, this claim can be taken in two senses. First, recognition can be *causally* constitutive of persons. This is widely accepted in psychological developmental studies. It is through early relationships to meaningful others that we build up our agency, sense of self, and security in the external world. Many key theorists – Hegel, Taylor, and Honneth for example – take the

[1] For Hegel recognition goes even deeper, as it is constitutive of self-consciousness and freedom (see master-slave dialectics in *Phenomenology of Spirit*, Hegel 1999).

[2] These positions are extensively argued for in the recognition literature, most notably in Honneth's work (1995, 2014, 2017) and in the works that further develop Honneth's contributions (e.g., McBride 2013; Ikäheimo 2014). Here these positions are thus taken mostly as given.

> constitution claim in a stronger sense. According to the *ontological* constitution claim, it is not just an empirical coincidence that human beings become persons through recognition, but that atomistic self-sufficient persons are actually impossibilities. In this line of thought, recognition is a necessary condition of becoming a person whether it be that personhood makes sense only in a recognitive community or – even more strongly – that self-consciousness cannot exist without recognition.

But what is it exactly that recognition constitutes? The usual answers are two-fold. The first is that recognition constitutes those normative and/or psychological features of personhood that it is also responsive to. Second, recognition is a necessary element in the constitution of those positive self-relations that enable us to relate to ourselves and to function in the actual world. Thus, recognition has a double function – it builds objective features of agents, but it also affects their attitudes toward themselves. Once developed, the objective features do not really disappear even if recognition were to cease. As such, recognition is a threshold concept in relation to the constitution of these features: enough recognition in the relevant phases of life helps to build up certain capabilities. This is not equally straightforward in self-relations, as social recognition in itself cannot guarantee a healthy self-relation. Here recognition should be understood as something that "upholds" or actively supports self-relation.

In addition to the *personal* side, recognition is also *political* and normative – human beings have expectations of recognition, and lack of recognition causes real harm (Taylor 1994; Honneth 1995). These harms are often formulated in languages of disrespect, misrecognition, non-recognition, and pathologies of recognition. There can be various structural, ideological, and personal reasons for withholding recognition. The general idea in the various recognition theories is that through the needs of recognition, we can set normative grammar for moral conflicts (Honneth 1995) and determine the conditions for a just society (Thompson 2006, 9).

However, it is not self-evident how recognition claims and justice are linked. Not all recognition claims can be justified, and not every claim requires a positive response from others. Recognition (or interpersonal attitudes that constitute personhood) is ambivalent, and interpersonal relationships are interlaced with power (McQueen 2015). There are intersubjectively constituted social statuses – very much related to immigration as well – which are harmful, denigrating, and exclusionary. Judith Butler is right in stating that sometimes it is better not to be recognized than to be recognized in a harmful way: "There are advantages to remaining less than intelligible, if intelligibility is understood as that which is produced as a consequence of recognition according to prevailing social norms" (Butler 2004, 3). Thus, especially in the case of immigration, the question is not merely about recognizing the migrant as having a status, but also about the quality of statuses and rights.

In the light of what has been stated above, an adequate analysis of recognition in the context of immigration will need at least a provisional account of what kinds of recognition can be justifiably expected. A promising starting point can be gleaned from Honneth's tripartite division of forms of recognition. According to him, love, respect, and esteem are the key types of recognition that attach to modern conceptions of the human lifeform and personhood. Without achieving sufficient recognition in all three spheres, one cannot achieve so-called full personhood or those positive self-relations that would enable fully-fledged agency in the social sphere.[3]

The three forms of recognition differ in their content, institutional realization, and the aspects of personhood that they are directed toward. a) *Love* or care is directed toward the physical and emotional needs of the other. Loving in this context means taking the other as a singular being whose ends and needs matter for me not because they benefit me in some fashion, but because they are her ends and needs. Honneth (1995, 107) states that love is limited to close "primary relationships" and it is usually thought to have its institutional realization in family – although there are clearly other institutions of care as well. Being loved enables one to form basic self-confidence, which is necessary for individual agency and acting in the world. b) *Respect* as a form of recognition is not limited to the closest sphere of meaningful partners and family members. Rather, it is based on the egalitarian ideal that every human being is in some sense similar to each other. Respect is universalizing recognition, which is best reflected in those legal institutions, democratic public spheres, and human rights statements where all individuals have the same status, based on their shared humanity (or personhood). What is relevant here is our shared status as co-authors of the normative realm (Ikäheimo 2007, 235). According to Honneth (1995, 120), respect is necessary for developing an understanding of oneself as a person who has an equal standing with others – self-respect. This is also a key element in the egalitarian theories of justice. The importance of respect is well reflected in the Kantian liberal tradition with the idea of dignity (see, e.g., Pinkard 2002, 53 for an explanation of Kant's idea). Also, republican political theory gives respect equally important standing, as exemplified by Philip Pettit's eyeball test: people should be able to "look others in the eye without reason for fear or deference" (Pettit 2014, xxvi). c) *Esteem*, the third form of recognition, is related especially to merits, achievements, and comparisons. It is institutionalized in the economic sphere (markets), which Honneth (2003, 140) describes as embodying the principle of "individual achievement" – or the "achievement principle."[4] Whereas respect referred to similarities between individuals, esteem is based on individuating features and distinction from others. It is nevertheless granted according to shared value horizons and varying cultures of esteem. The self-relation at stake with esteem is named self-esteem, denoting the capability of seeing oneself as a unique and valuable

[3] What is offered here is merely an overview of the Honnethian forms of recognition. The importance of these forms of recognition for modern humans is taken as granted. I also leave open the exact definition of "full personhood." In my view the concept of personhood is both political and historical in the sense that its exact contents and limits are constantly under debate. For example, it is not clear if there is any strict metaphysical standard for what rights or what opportunities (or freedoms) a person ought to have.

[4] It is of course contestable whether markets really function according to merit and achievements. However, even if the achievement principle is not an accurate empirical description, people still tend to understand the markets as if they should normatively be based on merit (Miller 1992).

member of a society (or group) whose contributions matter. In the modern societies, work is one of the central realms where esteem is distributed (Jütten 2017, 260).

There are two further conceptual clarifications, which are helpful in the context of immigration. First, recognition comes in two analytically distinct modes: *horizontal* and *vertical*. The horizontal recognition denotes intersubjective relations between agents. The vertical mode of recognition, on the other hand, refers to relationships between institutions and individuals where the institutions recognize the individuals and vice versa (Ikäheimo 2013, 17). In vertical recognition institutions – like state offices dealing with immigration – are active recognizers who have a license to bestow statuses like "refugee," "permanent resident," "guest worker," "citizen," and so forth.[5] Second, recognition is *institutionally mediated*. Honneth sees institutions as expressions of recognition relationships, but they also inform how we in practice arrange recognition and what acts count as recognitive acts (Honneth 2011, 403; also Deranty 2009, 232). Here institutions function as normative frameworks (and not agents), which inform and dictate what acts count as recognition and what counts as recognition-worthy. Thus, recognition is bound to normative frames, which are actualized in broader institutions like state, markets, and family. However, it is unclear if there is, to use Ikäheimo's (2013, 17) term, "purely intersubjective" recognition or if all recognition is institutionally mediated. It is plausible that some reference to a normative framework needs to be made for an act to count as recognition. However, it is also true that recognition does not need to always be tied to institutional roles or role-fulfillment.

9.3 Civic Selection, Civic Stratification, and Recognition

The main aim in this section is to see if the multi-faceted concept of recognition, as introduced above, provides a good theoretical apparatus for analyzing civic selection and civic stratification. At the core of civic selection is recognition as they both are fundamentally about giving and receiving social statuses. Civic stratification, in turn, focuses more specifically on recognition as a holder of certain rights. However, to get beyond the trivial acknowledgment of the centrality of recognition, we need to take a closer look at the different elements of civic selection and civic stratification relationships.

[5] It is debatable if this horizontal-vertical distinction should be taken literally or as a metaphor for role-fulfillment. It could be claimed that vertical recognition toward institutions does not make sense as recognition is supposed to refer to inter*personal* relationships – relationships between persons – and institutions are not persons. However, vertical recognition can also be taken to mean strongly role-bound and rule-mediated recognition. In this sense, vertical recognition denotes those cases where someone is filling a role or acting from the perspective of institutional reasoning. Horizontal recognition, in this interpretation, would be more "spontaneous" and not strictly tied to any institutional roles.

Focusing first on the institutionally recognized rights of movement, it is helpful to distinguish between pass-through rights and resettlement rights (see Hosein 2013, 33). Both of these are rights (or sets of rights) that enable and restrict the movement of individuals and as such function as background conditions for migration. The pass-through rights concern freedom of movement, whereas resettlement rights concern (alongside freedom of movement) the rights to choose a place of residence. The first applies most often to tourists and temporary, shorter-term visitors. Usually, pass-through rights do not imply demands for any participatory rights within the local society. However, what is expected is mutual respect: on the one hand, the "when in Rome" principle applies to pass-through rights as visitors are expected to respect laws and customs. On the other hand, visitors can often expect reasonable respect in the sense of hosting countries guaranteeing their security and individual human rights.[6] How the pass-through rights are distributed largely depends on the relationships between the originating and the receiving states. In certain cases, rights need not be specifically applied for (e.g., the Schengen area), but in others the possibility of getting even a temporary visa might be challenging – depending on the geopolitical situation and contracts (or lack of them) between states. There are also good reasons to deny pass-through rights, especially in matters related to security. This could be, for example, national security (war and terror or a pandemic) or environmental security (tourism and fragile ecosystems).

Although pass-through rights are important, the main issues surrounding immigration and civic selection are matters of resettlement rights. As with pass-through rights, security-based arguments apply for denying resettlement. However, in addition, resettlement rights are often highly contested. The positions in the debate range from open border policies and the abolition of borders to closed-border policies. Human rights (especially in the form of the right to movement and the right to choose a place of residence) are often invoked to allow resettlement. Adam Hosein summarizes one example line of thinking in this strand: "according to the democratic argument for open borders the right to participate in political decision making extends to people who live in other countries and according to the egalitarian argument governments ought to show equal concern even for non-residents" (Hosein 2013, 34). Here rights of movement are complemented by the egalitarian ideals of a right to participation and non-domination in the place of residence. Contra the egalitarian sentiment, there are also strong intuitions that groups do have a right to decide on or restrict their membership and that they should do this in order to protect their own socio-cultural legacy and the society's capacity to reproduce and uphold order.[7] Östen Wahlbeck (2016, 583) provides a case in point, stating that right-wing

[6] These normative expectations are not something that states necessarily commit to, but they seem to be the cornerstones of Western tourism and non-settling movement between different nations.

[7] This intuition comes up easily in the case of, for example, sports teams. It seems reasonable that my local ice hockey team does not have to accept me as a member – especially if the purpose of the team is to play at a competitive level. However, it is less clear how this applies to more encompassing groups and institutions like a nation-state. One line of thinking is that limiting membership is acceptable if it does not restrict opportunities too strongly and if there are alternative options. I

populism often takes the form of focusing on the unity of the people, a people that is somehow under a threat in the context of immigration. It does not require much imagination to see that seeing the other as an existential threat forecloses the possibility of giving rights to them. Of course, in the case of populist simplifications, it is quite unclear what exactly is threatened and even whether that threat is real.

Beyond the movement rights in the background of immigration, it is also helpful to distinguish the four relevant parties (or agent positions) in recognition relationships that are connected directly to immigration: namely, the emigrating individual, the receiving state, the residents of the receiving state, and the country of origin. Starting with the *emigrating individuals*, these are the moving persons who have varying reasons for their emigration. Obviously, in some cases – refugees and asylum seekers – physical and psychological harms or the threat of these, which can also be cast as violations of recognition expectations, are the reason for emigration. However, there are many other reasons for changing one's place of residence. When the emigrant enters or tries to enter the area of a sovereign nation state, there is an assumption of an unconditional recognition toward the institutions of the receiving state. Or, in other words, the emigrating individual does not really have a choice but to vertically recognize the receiving state and its institutions, rules, and so forth. Furthermore, the burden of proof for being recognition-worthy is on the emigrant. It is common that one needs to fulfill pre-set objective criteria to be recognized. This may take the form of having the right qualifications, right age, right occupation, or being threatened in the right way in the original context. These criteria are set by the receiving state, and while there are international conventions, the application of the criteria is in no way universal. The flip side of this coin is that the immigrant herself has been socialized in a different context, and thus while her needs for recognition might be universal, the actual expectations of their fulfillment might differ from the ones at place in the new context.

The second relevant party is the *receiving state*. In the current context, most nation states have their own categorizations for immigrants. These are mainly based on either meritocratic ideals (of needed workers or other suitable merits) or humanitarian commitments (like ratified human rights treaties for treating refugees). Again, it is impossible to give a general answer to how a state will or should respond to an immigrant. Their relationship is one of vertical recognition where an individual is related to an institution. The state sets a framework for the recognition relationships. It is notable that even if there are international contracts in place – like the commitments made by most liberal democracies – these contracts are not laws as such and the interpretation of their realization is always done in the receiving country, and the interpretations also differ.[8] Further, as the actual decisions are made by individuals and teams, the institutional recognition given by a state institution is

---

can play ice hockey in a different team (on a lower level of competition) or I can play football with my friends instead. With a state it is not as evident if there are as clear second options.

[8] For example, in Finland it was possible to tighten the screening of refugees from certain areas like Iraq and Syria through an administrative decision of interpretation of local rules, although the broader international commitments remained the same.

partly dependent on the personal interpretations (of the rules, regulations, and normative ideals) made by the officials and immigration workers.

Third, there is the horizontal level of recognition with the existing *residents of the receiving state*. These residents can work in the institutional roles and execute in practice the vertical recognition by the state, but they are also in horizontal interpersonal relations with the immigrants as individual persons (and not only through filling an institutional role). The existing residents are involved in their own recognition institutions, in their own lifeworld, or *Sittlichkeit* to use the Hegelian term. They have their own recognition expectations, own culture, and own habits, which, especially if left unreflected, can lead to challenges of multiculturalism.

Finally, the fourth relevant element is the *country of origin*, which has a role in relationships between nation states, which in turn have a large impact on the possibilities of migration. If, for example, an individual is migrating from a Schengen country to another, the challenges are lesser than migrating from a non-Schengen country to a Schengen country. Similarly, the internationally recognized status of the country of origin affects the possibilities of an individual obtaining refugee status. In short, the larger geopolitical challenges affect the individual-institution relations (challenges of getting a visa, working rights, residence rights, and so forth). However, they can also affect interpersonal relations as the reputation of the country of origin might impart negative or positive stereotypes to migrants. If recognition is a term that can be applied to relationships between states, then it is clear that recognition between states affects recognition and civic selection. And even if we would not conceptualize international relations using the language of recognition, their relevance for civic selection is clear.

With the multi-faceted view of recognition in mind, we can now look at how different forms of recognition manifest in the relationships between the different agents. In other words, if we formulate the resettlement rights from the perspective of Honnethian tripartite recognition, we can see that there are rights at stake which respond to all three forms of recognition. First, although states cannot guarantee love, recognition as care attaches to rights of security and the right to family that many refugees are after. The state as a guarantor of security is an idea that has been part of the liberal tradition at least since Hobbes's contract theory of the state, and it is still very much thought that the state should provide the basic security that its residents (or at least citizens) need in order to lead their lives. As with the other recognition-related rights, these do get different interpretations (compare, e.g., Nordic welfare states' conception of care and the libertarian night-watch states' conception of care), and the states might also very much fail to uphold these rights.

The second set of rights that concern resettlement are participatory rights and representation rights. In the language of recognition, these rights are related to respect, and their realization occurs – in the context of Western states – in the democratic public sphere. Voting rights are central to resettlement in the sense that they allow the individual to take part in the decision-making of the society she has become part of. However, this is often combined with an unconditional demand that she should respect the authority of local institutions.

The third patch of relevant rights (from the recognition theoretical perspective) concerns the rights to make a living and to have opportunities to make a contribution. In the modern context esteem-recognition is closely tied to institutions of work, and thus in practice this form of recognition requires the right to take part in the labor market and in the reproduction of social life.

Citizenship is the only category that comes with full institutional recognition in the sense of having the rights described above. Citizenship does not guarantee love nor esteem, but it guarantees the rights to pursue these forms of recognition. Regarding respect-recognition, citizenship is thus the "gold standard" of legal civic selection. Selective right-giving – civic stratification – on the other hand, limits the rights and thus also the opportunities of acquiring recognition. Having rights is a matter of legal recognition (respect), but these rights are related to other forms of recognition in the sense that they license the individuals to strive for love and esteem, as well as to express them, on equal ground with other citizens, without fear of persecution that others would not face. Whereas state institutions – especially in the legalized Western context – work much in terms of rights (freedoms, protections, and limits), this is not the whole picture of recognition nor civic selection. Inclusion in culture and markets is discussed more in the following section, but it is worth repeating here that while they are highly important, mere abstract rights cannot ensure that kind of recognition.

9.4 Desires, Needs, and Challenges of Recognition: Integration and Multiculturalism

The previous sections focused on the general features of recognition and recognizable rights in the context of migration. This section aims to get closer to the practical challenges that migrants face regarding their recognition needs and desires. Unsurprisingly, the position of a migrant has its own vulnerabilities, which manifest more or less strongly depending on the status and claims of the migrant. (For example, highly-skilled workers are more likely to get their esteem-related recognition needs met than refugees who might have to struggle even for institutional recognition of their basic human rights.) It is not claimed here that all migrants' claims are justified, but for the moment that is set aside until the justification of recognition claims is considered (in Sect. 9.5.). The focus is decidedly on the perspective of the migrants although the other previously introduced perspectives are still relevant. Similarly, the differentiation between forms of recognition is helpful in distinguishing the various kinds of claims at stake. From immigrants' perspective, recognition-related challenges come in two familiar forms which are expressed as the challenge of *integration* and the challenge of *multiculturalism.*

(A) The challenge of integration is, at its core, the challenge of obtaining recognition in a new cultural context, feeling at home, and being esteemed, respected, and loved in a new society.

Here it is helpful to look at the forms of recognition one by one. From the perspective of esteem-recognition, if esteem is tied to merit and achievement, no society can guarantee esteem for everyone. Unreasonable demands for esteem may well be disregarded, but the interesting question here is what counts as an achievement. Honneth (1995, 121) frames achievements in relation to contributions to the common good, in the light of a shared value horizon. When one moves from one society to another, one's reference point for the "common good" changes as does the evaluative framework. This shift can be challenging if there are drastic cultural differences between the country of origin and the receiving country. As Thompson (2013, 103) summarizes: "If the principle of esteem were to operate at the global level, it would have to be shown that institutionalized patterns of cultural value enable all individuals to make valued contributions to global goals." However, this seems unlikely. Lacking a universal standard for esteem, immigrants can easily fall outside of the relevant forms of social statuses and outside the potential to achieve them. Often this comes in the institutional form of not having one's capabilities (e.g., professional skills and degrees) acknowledged in the new context – especially not in the manner they were acknowledged in the original context.

However, we need to make at least one caveat with regard to the Honnethian picture. That is, we should not assume that any given state has a unified singular value horizon. Although often one can find a hegemonic culture, social settings are full of competing underground cultures where different kinds of contributions and claims are brought forth to compete. The hegemonic value horizon is also subject to change. Thus, the picture is not only one of a straightforward assimilation (as will be discussed more with multiculturalism).

The justice claims in the context of esteem take the form of counting something as a contribution or having an opportunity to contribute. The first is the sphere of cultural struggle, while the latter is in modern Western states closely tied to work rights and opportunities for work. In the recent literature on critical theory and work (see, e.g., Dejours et al. 2018; Jütten 2017), it has become clear that work is one of the main spheres of social esteem. This becomes evident when considering how a person's social worth can be closely tied to her work and, negatively, how unemployment can stigmatize and cause low self-esteem. In this light, the central claim that an immigrant can make in her new context is to have an equal access to work-life, with the same standards for merit and achievement as locals. Work-rights and providing opportunities for making a contribution are essential from the perspective of esteem-recognition. Forms of protectionism and double standards (closing job-markets from immigrants or making it difficult for them to prove their skills, having only a market of low-esteem jobs available for migrants) are straightforwardly in opposition with the immigrants' recognition claims and their possibilities of achieving recognition.[9]

[9]Although protectionism might be seen as recognition for the current citizens, here esteem has already shifted from achievement to belonging. As far as we think that the principle of desert or the principle of achievement is a good principle – or at least better than "inheritance" or "national-

As with esteem, it is clear that love is something that a society cannot guarantee.[10] However, if we take this form of recognition in its broader sense of care-recognition, in the context of immigration two rights rise up as central: rights to healthcare and rights to family unifications. Achieving meaningful personal relationships is not guaranteed by giving these (or any) rights, but is clearly dependent on institutional recognition and having the opportunities that come with these rights.

Integration in the respect-recognition sense consists of being taken seriously as a participant in the new society and seeing oneself as having that standing. The central point of respect is that one is a relevant co-constitutor of norms, and this ideal is based on an egalitarian ideal of universal humanity. Or, as Pettit (2014, 61, 99–100) formulated it, this kind of respect (and self-respect) means that one is on an equal footing with others, being able to look them in the eye. Here the core meaning of integration is to be taken as a person who can contribute to the norms and direction of the society they live in – namely, to be respected as part of the "us" of the society.

(B) The challenge of multiculturalism concerns fitting together the various "life-worlds" and recognizing diverging cultural practices and contributions as valuable. This can be formulated in the Taylorian sense of politics of difference, which is based on esteem and on different achievements. In this sense the struggle of multiculturalism shows itself as a struggle between different value horizons. I take it here that Taylor (1994) is quite right in stating that we have no clear criteria of judgement between the pluralities of values, although it would be a mistake to close out certain value horizons or cultural horizons *a priori*, especially if they have managed to provide meaning for lives for centuries.[11]

In the context of multiculturalism, struggles take a form that separates them from the struggles of integration. The struggle for esteem can also be seen as a struggle for a particular framework of esteem to be accepted or as a struggle of shifting the evaluative framework, whereas in the case of integration the esteem is striven for within a ready-set framework. In other words, immigration can cause two kinds of struggles for esteem: a struggle to be recognized in the new context and a struggle to make one's original achievements and practices count as recognition-worthy by shifting the value horizon of the new context.[12]

ism" – to distribute esteem, then we should be wary of protectionist lines of thinking and try to find solutions to the obviously harmful race to the bottom from other directions.

[10] In Honneth's model love as a recognitive attitude is limited to close interpersonal relationships, and perhaps it is indeed the case that we cannot be expected to feel unconditional sympathy for everyone. However, it is also clear that there are institutional solutions for providing fundamental care.

[11] Although here cultures are discussed as if they were unified entities, they should not be understood as too rigid or stationary. As Tariq Modood (2013, 90) points out, cultures are neither fictions nor essences but more akin to family-resembling collections. They consist of changing norms, practices, and recognition claims that require interpretation, affirmation, and acting-out on the part of their individual carriers.

[12] Shifting value horizons raises the question: what if the new value horizon is worse? Also, what normative benchmark should we use to judge value horizons? The fact that immigrants might want

The challenges of multiculturalism are not merely limited to the sphere of esteem. Different forms of expressing love and care and different understandings of political participation are also relevant for multicultural struggles. Whereas civic selection might assume that a new citizen integrates into a new society, in the context of a liberal society we should ask what this integration actually involves. Practical civic selection criteria often seem to assume assimilation – to become a citizen one needs to internalize and accept the local values – to the new context, but does this have to be so? This is a too broad question to properly analyze here, but it is worth noting that liberal societies are, in principle, supposed to be open for pluralities of conceptions of good, while still maintaining certain basic rights that foreclose those conceptions of a good life that would hinder others' possibilities for a good life. In short, universal respect and particularized esteem might be at odds, and in this case, as McBride (2013) argues, it is probably more just to emphasize the universalistic side of respect.

9.5 Grounding Demands for Recognition in the Context of Immigration

It has not been explicitly claimed that the receiving society has a duty to recognize immigrants or a duty to respond positively to their recognition claims. This naturally leads to questions about which recognition claims are justifiable and what, if any, normative import recognition theories offer in the context of immigration. To be a helpful theoretical approach, recognition-perspective should provide conceptual tools to demarcate justifiable stratification from unjustifiable stratification, as well as pinpoint recognition-related injustices more generally in civic selection processes.

We can start closing in on the issues by acknowledging that not all recognition claims are justified. This is clear with *strongly* unreasonable demands of recognition. These are claims for those kinds of views and practices which would undermine the person-constituting practices of recognition or even threaten the life of others. A classic example is the paradox of tolerance (see Popper 2013, 581) brought into practice: an intolerant person cannot justifiably expect others to tolerate his intolerance if that would undermine the whole way of life that the others have chosen.

More interesting in the context of immigration are the *weakly* "unreasonable" demands. That is, demands that are not fundamentally destructive to forms of life

recognition for their own cultural practices does not in itself guarantee that all these cultural practices would be morally acceptable. Following Honneth's (1995) ideas, we can state that recognition theories should be open to various ideals of a good life (and thus open to various value horizons). However, recognition does set a normative framework in the sense that moral progress can be identified with expansions of spheres of recognition as well as through eradication of non-recognition and misrecognition. In short, if any cultural practices lead to increased personal and social suffering, there seems to be good reasons to not accept them outright.

(or cultures), but which present positions that are rejected or frowned upon in the hegemonic culture.[13] How could one ground a claim for recognition – other than referring to experiences of lack of recognition – that challenges the norms of the existing recognition order?

From the perspective of esteem-recognition, not all claims for esteem are equally justified. Insofar as esteem is based on distinction and achievement, not all achievements should (or could) receive the same amount of esteem. If the standards of esteem (the value horizon) are acceptable to all affected parties, there seems to be no problem in denying claims for esteem that are not worthy of that esteem. For example, in a society that values hard work and honesty, it would be odd to demand esteem for shifty laziness. But how to ensure that the standards of esteem are just and not merely contingent cultural accidents? One way of doing this would be to ensure that those who are affected by the standards of esteem are also part of deciding those standards. Further, as Honneth (1995) argues, from the recognition theoretical perspective, it is important to expand the sphere of esteem to various kinds of contributions so that it enables and supports healthy socially achieved self-esteem for as many as possible, without fear of collective denigration. In this sense, justifiable exclusion from esteem would rely on shared acceptance (of instituted norms), ideally including discussion on the content and application of the norms.

Similarly, the perspective of respect-recognition allows that some claims for rights can be denied. In defense of stratification and partial rights, it can be acknowledged that some rights are positional and situational and thus not available for everyone. However, while rights are a central part of respect-recognition, the core of respect consists of, as stated by Ikäheimo (2007, 234–236), reciprocally admitting a status as a co-author of norms. Denying this status would go radically against the whole idea of respecting the other as a member of the abstract universal personhood – as an end in itself. This kind of respect is also tied to the psychological side of seeing oneself as a capable member of the civil society. Again, the Honnethian approach is to consider the expansion of respect as moral progress.

In short, the recognition-theoretical approach allows justifiable denial of recognition. However, there is a moral backstop that is grounded in the philosophical-anthropological roots of the need for recognition. The philosophical anthropology instructs the norms of interaction on the grounds of what is good for humans in general or good for the "lifeform of persons."

Although this moral ground in shared humanity is universalist in nature, recognition theories also include a strand of moral particularism. As Burns and Thompson (2013, 14) note, recognition is often understood as tied to an institutional setting and a particular lifeworld. Institutions are practical expressions of recognition relationships, and they are products of collective (often tacit) acceptance.[14] In other words,

[13] Here I focus only on esteem and respect as the claims for both of them are explicitly public claims.

[14] The collective acceptance model of institutions (as social facts) is part and parcel of contemporary social ontology (see, e.g., Searle 1995 for an early account or Epstein 2015 for an updated version). However, the details of the theories vary greatly.

institutions are construed around a normative core, and these norms are invoked (implicitly) in our demands for recognition as well as in our criticism of the institutional setting. Thus, as far as recognition is tied to actual institutions, recognition theorists must agree that standards and justifications of recognition are "up to us" and localized.

The philosophical-anthropological grounds are perhaps the same in every context, but the institutional setting is a historical achievement. Thus, in analyzing immigration from a recognition theoretical perspective, one main task would be to map out the actual historical normative commitments in immigration-related institutions. These commitments can, in turn, be used to determine what claims would be justified within those particular institutions and, more critically, if the institutions themselves live up to their normative expectations.

The first set of relevant institutions consists of the (state) institutions of the receiving country. These include official institutions like border control and immigration offices, and their functioning is tied to local immigration laws. These are often the first institutional actors that immigrants encounter, and they represent the receiving state in the relationship between the migrant and the new society. What are the normative promises and expectations in these institutions?

Though the guiding regulations differ, in the Hegelian story, the state institutions assume abstract personhood and equal respect. Everyone ought to get treatment as an equal legal person. However, this can be taken to apply only to citizens (or those who are part of the in-group of that particular institution), and the same treatment is not required with respect to external persons, or it might be conditional toward them. Thus, in the strictest interpretation, it seems that an immigrant might not be able to invoke the expectation of respect unless she is already a member of the society. However, if we follow Honneth (1995, 111–112), respect-recognition has a universalizing tendency that makes it apply to all persons in their abstract and universal personhood. Although the institutions are concrete manifestations of this abstract principle of right, the abstract principle could be used as a justification for equal treatment if the particular institution fails to realize it.

A more mundane argument is that the normative principles of an institution apply to all who come under its jurisdiction – all affected – and not only to those persons who are part of the "in-group" that created the institution. Thus, an immigrant has the equal right to demand respect within an institution that has the ideal of respect as its core. This is part and parcel of contemporary rule of law.

Moreover, most states and their immigration institutions make an explicit commitment to external normative sources, namely, human rights. These provide a ground for equal treatment and respect and elucidate what constitutes equal treatment.[15] However, as discussed below, these abstract rights are realized within the institutions in very different ways, and there are justifiable ways to limit rights.

---

[15] Being treated equally in the civic selection process is not the same as achieving full vertical state recognition. It is equal treatment in the ready-set normative framework and not respect in the full sense of all the affected parties being co-authors of the norms of the institution itself.

In addition to the commitments to broader normative frameworks like human rights, there are also international organizations and treaties in place that restrict how the local immigration institutions operate. Thus, alongside the local immigration institutions, the second relevant institutional sphere is the sphere of international institutions. It includes international organizations, contracts, and commitments, usually including membership in the United Nations (or the European Union in the context of Western liberal states), commitment to human rights, and commitment to Geneva convention on the treatment of refugees. These institutions represent explicit egalitarianism and include commitments to respect and individual rights to life as well as freedom of movement.

Immigrants can and often do appeal to explicit international normative commitments, such as the Geneva Convention Relating to the Status of Refugees. However, the challenge here is that states voluntarily commit to these principles (and thus can also – with some repercussions – choose to ignore them) and that the local application of abstract principles is not uniform or self-evident. The Australian interpretation of just implementation of the treatment of refugees might greatly differ from the German interpretation – even though both countries are part of the same international value community. Another example is the shift to stricter immigration rules in Denmark and Finland, although the background moral commitments have arguably remained the same.

Despite the challenges, the international community does have an important guiding role for the local institutions. Because the international community is first and foremost a value community which cannot directly enforce its values, its effects can be limited; but this does not mean that its effects are negligible. International treaties and institutions like the UN still function as guidelines and discussion forums on how more localized national systems should conduct themselves. International peer pressure in the globalized economy is not something that states can ignore. Becoming a pariah state is not a viable option for a liberal democracy in international politics.

However, whereas international institutions might have a guiding role in spelling out how universal respect ought to be realized, the local institutions can be based on similar normative commitments to respect. Therefore, the international institutions – while providing a normative guideline – are not strictly necessary for a good recognitive conduct at the local level.

In fact, local communities with local institutions (with universalizable principles as their normative core) seem to be central to immigration and civic selection issues. The global community is not the relevant forum for immigration issues if an institution is to be a practical solution for practical issues, which are perhaps globally shared but still necessarily locally instantiated.

This gets us to the third relevant social sphere: the sphere of everyday lifeworld that comes in the forms of the "original *Sittlichkeit*" and the "new *Sittlichkeit*." If we step beyond the spheres of state institutions and international commitments, there exists a varied range of different cultural expressions of the three forms of recognition. Given the variations of local cultures, practices, and habits, it is no wonder that the explicit expectations of recognition differ. Even if modern (more or less liberal) institutions would be based on the same expectations of recognition, their way of materializing these relationships is highly localized. This variation partly causes the everyday difficulties that an immigrant faces in her new everyday surroundings. There is a range of informal expectations that might be hard to fulfill even if formal institutional recognition has been achieved. The playing field might get more even, but the struggle for recognition does not end with residency or citizenship. Irene Bloemraad (2018, 20) notes that abstract bureaucratic citizenship is not as meaningful in everyday interpersonal interaction as, for example, stereotypes are. The fields of culture and markets might still harbor suspicion, xenophobia, and racism even if full legal rights were given to immigrants.

9.6 Conclusion

Talk of civic selection gives a partly misshapen picture of recognition. It gives an idea of one-sided rewards, recognition from above, and conditional status giving. It is also arguably too focused on nation states. One could say that the whole practice of civic selection only becomes an issue at the age of nation states and with the assumption of there being some people who are integral to a society and some who are foreign (Light 2013, 345; also Benhabib 2004).

A fuller view of recognition must instead focus on mutual reciprocal relationships (and their problems), and not only on granting rights. In this sense, combining the multi-faceted idea of recognition with civic selection gives a broader picture than focusing merely on granting rights or citizenship. To recognize someone is to invite them to be a part of a broader community, not merely about granting them rights that are limited to certain states. This perspective of recognition also highlights that although civic selection is a major issue in terms of restricted movement across nation states' borders, the same issues also apply at the scale of smaller communities. There is no reason to doubt that humans have lived in "we" and "them" groupings for a large part of our history. The recognitive issues remain even if states were to disappear.

As noted, in practice, the state recognition of migrants is problematic on many fronts. Documented cases of institutional mistreatment include unjust profiling and grouping (with country of origin as a defining feature), lack of case sensitivity, and cases of not relating to immigrants as individuals. Especially with potential refugees, individual self-assessment of the situation is often overlooked in favor of "more objective" selection criteria, which undermines the self-respect and agency of the migrant. On a more abstract level, immigrants are faced with a lack of reciprocity. They need to one-sidedly recognize the institutions, which, in turn, have full power to withhold recognition. Even as holders of global human rights, they are at the mercy of the local application of civic selection policies. These institutional challenges appear already before the civic selection processes, continue during them, and do not necessarily end after the process is completed. Even if the institutional status is granted, it does not guarantee interpersonal solidarity, esteem, or care.

It has been argued above that immigrants can provide normative grounds for their recognition claims from the normative commitments of liberal institutions and from the global moral community. The normative promises of liberal democracy seem clear in the case of respect-recognition. With that in mind, it could be said about many current immigration practices that if they were intended to be liberal, they are going about it the wrong way. Making physical borders (locking refugees in camps and centers) in order to limit the range of application of already accepted normative principles sounds odd at best. Although recognition is tied to particular institutions, it is unclear how one could defend a view that "our" freedom and equality (and the duties that they bring) are in principle – as normative commitments – different from freedom and equality in another geographical context.

Communities certainly can be expected to have some rights. It seems reasonable to expect that in most cases a group can decide its members and membership conditions. However, rather than being voluntary small-scale groups, states are over-encompassing institutional settings that one currently has to belong to. In many cases one cannot choose not to be a part of the state, and the exit options through moving are limited as well. When evaluating the reasonableness of recognition demands in immigration, asking to be recognized in general is certainly different from asking to be a member of a small community. Demanding that you let me play in your professional football team is different from asking you to let me escape danger and oppression and grant me the possibility to contribute to your broader society.

The unconditional one-sided expectation of vertical recognition goes against the basic principles of mutuality and reciprocity that are central for the Hegelian story. The apparent disparity in power creates the master-slave situation anew – one is forced to recognize while the other sets the terms – and, if Hegel is right, this in fact harms all the parties to the relationship. Recognition, to really count as recognition, should be freely given.

If membership of a state and citizenship are taken as natural properties that are automatically given at birth (and not at will), this obscures the crucial elements that are part of the modern egalitarian spirit of universal personhood. That is, if one is capable of giving and asking for reasons, one should be treated so and taken seriously as a co-author of norms – as a member of democratic will-formation. These egalitarian principles are part and parcel of the basic human rights that, in turn, are part of normative expectations that are built into our understanding of modern liberal-egalitarian states. These rights or normative expectations are also embedded in international institutions, and one could argue that they were set up to uphold those rights. Thus, with attention to misrecognition of migrants, one can conclude that liberal states are currently failing to keep true to their normative core.

References

Benhabib, Seyla. 2004. *The Rights of Others. Aliens, Residents, and Citizens*. Cambridge: Cambridge University Press.
Bloemraad, Irene. 2018. Theorising the Power of Citizenship as Claims-Making. *Journal of Ethnic and Migration Studies* 44 (1): 4–26.
Burns, Tony, and Simon Thompson. 2013. Introduction. In *Global Justice and the Politics of Recognition*, ed. Tony Burns and Simon Thompson, 1–22. Basingstoke: Palgrave Macmillan.
Butler, Judith. 2004. *Undoing Gender*. New York: Routledge.
Dejours, Christoph, Jean-Philippe Deranty, Emmanuelle Renault, and Nicholas H. Smith. 2018. *The Return of Work in Critical Theory*. New York: Columbia University Press.
Deranty, Jean-Philippe. 2009. *Beyond Communication. A Critical Study of Axel Honneth's Social Philosophy*. Boston: Brill.
Epstein, Brian. 2015. *The Ant Trap: Rebuilding the Foundations of the Social Sciences*. Oxford: Oxford University Press.
Hegel, Georg Wilhelm Friedrich. 1999 [1807]. Chapter IV: The Truth of Self-certainty. In *Hegel's Phenomenology of Self-consciousness. Text and Commentary*, eds. Leo Rauch and David Sherman (translated by Leo Rauch), 13–46. Albany: State University of New York Press.
Honneth, Axel. 1995. *The Struggle for Recognition*. Cambridge: The MIT Press.
———. 2003. Redistribution as Recognition: A Response to Nancy Fraser. In *Redistribution or Recognition? A Political-Philosophical Exchange*, ed. Nancy Fraser and Axel Honneth, 110–197. London: Verso.
———. 2011. Rejoinder. In *Social and Critical Theory. Vol. 12. Axel Honneth: Critical Essays. With a Reply by Axel Honneth*, ed. Danielle Petherbridge, 391–421. Leiden/Boston: Brill.
———. 2014. *Freedom's Right*. Cambridge: Polity Press.
———. 2017. *The Idea of Socialism*. Cambridge: Polity Press.
Hosein, Adam. 2013. Immigration and Freedom of Movement. *Ethics & Global Politics* 6 (1): 25–37.
Ikäheimo, Heikki. 2007. Recognizing Persons. *Journal of Consciousness Studies* 14 (5–6): 224–247.
———. 2013. Hegel's Concept of Recognition – What is it? In *Recognition – German Idealism as an Ongoing Challenge*, ed. Christian Krijnen, 11–38. Leiden: Brill.
———. 2014. *Anerkennung*. Berlin: DeGruyter.
———. 2019. Personhood and Recognition. In *Handbuch Anerkennung. Springer Reference Geisteswissenschaften*, ed. Ludwig Siep, Hekki Ikäheimo, and Michel Quante. Wiesbaden: Springer VS.
Jütten, Timo. 2017. Dignity, Esteem, and Social Contribution: A Recognition-Theoretical View. *The Journal of Political Philosophy* 25 (3): 259–280.
Laitinen, Arto. 2002. Interpersonal Recognition: A Response to Value or a Precondition of Personhood? *Inquiry: An Interdisciplinary Journal of Philosophy* 45 (4): 463–478.
———. 2006. Interpersonal Recognition and Responsiveness to Relevant Differences. *Critical Review of International Social and Political Philosophy* 9 (1): 47–70.
Light, Matthew. 2013. Regulation, Recruitment, and Control of Immigration. In *Routledge International Handbook of Migration Studies*, ed. Steven J. Gold and Stephanie J. Nawyn, 345–354. Oxon: Routledge.
McBride, Cillian. 2013. *Recognition*. Cambridge: Polity Press.
McQueen, Paddy. 2015. Honneth, Butler and the Ambivalent Effects of Recognition. *Res Publica* 21 (1): 43–60.
Miller, David. 1992. Distributive Justice: What the People Think. *Ethics* 102 (3): 555–593.
Modood, Tariq. 2013. *Multiculturalism. A Civic Idea*. 2nd ed. Cambridge: Polity Press.
Morris, Lydia. 2003. Managing Contradiction: Civic Stratification and Migrants' Rights. *The International Migration Review* 37 (1): 74–100.
Pettit, Philip. 2014. *Just Freedom*. New York: W.W. Norton & Company.

Pinkard, Terry. 2002. *German Philosophy 1760–1860. The Legacy of Idealism*. Cambridge: Cambridge University Press.
Popper, Karl. 2013. *The Open Society and Its Enemies*, New one-volume edition. Princeton: Princeton University Press.
Searle, John. 1995. *The Construction of Social Reality*. New York: The Free Press.
Taylor, Charles. 1994. The Politics of Recognition. In *Multiculturalism. Examining the Politics of Recognition*, ed. Amy Gutmann, 25–74. Princeton: Princeton University Press.
Thompson, Simon. 2006. *The Political Theory of Recognition*. Cambridge: Polity Press.
———. 2013. Recognition Beyond State. In *Global Justice and the Politics of Recognition*, ed. Tony Burns and Simon Thompson, 88–107. Basingstoke: Palgrave Macmillan.
Wahlbeck, Östen. 2016. True Finns and Non-True Finns: The Minority Rights Discourse of Populist Politics in Finland. *Journal of Intercultural Studies* 37 (6): 574–588.

Chapter 10
Managing Invisibility: Theoretical and Practical Contestations to Disrespect

Benno Herzog

Abstract By aiming at the recognition of normative claims contained in affective reactions to disrespect or social suffering, Recognition Theory points—at least implicitly—also toward the visibilization of groups and individuals who suffer from disrespect. Social invisibilization is therefore understood as hindering recognition and impeding even the perception of legitimate normative claims. However, since Foucault we also know that visibilization could be a mechanism of control and domination, thus pointing to a form of disrespect as the opposite of recognition. In this chapter the diverse forms of migrants struggle for recognition will be analyzed with regard to the question how these struggles negotiate processes of public visibility. If visibilization can be a mechanism of control that in some cases can even lead to detention and deportation, then self-invisibilization, e.g., of undocumented migrants, could help to escape this kind of disrespect. In the first part, and with the help of Axel Honneth's Theory of Recognition, I will draw the significant key elements of a theory of invisibilization (I). I will then describe direct struggles for the recognition of migrant's normative claims (II). Using the approach by Dimitris Papadopoulos et al. in a second part, subversive struggles of escaping mechanisms of visibilization and control will be shown, and proposals of derecognition, as an emancipatory strategy, will be discussed (III). Not granting certain social institution the right to recognize also means not granting them the power to promote disrespect.

Keywords Recognition · Disrespect · Migration · Racism · Invisibilization

---

B. Herzog (✉)
Department of Sociology and Social Anthropology, University of Valencia, Valencia, Spain
e-mail: benno.herzog@uv.es

G. Schweiger (ed.), *Migration, Recognition and Critical Theory*, Studies in Global Justice 21, https://doi.org/10.1007/978-3-030-72732-1_10

10.1 Introduction: Managing Invisibility—Theoretical and Practical Contestations to Disrespect

In his seminal work "Invisibility: On the Epistemology of Recognition," Axel Honneth (2001) sketches out a theory of social invisibilization. Summing up his argument, social invisibility is a kind of disrespect in which one does not view another as relevant partner of interaction. Honneth develops some elements of his theory of invisibilization with the help of Ralph Ellison's novel *The Invisible Man* where a Black protagonist struggles against being ignored. This "looking through" has been exemplarily shown by Rebeca Solnit (2014) in the case of mansplaining, i.e., the case where a communicative offer of a woman is ignored or not given the same importance as that of a man. In migration research, media pieces, and even art works on migration, "invisibility" is an often-heard complaint about the situation of groups of migrants.

Honneth then goes on to frame his thoughts on the epistemology of social invisibility with the help of his well-known Theory of Recognition (Honneth 1995). Social invisibilization, similar to disrespect, seems to be the opposite of recognition, i.e., the struggle for recognition is also a struggle against social invisibility. Honneth insists that this kind of invisibility has a social and not a physical character. Furthermore, in order to be socially invisible, one has to clearly be physically perceptible. Only when it is clearly identified can the ignoring be perceived as a form of disrespect by the invisibilized people and bystanders.

However, in a footnote of the slightly modified German version of the article on invisibilization, Honneth addresses the possibility of a far more dialectical approach on invisibilization by stating that "[o]f course, there is also another form of power exercise, the strategy of visibilization that could range from the communicative revelation to the visual control in the panoptic system researched by Foucault" (Honneth 2015: 11). In other words, both invisibilization and visibilization can be or can lead to disrespect. Since Foucault (e.g., 1975, 1990), we know that the eye of a prison guard, a priest, a teacher, or a medical practitioner is a powerful mechanism turning the other into an object of unequal power relations. These unequal power relations and regimes of visibilization and control (see also Hempel et al. 2010) could be understood as subordination and often also as forms of disrespect.

The aim of this chapter is to explore the theoretical potential of the concept of invisibilization for the analysis of migration and racism and to analyze the struggles of negotiating (in)visibility as struggles for recognition.

Therefore, I will first mark out the key elements of a theory of invisibilization, drawing mainly on the logic underlying Axel Honneth's Theory of Recognition. On an empirical level, this leads, in a second step, to the analysis of migrants' struggle for the direct recognition of their normative claims. In the third part of this chapter, I will use the approach of Dimitris Papadopoulos et al. (2008) in order to understand the logic of escape and self-invisibilization. Here, I will present the subversive struggles of migrants against the control regimes and interpret them as indirect struggles for a new recognition order. In this section, we will also address the proposals of withdrawal and derecognition as radical emancipatory strategies. These strategies will be discussed with regard to the overall framework of the struggle for recognition and against social invisibility. If visibilization can be a mechanism of control that in some cases can even lead to detention and deportation, then self-invisibilization, e.g., that of undocumented migrants, could help to escape this kind of disrespect. Not granting certain social institutions the right to be recognized also means not granting them the power to promote disrespect.[1]

10.2 Elements of a Theory of Invisibility

10.2.1 Different Spheres and Modes of Invisibilization

When we understand invisibilization as the opposite of recognition, we can differentiate invisibilization by the elements of one's personhood that are denied or undervalued (see Herzog 2020). Following Axel Honneth's Theory of Recognition, there are three different spheres of recognition in western bourgeois society where different modes and forms of recognition are institutionalized—at least as normative claims. These forms of recognition refer to different dimensions of one's personality, namely, one's needs and emotions, one's moral responsibility, and one's traits and abilities (Honneth 1995). If people are not recognized, their practical self-relation gets hurt. In this case, individuals are unable to develop or maintain basic self-confidence, self-respect, or self-esteem.

In our words, we could say that if one dimension of one's personality is systematically invisibilized, one cannot develop or maintain a non-pathologic self-relation. With regard to migration, we can find clear structural invisibilization of all three forms. Structural here means that these forms of invisibilization do not depend on the will of other individuals, but they are the result of a certain social order. One example is the fact that the material and emotional needs of migrants are often not seen as of equal importance as those of natives. Often, this structural distain comes along with other forms of discrimination such as those of class. Therefore, while the family problems of the rich and famous can move millions to tears, people often are quite unmoved by the urgent needs of the migrants in refugee camps all around the world. The perception of the needs of others frequently follows lines of social hegemony, creating a structural invisibility of the need for the physical and emotional integrity of non-hegemonic groups such as migrants.

An example of the invisibilization of the other as a morally responsible person would be treating the other as having fewer rights. In most western societies, democratic participation rights are clearly limited for migrants, following a hierarchy of status (e.g., EU citizens, legal migration, and undocumented migration). In these

[1] The general reflections of this work draw on my book *Invisibilization of Suffering: The Moral Grammar of Disrespect*.

cases, their capacity to take part in decision-taking procedures is denied. Instead of being part of the people as democratic sovereign citizens, they are treated as subjects in its literal meaning as "thrown under."

Finally, we can find a systematic devaluation of the traits and abilities of the other. The contributions of migrants to society are seldom seen as positive. Often, in public debates, migration is even framed as a problem, or more positively as a challenge, but seldom as a gain for society. This devaluation includes two elements. First, the same capacities are not recognized in the same way as for natives. We just have to think about the difficulties of the official recognition of foreign formal education. Second, different capacities are seen as "unwieldy" knowledge and abilities that do not fit in the preformed hierarchy of social evaluation. Therefore, for example, migrants' proven mobility, intercultural capacities, and language skills are usually not valued since these skills are usually not positively assessed. Being able to speak Arabic or Romanian is seen in central Europe more often as a defect than as a virtue. These capacities are not visibilized in the sense of socially valued or recognized.

Even seemingly positive evaluations are often framed in a system of general devaluation. Therefore, for example, when supposed original products of handicraft or cuisine are praised, this is similar to the appraisal of the ability of the real or supposed capacity of Black people to play jazz music or basketball. This type of "recognition" is located in a social space that assigns migrants or Black people a lower status in the social hierarchy. While complex, abstract, and intellectual labor is reserved for the hegemonic group, Blacks or migrants are pushed to look for recognition in the entertainment industry.

All these forms of invisibilization have negative consequences for the development or maintenance of a positive self-relation. The lack of self-confidence, self-respect, and self-esteem can be a result of systematic invisibilization. Furthermore, in a society that positively values a certain self-confident being in the world, a damaged self-relation can be the cause for further discrimination instead of leading to actions and policies of mitigation.

10.2.2 Social and Physical Invisibilization

Until now, we could ask the following question: what is the difference between invisibilization and disrespect? It seemed as if both concepts could be used as synonyms. In fact, disrespect is a form of social invisibilization in so far as disrespect underrates and invisibilizes systematically dimensions of the personality of the other. However, the concept of invisibilization allows us to explore some aspects and dialectical relationships of disrespect that otherwise would remain in the dark. When Honneth speaks of *social* invisibility, we can raise questions of the relation of social invisibility as disrespect and physical (or material) invisibility. In other words, we can use invisibility in a metaphorical and in a literal way to explore the relation to non-hegemonic social groups.

Physical invisibilization could be a result and a reason for social invisibilization. On the other hand, Honneth (2001) argues that for social invisibilization, the physical *visibility* of the other is important. Here, it seems as if social invisibilization for Honneth refers only to a direct disrespect of the other. However, we could also speak of structural, institutionalized invisibilization as an impersonal form of disrespect, similar to the case of the non-recognition of foreign educational qualifications. In this case, it is a distant, impersonal, and even anonymous institution that denies recognition, often without ever having personally seen those who want their qualifications recognized.

The differentiation between social and physical invisibilization could help us to clarify the different types of disrespect and their relations. This differentiation can help us to open up the topic to analyze the struggles against invisibilization and for recognition in a next step. We could find material or physical invisibilization in all three spheres of recognition. Therefore, in the case of migrants, immigration laws may impede family reunifications. This means that migrants could be physically prevented from living with their families and offer all the forms of emotional support that family members are supposed to give each other in the social institution of a family. In other words, here we have physical invisibilization of migrants and their families, leading to the lack of recognition and therefore to a lack of social visibilization. In the sphere of the recognition of the moral responsibility through equal rights, especially illegal migrants are prevented from physically showing up in the public space. When there is no right to reside and work, there is also no right to assembly and no right to demonstrate. In other words, undocumented migrants are hindered to be perceived as morally responsible political actors in the public space. Finally, there often is workplace segregation where those works that are economically and symbolically well recognized are performed in places physically separate from low-paid work in precarious conditions with little recognition. This physical invisibilization of the specific contributions of the different workers for the whole work-process occurs on several levels at the same time. Internationally, products and their marketing campaigns are often designed (and bought) in countries different from where they are produced. This separation impedes the direct contact through the maximal physical distance and thus perceiving and recognizing the work of those on the other side of the globe. On a local level within a metropolitan landscape, economically and symbolically little recognized social groups often find themselves in precarious housing conditions in the suburbs where the urban elite cannot physically perceive how little these groups are valued. This segregation often exists even in the very same workplace with clean offices and a separate canteen for some highly recognized workers, physically invisibilizing the destiny of many others.

Since Foucault (2007), we also know other forms of physical separation. In his inaugural speech on the order of discourse, he mentions social, not physical, reasons for others to be invisible. He cites the privileged or exclusive right to speak on a particular subject. Only some people, usually natives with high economic and cultural capital (i.e., competences considered capital by the hegemonic society), become visible in the public arena. One example is the public discourse on migration in Spanish newspapers. This discourse is mainly *about* migration and migrants; it is only very rarely a discourse on the migrants themselves. The discourse is usually produced by politicians, NGOs, journalists, and the police (Herzog 2009). This privileged right to speak also means a privileged physical visibility in order to claim love, rights, and respect or, on the other hand, a physical invisibilization of the needs, moral responsibility, and specific capacities of the rest.

Additionally, according to this physical invisibilization, there is social invisibility. Axel Honneth is referring to this kind of "looking through" with regard to the Black protagonist of Ralph Ellison's novel *The Invisible Man*. Honneth uses invisibilization here in a metaphorical sense, referring to those who are clearly perceptible. Examples are the famous mansplaining where men (and also women) socially invisibilize women who are physically present (Solnit 2014). What is true for women is also true for other, non-hegemonic social groups such as migrants and refugees. Although it was said that the discourse on migration is mainly a discourse *about* and not by migrants, even in the cases where migrants are present, often the confirming voice of a native "expert" is needed in order to provide validity to the speech act of the migrant. In other words, the moral responsibility to take part in the public discourse is socially invisibilized for migrants. Equally, other dimensions of the personality of migrants are not socially perceived. Gutiérrez-Rodríguez (2007) shows how employers treat migrant household workers as invisible and as socially irrelevant, similar to Honneth's example that "the nobility were permitted to undress in front of their servants because the latter were simply not there in a certain sense" (Honneth 2001: 112).

Now we have all the analytical elements to understand the relations between social and physical invisibilization. Sometimes it is physical visibility and not invisibility that is the condition for social invisibilization as a form of disrespect, similar to the case of the invisibilized household workers. However, sometimes the physical invisibility can also be a form of disrespect. If we understand the material conditions as forms of recognition/disrespect (see the debate between Fraser and Honneth 2003), then, for example, housing segregation—i.e., spatial invisibility—can clearly be understood as a form of structural disrespect, i.e., as social invisibility. Social invisibility can lead to physical invisibilization and vice versa. The "looking through" the other expresses a lack of interest for the situation of the other and that can be reinforced by physical segregation. The white middle class knows little about the living conditions of their household and care workers physically separated in other urban areas with their children attending different schools. On the other hand, this physical distance, at the same time, can reinforce the lack of communication when sharing a common space, thus reinforcing the "looking through." Within the existing power hierarchies, those on the top can allow themselves to be uninterested in the lives of those on the bottom. For the servants, on the contrary, there is a vital interest in the conditions of those on top. The anticipation of the moods and mental states of authorities, their employers, and their superiors is a social survival strategy (Scott 1990).

10.2.3 Silencing, Evidentiality, and Self-Invisibilization

Until now, we have used the term invisibilization in two senses: literally, as physical invisibility and metaphorically, in a social way, i.e., as a form of ignoring, looking through or disrespect. Both usages can be analytically split into two forms that, taken together, allow for new, revealing analytical perspectives.

When using "invisibility" as a form of physical imperceptibility, then the question that is raised is why this imperceptibility should be limited to the sense of vision. Often, when referring to physical imperceptibility, invisibilization is used in a second metaphorical sense, also referring to forms of silencing. We would then have the case of someone being physically present but as a muted subject. Unlike in the case of mansplaining where the other is speaking but not taken seriously, here the other has no right to speak or is ashamed to raise his/her voice.

Since Goffman's seminal work on *Stigma* (1986) and *The Presentation of Everyday Self* (1990), we know that stigmatization does not only depend on the mere physical perceptibility, but it also depends on the evidentness or evidentiality. These terms refer to the fact that a physically perceptible characteristic has to be correctly identified in order to lead to discrimination and disrespect. Therefore, for example, only those who identify clearly visible puncture holes in the forearm of the other as stemming from intravenous drug consumption can identify the other as a drug user who then can be stigmatized. Clues to stigmatizable characteristics therefore often undergo a careful impression management. Wilcke (2018a, b) offers the example of a Black, undocumented migrant in Berlin who always carries a tourist guide with her. Her skin color that otherwise would be interpreted as related to her probable residential status and lead to a higher probability of police control now is interpreted as part of her status as a tourist. The Black woman cannot hide her skin color in the public space. What she can do is modify the evidentness of this visible sign in a significant way. Another interesting combination of these strategies of impression management with physical invisibilization is mentioned by Goffman himself when talking about a Black person who prefers doing business by phone so that his Black skin cannot be perceived and not lead to stigma.

This example also shows that invisibilization does not have to come directly from those who discriminate but can also stem from the disrespected themselves. This leads us to a final, analytical distinction, i.e., that between self-invisibilization and invisibilization by others/society. From a strict sociological perspective, there is no such thing as an individual that freely decides to invisibilize itself. Self-invisibilization and hiding is part of a learned behavior. Especially, members of non-hegemonic groups have learned to socially hide themselves and to become muted subjects. They have learned that their claims for recognition are considered socially less important or less urgent.

Having said that the differentiation between self-invisibilization and invisibilization by others can only be an analytical distinction, we can now proceed to use this differentiation and combine it with all the other analytical forms of using the concept of invisibility. Multiple analytical combinations of these different forms of invisibility and usages of the term can shed light upon reality. Physical and social invisibility, evidentiality and perceptibility, visual imperceptibility and silencing, and self-invisibilization and invisibilization of others can be combined with all three spheres of recognition and disrespect described by Honneth. Therefore, we can understand muted subjects in the classroom who, due to their socially learned shame, do not dare to speak up and therefore have difficulties having their specific traits and capacities recognized. These students, who are more likely to be from non-hegemonial groups, are physically visible but muted. They invisibilize themselves, or, at least, in this sphere where it is mostly about the recognition of specific characteristics, they invisibilize their capacities. In practice, this form of invisibilization would probably mix with other forms of invisibilization. When finally daring to speak up, we could imagine social invisibilization in the form of mansplaining.

Armed with this analytical toolbox, we can now come back to practices of invisibilization in the case of migration.

10.3 Struggles Against Social Invisibilization

10.3.1 First-Order Struggles

What Axel Honneth seems to have in mind when writing his essay on invisibilization is mainly the struggle for recognition as a struggle against social invisibility. In the case of migration, we can find all kinds of struggles for recognition in the three spheres. We can find the struggle for family reunion in order to provide effective care in the institution of the family. Second, we can identify various forms of the struggle for the equal rights of migrants including regarding residency status, health care, voting rights, etc. Finally, there are struggles to recognize the specific traits and abilities of migrants and other minorities. Examples here are the struggle for the formal recognition of specific educational qualifications or the informal campaigning for the perception of migration as social capital and not as a problem. In all cases, these struggles aim at the visibilization and therefore recognition of the different dimensions of the personality of migrants, thus allowing them to create or maintain a non-pathologic self-relation.

We have seen a relation between this form of social recognition and physical visibility. Only when correctly identified and when physically perceptible can migrants make their claims for recognition. These claims can be made openly and directly, or they can show "affective reactions of disrespect" (Honneth) that can be understood intersubjectively and indirectly by society and turned toward positive formulated claims of recognition. In that sense, the claim for social visibility is often directly related to the claim of physical perceptibility. Following this argument, if more migrants would be seen as experts in newspapers, if political parties had more migrants in elevated positions, and if people would speak more to migrants, then, with this grown physical visibility, there would be more social visibility, more recognition.

With the classical theory of the three forms of recognition, we can now differentiate three basic modes of struggles against invisibility. The first is a struggle to recognize basic material and emotional needs. This struggle tries to physically visibilize migrants as human beings with the same needs. For example, it denounces the infrahuman conditions in migration camps, the deathly risks migrants often have to face when crossing the border or especially the inhuman treatment and abuses that undocumented migrants are facing in the so-called host societies. Second, we can identify the struggles for the same rights for migrants. Again, these struggles often are related to making unequal situations visible. The argument here is that only if society is aware of the situation, i.e., if migrants and their lack of formal or informal access to rights are made physical visible (or audible or readable), can social visibility be achieved. Finally, in the public perception, there is often a relation between rights and contributions as if having rights would be the reward for contributing, especially to the economic well-being of a nation. Therefore, struggles to visibilize the work done by migrants, their traits, or their impacts on GDP not only aim at recognizing their capacities. This example shows how recognition in one sphere can have simultaneous positive (i.e., visibilizing) effects in other spheres.

10.3.2 Second-Order Struggles

All these forms of visibilization depend on social institutions able to provide recognition. However, the existence of these institutions should not be taken for granted. Elsewhere (Herzog 2015), I have developed the notion of second-order recognition, i.e., the recognition of institutions or groups that can grant first-order recognition. For example, the institution par excellence that provides emotional support is the family. However, families themselves have to be recognized by the state and civil society in order to fulfill this role. In societies in which same sex couples are neither legally nor socially recognized as families, i.e., where they lack second-order recognition, they cannot provide (first-order) recognition in the form of love and care in the same way as socially and legally recognized families. These struggles for the recognition of institutions are called struggles for second-order recognition because they are not direct struggles for recognition but are struggles for an institutional framework able to provide adequate recognition.

In relation to migration, we can find struggles for alternative institutions of recognition or for alternative interpretations of the given ones. The struggle for the recognition of individual migrants and for migration as a family project that requires family reunification would be a struggle for the second-order recognition of institutions that could then provide first-order recognition in the form of love and care. Additionally, there are also struggles for alternative legal institutions. These struggles are not aiming primarily at modifying the existing legal institutions, such as the state, but aim at introducing alternative ones. Here, we can find approaches aiming at a global community in order to grant something similar to global citizenship instead of national citizenship and also the struggle for the recognition of particular legal institutions such as sharia courts in western societies that point toward legal, i.e., moral, recognition different from the established state structures. Finally, we can find struggles for the second-order recognition of institutions that value the traits and abilities of their members. Here again, we could think of the legal recognition of certain religious minorities that value their member-specific capacities and contributions different from those recognized in the broader society.

In all these cases, the main struggle would be about visibilizing the contribution to the recognition of these *institutions*. In the classical approach that understands physical visibility as a prerequisite for social visibility, this would mean to visibilize the virtues of certain institutions in terms of their capacity to organize first-order recognition.

The idea presented here is very similar to the reflection by Honneth on "countercultures of compensatory respect" (Honneth 1995: 124). This term refers to a situation in which the given institutions are not able to provide the desired recognition (love, rights, or social esteem) of individuals or whole groups. Therefore, these individuals, instead of struggling for recognition in the given institutions following the established paths in order to deserve love, rights, and respect, are creating a new whole system, a counterculture. We can understand these countercultures as alternative institutions that try to compensate for the lack of recognition in one or several social spheres. Therefore, where there are dysfunctional family structures that are unable to recognize basic needs, alternative structures, even some calling themselves a "family," often serve to compensate for the lack of recognition. This compensatory recognition sometimes is an additional recognition, but sometimes it is a whole system of counterculture. We can understand this search for alternative institutions as a process of derecognition in which the given institutions such as the family, the state, or the labor market are considered inadequate for providing recognition.

Taking Ikäheimo's (2002) definition of recognition as "a case of A taking B as C in the dimension of D, and B taking A as a relevant judge" (p. 450), we can see that recognition depends on mutuality. This means that if B is not considering A to be a relevant judge anymore, if an individual or a group does not think the family, state, or labor market are relevant providers of recognition, then we can speak of derecognition. Using our conceptualization of invisibility, we can say that if people feel invisibilized by one or several institutions, they can seek or create alternative cultures/institutions where the dimensions of their personality are recognized, i.e., physically and socially visibilized.

All these examples of the struggle for recognition, second-order recognition, compensatory respect, etc. stem from the classical approach of the struggle for social visibilization through physical visibilization. It is, ultimately, due to the communicative approach immanent in the Theory of Recognition that the claims for recognition are thought to require physical or public visibility in order to become effective.

10.4 Invisibilization as a Strategy Against Disrespect

As we have seen at the beginning, visibility is not only a precondition for recognition but also for disrespect. In order to ignore, insult, or dishonor the other, one has to correctly identify the other in the first place. Furthermore, the act of correct identification itself has to be perceived by the disrespected and by possible bystanders. Therefore, if disrespect requires visibility, then perhaps a way to overcome disrespect does not have to be developed by exposing personal dimensions, but almost by its opposite, by invisibilization. A lot of cultural products such as Harry Potter's invisibility cloak, the ring that turns its bearer invisible in the Lord of the Rings, and the extraordinary capacity of the Invisible Woman from the Fantastic Four comics show that invisibility can also protect people from all kinds of danger. Physical invisibility can protect one from direct disrespect. In what follows, I will explore the possibility of invisibility and invisibilization as a means to overcome disrespect.

In their book *Escape Routes: Control and Subversion in the 21st Century,* Papadopoulos et al. (2008) (see also the more focused article by Papadopoulos and Tsianos 2008) develop the concept of *imperceptibility*, pointing toward becoming imperceptible as the aim of all. Their approach is not about physical invisibility, but rather it is about becoming impersonal, indiscernible, and therefore able to escape the control mechanisms of the state. In their work on undocumented migrants, they show how becoming indiscernible as an individual behind the visible phenomenon of mass migration can be a conscious strategy of resistance. Breaking the relation between one's life or body and one's name can be a step toward resistance to being trapped in a stigmatized category. Similar to how queer theory tries to escape fixed gender categories, aiming at the rupture of a general social order based on gender and sexual descriptions, Papadopoulos et al. argue that becoming imperceptible immanently aims at a general critique of the social order based on nations, borders, and state control. One example is the "herraguas," which is Moroccan for "the burners," that is, those who burn their legal documents as a strategy of voluntary dehumanization, deindividualization, and disidentification in order to become difficult for the European authorities to manage.

The imperceptibility of migrants does not mean that migration itself is imperceptible. On the contrary, migrational flows become highly visible and turn to be the most prominent targets for registration, regulation, and restriction by a sovereign power. Becoming imperceptible is an immanent act of resistance because it makes it impossible to identify migrations as processes that consist of fixed collective subjects. Becoming imperceptible is the most precise and effective tool migrants can employ to oppose the individualizing, quantifying, and representational pressures of the settled, constituted geopolitical power (Papadopoulos and Tsianos 2008: 4).

In this case, and although migration as a mass phenomenon is quite visible, migrants as individuals can tend to become invisible. Similar to Axel Honneth's description of the invisibility of a Black individual, the visibility of mass migration is the reason for the creation of an enormous state and supra-state apparatus of power, control, and bordering that we can now understand as a process of disrespect or social invisibilization toward individual migrants. As we have seen, instead of struggling to become a political subject recognized by the state, an alternative strategy can be to refuse to become a political subject or to become a subject at all. The term "subject" comes from the Latin subjectus, which means "brought under" in the sense of power and obedience, literally being the combination of *sub-* (under) and *jacere* (throw). Thus, a subject is someone who is thrown under the power or authority of others. When becoming a subject means being thrown under the domination and ultimately the exclusionary rules of others, refusing to become a subject can be one way to escape this domination. Perhaps this is also what Foucault had in mind when he understood critique as the process of "desubjugation" (Foucault 2007; see also Butler 2001).

The strategy of invisibility is related to a practical crisis of representation. The strategy of invisibilization means that individuals and groups do not struggle (any longer) for more visibility and participation in the public sphere. If migrants do not perceive the existing power as representing the individual or collective experience or when they perceive the official apparatus even as hostile to one's own goals, then entering the game of representation via visibility appears to be social and political suicide. Whereas representation means subjugation, invisibility becomes resistance or, in the words of Papadopoulos et al. (2008), "the decline of representation as the core politics of resistance and subversion means simultaneously the end of the strategy of visibility" (p. 218).

This strategy is only a negative one protecting one from direct disrespect. Still, the very fact that people hide themselves in order not to be disrespected points itself to a form of contempt. The self-invisibilization as a strategy to struggle against disrespect can, thus, lead to a damaged self-relation, i.e., to a lack of self-confidence, self-respect, or self-esteem. This mere negative strategy of imperceptibility is therefore often combined with others, aiming not only at avoiding disrespect but also toward forms of recognition.

One of these strategies is the already mentioned counterculture of compensatory respect, i.e., alternative forms of finding recognition apart from hegemonic institutions. Another strategy consists of maintaining the individual invisibility while trying to gain public visibility in an impersonal way. Wilcke (2018a, b) presents the example of the struggle of undocumented migrant workers in Germany to join labor unions. In a first instance, since being undocumented means not being able to have a legal work contract, unions refused to accept illegal migrants as members. Since it is very dangerous for undocumented migrants to protest publicly due to the risk of detention and deportation, a group of migrants organized protests in the meetings of labor unions that they considered a safe space. Through this pressure, they reached their goal so that now German trade unions accept undocumented migrants as members and represent them in labor conflicts. Through representation, it is now possible to struggle for social visibility. Trade unions can make the situation of undocumented migrants visible in general without exposing and endangering the individual migrant. In other words, there is personal physical invisibility that avoids direct disrespect (social invisibility). In addition, this strategy aims at public social visibility (recognition) through physical visibilization, i.e., the public exposure of unjust social situations. A similar strategy is making anonymous but public claims, e.g., by writing newspaper articles under a pseudonym.

This episode also points toward another negotiation of visibility. Even physical visibility is not an all-or-nothing game. We always have to ask for whom someone is visible. Being visible for those who are likely to show disrespect is not the same as being visible for those where there is a reasonable possibility to find solidarity. In the mentioned example, the group of undocumented migrants carefully chose the place of struggle for social visibility through physical exposure. Here, we are again in a form of impression management. Migrants have to constantly evaluate whether a space is safe or not and whether the school director of the children, their neighbor, or even their friend could be considered a space where their identity as an undocumented migrant can be laid open.

In the same sense, we can also read the already mentioned impression management of the Black undocumented migrant carrying a tourist guide with her in order to falsely visibilize herself as a tourist and not as a possible undocumented migrant. In this case, she opted for a kind of social invisibility, of the invisibility of a certain social status, reclaiming a very basic recognition in the public space: the recognition of the right to move freely. This person could only achieve this factual recognition through a specific form of invisibility. Here, it is not the physical invisibility of the person as a whole, but rather the imperceptibility of a social status through the negotiation of what can be seen for whom.

All these examples show that when physical visibility is a precondition for recognition and for disrespect, then what is seen, how, and for whom is a crucial question, especially for the most vulnerable.

Different from these strategies—that although relying on personal invisibility still depends on certain forms of visibility, such as misleading visibility through impression management, collective (impersonal) visibility, or visibilization through representation by others—we can find strategies that try to deny any kind of visibility. Smith et al. (2018) develop the notion of "stoic withdrawal" for cases where not only visibility is denied, but also no explanation for this invisibilization is given. The best-known literature figure here is perhaps Bartleby the scrivener (Melville 2017 [1853]) who refuses not only to do his work but also to give any explanation. Always using the same words "I would prefer not to," he becomes invisible and not understandable for the others. His superior is therefore lacking any access to Bartleby and unable to develop strategies to make him do his job. Invisibilization can thus be seen as the opposite of the permanent self-presentation, as described, for example, by Erving Goffman (1986, 1990). Goffman contemplates invisibilization only in the case of a stigma, which is a negative ascribed or acquired aspect that people want to hide. However, in the stoic tradition, invisibility can refer to the refusal of a general social norm, will, or obligation to always present oneself to others in a specific socially favorable way. If we translate the concept of "loneliness," which is widely used as a positive value in the stoic tradition, into the modern vocabulary of social philosophy, then we could perhaps speak of independence or autonomy as the positive ethical attitudes behind self-invisibilization.

This withdrawal is not the same kind of self-invisibilization that occurs due to shame and that can be understood intersubjectively. Perhaps we could say that we are here in front of a double invisibilization because even the reason for invisibilization is deliberately hidden away. This type of invisibilization is perhaps the most radical form since the invisibilized itself denies recognition to the given social institutions. We could say that through stoic withdrawal, the existing institutions are derecognized.

Since there is still a need for self-confidence, self-respect, and self-esteem and since this need can only be met intersubjectively, the stoic withdrawal from the institutionalized order of recognition can lead to parallel organizations of recognition. Here we can find partial recognition through the already mentioned countercultures of compensatory respect where only one part of the institutional order is replaced by the other. For example, the family can be replaced by a gang, or the workplace can be replaced by a religious community. Those who do not recognize the family or the labor market as institutions of legitimate recognition are finding the social visibilization of some of the dimensions of their personality in these institutions. This recognition is needed for a satisfactory self-relation. The complete stoic withdrawal would mean the creation of a whole parallel society with different institutions that provide non-pathologic self-relations. However, when there is the warning of parallel societies in public discourses, this term usually refers to the withdrawal only from some—and not all—of the institutions of recognition.

10.5 Conclusion

We have seen three different forms of managing invisibility. These are not only different strategies in the struggle for recognition. They also differ regarding the depth of social change required in order to achieve acceptable recognition. This means that the strategies differ in their judgments of how deep the actual situation of disrespect is engrained in the social order. Therefore, struggles for recognition that aim at direct unmediated physical visibility in order to achieve social recognition engage in a kind of communicative action, making a claim directly through speech acts or indirectly through the visibilization of their suffering. These struggles directly address broader spheres, mainly the institutions to which they reclaim recognition. In the case of individuals or groups seeking social visibility through physical visibilization, we could say that they recognize the given order of recognition. They just claim (more) recognition for themselves within this order.

The situation changes sensibly if the institutionalized recognition order is not considered adequate for granting recognition. When opting for physical invisibilization as a strategy to defy disrespect, individuals and groups do not assume that it is just a lack of information (or of physical visibility) that leads to disrespect. On the contrary, these groups assume that if the other had more information about their own group, the direct consequence would be more, not less, disrespect. However, often there is a struggle for *another* visibility or recognition in other *categories*.

Instead of being perceived as (undocumented) migrants, groups of people want to be recognized as citizens, tourists, human beings, etc. This means that this type of invisibility still implicitly recognizes the existing institutions of recognition.

Finally, in the case of the stoic withdrawal, we are not facing only a questioning of the categories of recognition but also a questioning of the very institutions of recognition or even the whole recognition order. Bartleby does not claim to be recognized in a different way, say as a broker instead of a scrivener. He does deny making even implicit claims of recognition, therefore factually not recognizing the recognizer. In real life where people depend on positive self-relations, i.e., on self-confidence, self-respect, and self-esteem, we could assume that certain forms of recognition are indispensable. However, it does not necessarily have to be *these* institutions that distribute *this* type of recognition in *that* way and for *that* price.

What should be clear by now is that we cannot speak of visibility or invisibility as if they were by definition positively or negatively related to recognition. There are not only different forms of visibility, ranging from social invisibility to physical invisibility or ranging from self-invisibilization to invisibilization by others. Furthermore, there are different uses of this invisibilization. All types of invisibilization can be used as part of disrespect or as its contrary as a form of avoiding disrespect. Whether invisibilization has positive effects (or prevents negative ones) or is part of discrimination depends on the social conditions and power relations surrounding the concrete situation. As a consequence, the analysis of invisibilization has to always consider the social power relations. Further questions that can be raised in this context are the following.

First, what is visible and what is invisibilized *before and after* the visibilizing practice of the researcher? There is never a practice that sheds light on all aspects of an object. When focusing on a specific aspect, others necessarily become less perceptible. By using certain words to describe injustices, many others that could also have been pronounced remain silent. It is impossible to escape this dialectic of visibility and invisibility. However, we can be aware of this problem and make conscious decisions about what we want to focus on. Here is the place to also ask about the (in)visibility of the amorphous processes of power relations and the production of suffering: how far does the visibilization of the *situation* of the invisibilized itself impede the perception of certain *processes*?

Second, we can ask to whom is someone or something (in)visible. It is not the same to be visible to the government as to civil society. The visibility of my suffering in the public sphere is not the same as its visibility in the private sphere. It is not that one sphere is inherently better than the other; rather, they have different implications. While the public is almost always related to the political, the private is often perceived as a space of special protection from intromission. In addition, the very division between political and private is also a socio-political one. This question again is clearly a question of power.

Third, who can rightfully claim (in)visibility for oneself or transparency for the other? It is not the same to expose oneself as it is to be exposed. A critical theory of invisibility and of suffering must always identify the intervening power. Who has the power to visibilize or invisibilize which type of suffering?

All these questions are also relevant for us as researchers. With our research, we engage in processes of visibilization while at the same time contributing to the irrelevance of the invisible. While being unconscious about this question almost always leads us to being part of the given power structure, conscious use of academic strategies of visibilization and invisibilization can lead to both the maintenance of structural disrespect and its subversion. As researchers, we can help in the impersonal visibilization for those migrants to whom the direct process of visibilization is closed. We can furthermore point toward the *processes* of disrespect themselves instead of focusing directly on the migrant population as objects of visibilization, power, and management. The visibilization of dominating power that prefers being invisible can have a more emancipatory effect than exposing the invisibilized.

References

Butler, J. 2001. What Is Critique? An Essay on Foucault's Virtue *EIPCP – European Institute on Progressive Cultural Policies*. Retrieved from http://eipcp.net/transversal/0806/butler/en.

Foucault, M. 1975. *Discipline and Punish: The Birth of the Prison*. New York: Random House.

———. 1990. *The History of Sexuality. An Introduction*. Vol. 1. New York: Random House.

———. 2007. *The Politics of Truth*. Los Angeles: Semiotext(e).

Fraser, N., and A. Honneth. 2003. *Redistribution or Recognition? A Political-Philosophical Exchange*. London: Verso.

Goffman, E. 1986. *Stigma: Notes on the Management of Spoiled Identity*. New York: Simon & Schuster.

———. 1990. *The Presentation of Self in Everyday Life*. London: Penguin.

Gutiérrez-Rodríguez, E. 2007. Reading Affect—On the Heterotopian Spaces of Care and Domestic Work in Private Households, *Forum: Qualitative Social Research*, 8(2). Retrieved from http://nbn-resolving.de/urn:nbn:de:0114-fqs0702118.

Hempel, L., S. Krasmann, and U. Bröckling. 2010. *Sichtbarkeitsregime. Überwachung, Sicherheit und Privatheit im 21. Jahrhundert*. Wiesbaden: VS Verlag.

Herzog, B. 2009. *Exclusión discursiva – el imaginario social sobre inmigración y drogas*. Valencia: PUV.

———. 2015. Recognition in Multicultural Societies. Intergroup Relations as Second-order Recognition. *Revista Internacional de Sociología* 73 (2).

———. 2020. *Invisibilization of Suffering. The Moral Grammar of Disrespect*. London: Palgrave.

Honneth, A. 2001. Invisibility: On the Epistemology of Recognition. *Aristotelian Society* 75 (1): 111–126.

———. 1995. *The Struggle for Recognition: The Moral Grammar of Social Conflicts*. Cambridge: Polity Press.

———. 2015. Unsichtbarkeit. Über die moralische Epistemologie von "Anerkennung". In *Unsichtbarkeit. Stationen einer Theorie der Intersubjektivität*. Frankfurt/Main: Suhrkamp.

Ikäheimo, H. (2002). On the Genus and Species of Recognition. *Inquiry*, 45: 447–462.

Melville, H. 2017. Bartleby, the Scrivener. In *A Story of Wall Street*. Edinburgh, Mockingbird.

Papadopoulos, D., and V. Tsianos. 2008. The Autonomy of Migration – The Animals of Undocumented Mobility. In *Deleuzian Encounters. Studies in Contemporary Social Issues*, ed. A. Hickey-Moody and P. Malins, 223–235. Basingstoke: Palgrave Macmillan.

Papadopoulos, D., N. Stephenson, and V. Tsianos. 2008. *Escape Routes: Control and Subversion in the 21st Century*. London: Pluto Press.

Scott, J. 1990. *Domination and the Arts of Resistance: Hidden Transcripts*. New Haven/London: Yale University Press.
Smith, W., M. Higgins, G. Kokkinidis, and M. Parker. 2018. Becoming Invisible: The Ethics and Politics of Imperceptibility. *Culture and Organisation* 24 (1): 54–73.
Solnit, R. 2014. *Men Explain Things to Me*. Chicago: Haymarket Books.
Wilcke, H. 2018a. *Illegal und unsichtbar? Papierlose Migrant*innen als politische Subjekte*. Bielefeld: Transcript.
———. 2018b. Imperceptible Politics: Illegalized Migrants and Their Struggles for Work and Unionization. *Social Inclusion* 6 (1): 157–165. https://doi.org/10.17645/si.v6i1.1297.

Chapter 11
A Quest for Justice: Recognition and Migrant Interactions with Child Welfare Services in Norway

Alyssa Marie Kvalvaag and Gabriela Mezzanotti

Abstract Norwegian Child Welfare Services (NCWS) has faced intense criticism regarding their interactions with migrant families, with international human rights monitoring mechanisms expressing concern regarding ethnic discrimination over the past decade. Our aim is to contribute to the academic discussion around migrant interactions with NCWS through exploring the suitability and relevance of Nancy Fraser's theory of social justice, with a particular focus on recognition. We utilize the narratives of two migrant parents and two child welfare practitioners supplemented by critiques from international human rights monitoring mechanisms to bridge the gap between the theoretical level, institutions, and daily practices. Three areas regarding the suitability of recognition in the case of NCWS are discussed: misrecognition as institutionalized subordination; equality, sameness, and difference in the Nordic welfare state; and the dynamic nature of culture. While we find recognition to be an essential element to be considered in the case of NCWS, we emphasize recognition must also be considered within Fraser's larger understanding of social justice, alongside redistribution and representation.

Keywords Recognition · Child welfare · Migrant · Social justice · Redistribution

A. M. Kvalvaag (✉)
Nord University, Bodø, Norway
e-mail: alyssa.m.kvalvaag@nord.no

G. Mezzanotti
University of South-Eastern Norway, Drammen, Norway
e-mail: gabriela.mezzanotti@usn.no

G. Schweiger (ed.), *Migration, Recognition and Critical Theory*, Studies in Global Justice 21, https://doi.org/10.1007/978-3-030-72732-1_11

11.1 Introduction

Norwegian Child Welfare Services (NCWS) has faced intense criticism regarding their interactions with migrant families, with international human rights monitoring mechanisms highlighting concerns regarding ethnic discrimination over the past decade; with this chapter, we hope to contribute to the academic discussion by examining what Nancy Fraser's understanding of social justice and recognition contribute in the case of NCWS. In its Universal Periodic Review from May 2019, the Human Rights Council recommended Norway review current practices relating to out-of-home placements, deprivation of parental rights, and limitation of contact rights (UN Human Rights Council 2019). In general, migrants are overrepresented – compared to the majority population – in interactions with NCWS, particularly in voluntary assistance including advice, guidance, and poverty reduction measures; however, overrepresentation varies greatly between migrants from differing countries of origin, with some groups being strongly overrepresented[1] in voluntary assistance measures and some migrant groups being underrepresented[2] (Berg et al. 2017; Christiansen et al. 2019; Dyrhaug and Sky 2015; Staer and Bjørknes 2015; Thorud 2019). The issue of migrant family interactions has been raised previously by the UN Committee on the Rights of the Child in 2010 and 2018, when concerns that "child welfare assistance for children from ethnic minorities is of a much lower standard" were raised (UN Committee on the Rights of the Child 2010: 14). Such critiques relate to a context of complex migration processes and increased public hate speech, whether by politicians or media professionals and social media toward different targeted populations including asylum-seekers, Muslims, the indigenous Sámi population, recognized national minorities including Jews and Roma, and others (UN Human Rights Council 2019).

Research in different fields has addressed growing criticism of NCWS; our aim is to contribute to this current debate by utilizing an interdisciplinary perspective focused on the interplay between critical theory, child welfare, and human rights. From this angle, our chapter presents the context of NCWS and the extent to which Nancy Fraser's critical theoretical approach to recognition, redistribution, and representation (Fraser 2000; Fraser 2001; Fraser 2007; Fraser 2008a; Fraser 2008b; Fraser 2010; Fraser and Honneth 2003) provides a suitable analytical framework to examine sociocultural, socioeconomic, and sociopolitical aspects of injustice. Thus, in this article, we ask: to what extent may Fraser's understanding of social justice – encompassing recognition, redistribution, and representation – be suitable to provide insights in exploring (mis)recognition of migrant families in the case of NCWS? To answer this question, we utilize the narratives of migrant parents and

[1] Migrants from Latin America, Africa, Asia, and Turkey and Eastern European countries that are outside of the EU as well as refugees are overrepresented in voluntary assistance measures (Berg et al. 2017: 44).

[2] Some groups that are underrepresented in voluntary assistance measures include migrants from Poland, Russia, and India (Berg et al. 2017: 44).

child welfare practitioners and relate to international human rights law documents and domestic legislation regarding migrant interactions with NCWS.[3] The narratives of two parents and two child welfare practitioners will be utilized to bridge the gap between daily social interactions and large-scale social structures (Ewick and Silbey 1995). While the individual narratives shared in this chapter cannot be generalized in and of themselves, they are essential to construct an understanding about (mis)recognition in NCWS as it is understood and lived by the people directly involved with different aspects of the child welfare system.

In this article, we will share the experiences of John, Emma, Anna, and Ashley.[4] John is a migrant parent who has lived in multiple countries and has obtained Norwegian citizenship; his experience with NCWS comes through work as a translator for the municipality, where he has translated for migrant families in NCWS cases. His concerns were wide, including a lack of knowledge migrant families have of NCWS, the power of NCWS as an institution, and a Norwegian normative framework which he understands as guiding NCWS. He suggests greater access to information about NCWS and more room in legislation for cultural nuances would improve the relationship between migrant parents and NCWS. Emma is a child welfare practitioner who works for NCWS – she has lived in multiple countries and has an international family. A primary concern of Emma was (lack of) cultural competency and how the child welfare system is designed considering the Norwegian norm, with seemingly little room for difference. In her opinion, NCWS is in need of reform to better accommodate the families residing in Norway today. Anna is a child welfare practitioner who works for NCWS; she is a migrant to Norway and expresses the time needed to become established in a new place. She emphasizes building trust as an important factor in enhancing the relationship between migrant families and NCWS. Ashley is a migrant parent who was initially very afraid of NCWS due to the large negative publicity the institution has received in her home country; however, the longer she has lived in Norway, the more positive things she has heard about NCWS. She emphasizes trying to adapt to Norwegian society and navigating "a perfect middle way" of combining aspects from Norway and her home country. With the hope of exploring more in-depth some of the nuances relating to recognition, these narratives will be integrated throughout the following sections.

First, we examine the Norwegian context to provide a brief overview of migration in Norway, migrant trust in NCWS, and child welfare legislation. Second, we discuss our conceptual framework by setting out Fraser's theory of social justice, recognition, and participatory parity which will guide and inform our analysis. Third, we present data from the selected narratives and analysis in relation to the suitability of recognition, focusing on (a) misrecognition as institutionalized subordination; (b) equality, sameness, and difference in the context of the Nordic welfare state; and (c) the dynamic nature of culture, with consideration for narratives from

---

[3] These narratives were collected by Kvalvaag in 2017 and 2018 as part of a larger project.

[4] Pseudonyms.

parents and practitioners and Fraser's theory of recognition. Fourth, we focus on the importance of redistribution and representation in the case of migrant interactions with NCWS. Finally, we conclude with final thoughts relating to social justice, recognition, and migrant interactions in the case of NCWS.

11.2 The Norwegian Context

Norway is recognized as being an international leader in human rights, notably children's rights and child protection. The first law addressing child welfare in Norway came into force in 1896, and in 1953, new children welfare legislation emphasized preventative measures, counseling, guidance, and in-home treatment (Folleso and Mevik 2011). Current legislation on child welfare in Norway – the *Child Welfare Act* (Norwegian: *Barnevernloven*) – was established in 1992, and the most recent amendments were passed in 2018. The Nordic countries are often portrayed or described as child-centered paradises; however, the region is not immune to issues of social exclusion and poverty – particularly for those in ethnic minority communities (Forsberg and Kröger 2011). This may be understood within the larger frame of what Langford and Schaffer (2015) termed the *Nordic human rights paradox*, where social pluralism is restructuring domestic implementation of human rights.

As we begin with an assumption that NCWS does not operate within a societal vacuum, it is important to briefly consider the wider context of migration in Norway. In the aftermath of the so-called 2015 migration crisis – where more than 30,000 asylum-seekers applied for asylum in Norway – stricter asylum policies were implemented and broadly publicized. As a consequence, the number of asylum-seekers arriving in Norway today is one of the lowest in many years; in the last 2 years, only around 5000 persons applied for asylum (NOAS 2019). Stricter policies have led to the adoption of many state-driven actions toward *non-entrée*, and important restrictions to family reunification and access to resident permits were implemented. At the same time, Norway's international discourse advocates for "genuine" solidarity toward refugees elsewhere (Mezzanotti 2018). These new policies were carried out within the context of concerns relating to the rise of right-wing discourses, hate speech, and xenophobia against minorities, particularly migrants (including Norwegian born to migrant parents) and national minorities (UN Human Rights Council 2019). As perceived discrimination can influence the lack of trust between migrants and institutions, these discourses and stricter policies may be interpreted as barriers to the participation of migrants in the political sphere and contributing to a lack of trust between migrants regarding public services, such as NCWS (Korzeniewska et al. 2019).

The issue of trust between migrant families and NCWS has received considerable attention in Norway – both in the media and as a subject of public debate (Korzeniewska et al. 2019). Success factors in building trust between migrants and public institutions include being familiar with the institution, understanding and interpreting an institution, and predicting, with reasonable certainty, an institution's actions (Korzeniewska et al. 2019). A common theme in Anna's narrative illustrates a concern about the lack of trust of migrants to NCWS, particularly in initial stages of contact:

> *People know or they've heard a lot about us and they come here [to NCWS] very scared. So much of the time you spend, like trying to make them understand, to calm down, to that level where you can actually work and interact properly... There was one mom that told me that, you know, when she goes into the door there [at NCWS] and people see her and start thinking "oh, they are going to take your kid very soon"... Well, at the end of the day, we didn't do that. And it kind of made her realize that we do more than taking the child, which is the general belief of many people really. So we do other things. We actually give counseling, we help with other small things too, like trying to find the right body, or person, or organization to help them, you know, depending on the kind of problem. So we do more than taking custody of people's children. But it just that, when people come in and are afraid already, then you just know, you have to be very patient and you have to spend up to like a year trying to convince them before they actually see that we all want the same thing. We want what is best for the child – you want what is best for the child and so do I.* (Anna)

Correspondingly, the UN Committee on the Rights of the Child has expressed concern over "insufficient communication and information exchange between child welfare services and families, in particular migrant families" (UN Committee on the Rights of the Child 2018). Current research also suggests a general lack of trust in NCWS, and many migrant parents view NCWS as a threat rather than a helpful service (Ipsos 2017; Korzeniewska et al. 2019; Øverlien 2012). For example, Fylkesnes et al.'s (2015) findings suggest that fear of NCWS is generally related to general perceptions or generalizations that NCWS primarily takes children away from their parents, that NCWS does not go into a dialogue with parents, and that NCWS discriminates against migrant families; these perceptions are not based on their interviewees own experiences, but rather what they describe as common representations among migrant families in general (Fylkesnes et al. 2015). Christiansen et al.'s (2019: 11, 239) findings indicate a larger number of families with migrant background, than nonmigrant background, experience risk assessments and investigations early on in interactions with NCWS, suggesting a more intervention-based mode of child protection; they suggest this more intervention-based experience may also influence some of the skepticism and fear which exists between migrant families and NCWS, not only cultural-related perceptions about the relationship between governmental authorities and families.

As a response to the critique and lack of trust, revisions were made to child welfare law in 2018 (Thorud 2020). The purpose of these amendments is to adapt the child welfare legislation to be more appropriate for today's society and to strengthen children's rights (Thorud 2020). There is also a draft for amendments intended "to strengthen the consideration of the child's religious, cultural and linguistic background" (Thorud 2020: 81). Further, a *Competence Strategy for the Municipal Child Welfare Services (2018–2024)* has been implemented, which includes an educational aspect to increase "greater understanding and sensitivity in the follow-up of children and families with minority backgrounds" (Thorud 2020: 81). However, it is unclear how much the voices of migrants have been included and considered in these reforms. Between 2017 and 2018, children and youth with a migrant background constituted 18% of the population and 28% of the children and youth who received assistance from NCWS (Thorud 2019). It is interesting to note that migrants are not overrepresented in out-of-home placements or child removals as migrant families seem to fear, but rather in voluntary assistance measures (Berg et al. 2017); this may be related to the emphasis the system places on preventative measures. Staer and Bjørknes (2015: 30) investigate NCWS involvement among families with children ages 6–12, finding that children with non-Western migrant parents have double the likelihood of voluntary assistance measures being implemented by NCWS when compared to children with nonmigrant parents; however, migration background was not found significant when controlling for socioeconomic conditions, suggesting the essential role of socioeconomic background in NCWS cases with migrants (Christiansen et al. 2019; Staer and Bjørknes 2015: 30). Removing a child from his or her family and resorting to alternative care without parental consent is deemed a measure of last resort (Thorud 2020).

There are several who argue the current interactions and general overrepresentation of migrants in NCWS may not be as problematic as it appears on the surface. For example, Skivenes (2014) suggests an overrepresentation of migrants in voluntary assistance measures by NCWS is not alarming as welfare state ambitions are in line with poverty reduction among migrant families (see also Christiansen et al. 2019; Staer and Bjørknes 2015). Yet, migrant interactions with NCWS and concerns about discrimination against minorities have received extensive attention and critique from international human rights monitoring mechanisms, including the UN Human Rights Council during the Universal Periodic Review,[5] the UN Committee on the Rights of the Child,[6] the UN Committee on the Elimination of Racial Discrimination,[7] the UN Committee on Economic, Social and Cultural Rights,[8] and the UN Human Rights Committee.[9] Further, as indicated by the previously mentioned amendments to domestic law, critiques from international human rights

[5] See Report of the Working Group on the Universal Periodic Review: Norway (Report No. A/HRC/42/3); Report of the Working Group on the Universal Periodic Review: Norway (Report No. A/HRC/27/3); and Report of the Working Group on the Universal Period Review: Norway. Addendum (Report No. A/HRC/27/3/Add.1).

[6] See Concluding Observations on the Combined Fifth and Sixth Periodic Reports of Norway (Report No. CRC/C/NOR/CO/5-6); Consideration of Reports Submitted by States Parties Under Article 44 of the Convention (Report No. CRC/C/NOR/CO/4); and Consideration of Reports Submitted by States Parties Under Article 44 of the Convention (Report No. CRC/C/NOR/5-6).

[7] See Combined Twenty-Third and Twenty-Fourth Periodic Reports Submitted by Norway Under Article 9 of the Convention, Due in 2017 (Report No. CERD/C/NOR/23-24) and Consideration of Reports Submitted by States Parties under Article 9 of the Convention (Report No. CERD/C/NOR/CO/19-20).

[8] See Concluding Observations on the Fifth Periodic Report of Norway (Report No. E/C.12/NOR/CO/5).

[9] See Consideration of Reports Submitted by States Parties Under Article 40 of the Covenant (Report No. CCPR/C/NOR/CO/6) and Concluding Observations on the Seventh Periodic Report of Norway (Report No. CCPR/C/NOR/CO/7).

monitoring mechanisms, aligned with transnational and domestic mobilization, appear to have an effect on domestic legal reforms. States comply with international recommendations for different reasons (Krommendijk 2015), but Norway's general conduct on the matter suggests a dichotomy not easily overcome. The country seems to be committed to maintaining its reputation for abiding their international law obligations, thus maintaining its reputation in the international arena, but at the same time approves a series of stricter regulations related to migration (Mezzanotti 2018). The country's self-characterization as a human rights-friendly society, a "moral superpower" (Langford and Schaffer 2015), contradicts its current policies toward different aspects of a migrants' life.

As mentioned previously, there have been amendments to the current *Child Welfare Act*, and a further revision to strengthen the consideration for the child's religious, cultural, and linguistic background has been proposed (Thorud 2020). This recent reform of the NCWS will only prove itself effective from the perspective of migrants' recognition if NCWS' further work, in practice, enables the participation of migrants in its processes as peers and with parity in relation to Norwegians. Legal or policy change will only lead to social justice if institutionalized value patterns that impede parity of participation are replaced with ones that enhance it (Fraser 2000). In the sections to come, we aim to demonstrate how multifaceted and institutionalized misrecognition has been.

11.3 Social Justice, Recognition, and Participatory Parity

In our particular case, it is important to understand recognition within Fraser's larger theory of social justice. Social justice is understood as multidimensional, including spheres of redistribution, recognition, and – in Fraser's more recent works – representation. Participatory parity is the normative core of social justice, where "justice requires social arrangements that permit all (adult) members of society to interact with one another as peers" (Fraser 2008b: 16; Fraser and Honneth 2003: 36). There is a focus on the macro level, where overcoming barriers to justice "means dismantling institutionalized obstacles" that create barriers to the parity of participation (Fraser 2008b: 16). Fraser acknowledges that recognition and redistribution philosophically seem paradoxical; however, she integrates the relationship between recognition and redistribution through perspectival dualism, reconciling their seemingly contradictory nature with a single normative standard – participatory parity (Fraser 2007; Fraser and Honneth 2003). Like Fraser, we understand misrecognition as institutionalized status subordination (Fraser 2000; Fraser and Honneth 2003; Olson 2008). Status subordination prevents individuals from participating as a peer in social life, resulting in misrecognition; the remedy then lies in "establishing the misrecognized party as a full member of society, capable of participating on a par with other members" (Fraser 2004: 129).

Further, *two-dimensionally subordinated* groups face struggles with both misrecognition and maldistribution; in these cases, "neither a politics of redistribution alone nor a politics of recognition alone will suffice" (Fraser and Honneth 2003: 19). There is a need for child welfare systems to pay attention to both the cultural and material dimensions associated with social class, which seems to reinforce the relevance of Fraser's perspectival dualism "treating every practice as simultaneously economic and cultural, albeit not necessarily in equal proportions" (Fraser and Honneth 2003: 63). Lazzeri (2009) argues that recognition alone may appear to present culture as a main factor behind claims and conflicts while ignoring the significant role of economy. In the case of NCWS, this is important as culture is often recognized as a primary contributing factor to the overrepresentation of migrant families; however, when controlling for economic status, there is a reduction in overrepresentation of migrants in NCWS (Berg et al. 2017; Staer and Bjørknes 2015). Therefore, while recognition is one crucial element of Fraser's theory, it should be considered in relation to redistribution as a matter of social justice.

Again, participatory parity is crucial – the evaluative standard – in Fraser's theory when distinguishing struggles for recognition which are legitimate and necessary; only practices which promote parity are justified (Fraser 2001, 2007; Fraser and Honneth 2003). The norm of participatory parity highlights injustice and institutional hierarchies, focusing on structural factors including marginalization and exclusion (more related to misrecognition) and poverty and unemployment (more related to maldistribution) (Fraser and Honneth 2003). Misrecognition is understood as being "relayed through institutions and practices that impede parity" and cannot be solved solely through eliminating prejudice (Fraser 2001: 32–33; Fraser 2007: 309; Fraser and Honneth 2003: 38).

11.4 Participatory Parity and the Subject(s) of Justice

Does the norm of participatory parity actually fit a context like NCWS where there are inherent power imbalances between families and practitioners? And what kind of participation may be demanded by justice? Social workers are, by the nature of their work, in a position of power. When interviewing parents about their experiences with NCWS, Havnen et al. (2020: 142, 167) describe that while the experiences were largely positive, many parents described a fear of NCWS due to the power inherent in such an institution. In this position of power, practitioners can sometimes – whether deliberately or unconsciously – display cultural superiority in their work with clients from other cultures (Christiansen and Anderssen 2010; Fitzsimmons 1997; Piña and Canty-Swapp 1999; Zavirsek 2001). For example, Hennum (2011) describes the power of practitioners in document writing and how this strengthens ruling definitions of normality by reinforcing norms of cultural consensus on familial life and parenting. However, participation and dialogue seem to be emphasized as ideals to strive after in Nordic social work, placing participatory parity as an ideal for family-practitioner interaction.

In regard to parent(s) ability to participate, the 2018 *Child Welfare Act* is clear that NCWS has a duty to show respect for and, as long as possible, work together with parents (§1–7) (Havnen et al. 2020: 25). What can be seen as an attempt to equalize the power relations between practitioners and parents, there has been a recent movement[10] by the County Social Welfare Boards[11] toward a "dialogue process": here, the focus is on solutions and coming to an agreement through dialogue and between NCWS, parent(s), and – when the child is old enough – the child(ren) (Andersen 2020). In June 2020, the Norwegian parliament decided this dialogue process would be implemented as a permanent solution in all County Social Welfare Boards across the country (Andersen 2020). In this way, parent(s) ability to participate within the larger structure of NCWS has been enhanced. Depending on implementation, this may be a move toward increased parity of participation for migrant families and should be examined in future research. In addition, we understand the normative core of social justice – participatory parity – would require migrant families to be able to participate on par with nonmigrant families in regard to interactions within the larger children welfare system; this may, in some instances, require differential treatment (Fraser and Honneth 2003: 47; Olson 2008: 136–137). Therefore, justice – from a child right's perspective – may require that migrant parents are treated with respect, are informed about the process of NCWS, and are able to participate as full partners in interaction where assimilation to dominant norms is not a criterion for access to participation in the institution at large.

In addition, there are inherent power imbalances between children and adults. NCWS is often heralded for taking a child-centered approach; however, there remain limitations and challenges regarding participation of the child in NCWS. Havnen et al. (2020) examine the ability of children to participate and express their views in child welfare investigations. In 1123 NCWS investigations involving children from 0–17 years old, only 60% of children had conversations with NCWS; this percentage increased to 76% of children over 6 years old (Christiansen et al. 2019: 5; Havnen et al. 2020: 3, 8, 61, 158, 169). The 680 cases where there was a dialogue with the child were organized into themes based on the journal notes: only 31.5% of cases had themes which included a child's perspective or opinion (Havnen et al. 2020: 4, 9, 70–71, 161). While the content of conversations suggested limited participation of the child in regard to expressing his or her views, Havnen et al.'s (2020: 75, 162, 169) findings suggest conversations with the child had influence on which themes NCWS became worried about during the investigation.

How can, and should, children participate in decision-making processes in NCWS considering justice?[12] From a children's rights perspective, the Convention

---

[10] Five of the ten County Social Welfare Boards began to offer this dialogue process 4 years ago (Andersen 2020).

[11] If children are to be placed outside of the home against the wishes of the parent(s), a proposal must be brought before the County Social Welfare Board (Christiansen and Anderssen 2010).

[12] For more on participation of the child in child welfare services, see Archard and Skivenes (2009a); Archard and Skivenes (2009b); Havnen et al. (2020); Kosher and Ben-Arieh (2020); Skivenes and Strandbu (2004, 2006); Steinrem et al. (2018); Strandbu and Vis (2008); van Bijleveld et al. (2015); Vis (2014); Vis et al. (2012); Vis and Thomas (2009).

on the Rights of the Child promotes the right to express views on issues affecting the child's interests (Article 12.1) and the best interests of the child (Article 3.1) (Archard and Skivenes 2009a; Havnen et al. 2020: 18). In addition, the 2018 changes in the *Child Welfare Act* emphasize the child's right to participate or contribute (§1–6) (Havnen et al. 2020: 18). In Norway, children over 7 years of age have the right to express their meanings, perceptions, and opinions (Havnen et al. 2020: 8, 157). A promising tool for children's participation in practice includes a process model by Skivenes and Strandbu which specifically addresses children's participation in the context of NCWS, which includes (1) information, (2) forming of opinions, (3) expressing opinions, (4) inclusion of opinions, and (5) follow-up (Havnen et al. 2020: 34–35; Skivenes and Strandbu 2004, 2006; Strandbu and Vis 2008). In relation to the right of the child to be heard, Follesø and Mevik (2011) address the potentials of a participatory approach, with an appreciative attitude and understandings of dialogue inspired by Freire (2003) as equalizing measures between children and adults in NCWS. This is dependent on addressing individuals with the same respect "regardless of age, feelings, and understandings" where the experiences and opinions of children are taken seriously (Follesø and Mevik 2011: 110). Therefore, justice – from a child right's perspective – may require participation from the child in terms of having that opportunity to express his or her understandings, preferences, and choices through verbal and nonverbal communication; this participation should take into consideration the child's ethnic, religious, cultural, and linguistic background.[13]

This begs the larger question of who is the subject of justice? In social work discourse, the subject is often divided into three entities – the child(ren), the parent(s), and the family. For this chapter, we are primarily concerned with the interactions of the migrant family, although the voices of migrant parents have more representation within our chapter than the voices of children. Participatory parity has a focus on an individual's ability to participate on par; in this way, we understand the family as comprised of individuals who, ideally, have the opportunity to participate as peers in social arrangements (for children, this is according to their age and maturity). In this way, our understanding is that 3-year-old children would not participate as peers with their parents or child welfare practitioners, for example, but that migrant children and nonmigrant children of the same age have equal possibilities for participation. While our focus is on the larger societal context of the institution of NCWS and the subject of justice being migrant families, it is important to consider how Fraser's framework of social justice operates when the child or parent is the subject of justice, particularly how the normative core of social justice – participatory parity – operates (1) in the case of children of varying ages, (2) in the relationship between the child(ren) and parent(s), and (3) in the interactions between the child(ren) and NCWS. This is something which should be explored in

[13] This definition of participation is established in §3 of the 2014 Regulations on Participation and Child's Advocate (Norwegian: *Forskrift om medvirkning og tillitsperson*).

further research, in dialogue with the current literature on participation of the child in NCWS.

The case of NCWS is compatible with the all-subjected principle (Fraser 2008a, b, 2010; Fraser and Honneth 2003); that is, "all who are jointly subjected to a given governance structure have a moral standing as subjects of justice in relation to it" (Fraser 2008a: 411; Fraser 2008b: 96; Fraser 2010: 292–293). As the mandate of NCWS applies to all children who reside in Norway, regardless of their background, residency status, or citizenship, migrant families are subjected to the governing structure of NCWS and, thus, have relevant claims as subjects of justice in relation to it. In addition, Norway is a signatory of the Convention on the Rights of the Child, and the Norwegian national legislation explicitly states that if there is conflict between the national legislation and the Convention on the Rights of the Child, that the Convention on the Rights of the Child will have precedence.[14] The Convention grants the rights to each child within a State's jurisdiction, "irrespective of the child's or his or her parent's or legal guardian's race, color, sex, language, religion, political or other opinion, national, ethnic or social origin, property, disability, birth or other status" (Article 2). Therefore, regardless of legal status, migrant families have a well-established position as subjects of justice and claimants of rights in this case.

In examining the suitability of Fraser's theoretical approach to recognition, we suggest three particular areas a critical theory of recognition may be useful in regard to the social relations and processes in NCWS interactions with migrants, which will be further explored below.

11.5 Misrecognition as Institutionalized Subordination

How exactly does misrecognition occur through institutions – in rules, expectations, and practices? According to Fraser, misrecognition occurs through many different modes – codified in formal law; institutionalized via government policies, administrative codes, or professional practice; or institutionalized *informally* in associational patterns, long-standing customs, and social practices; in each of these cases, "an institutionalized pattern of cultural value constitutes some social actors as less than full members of society and prevents them from participating as peers" (Olson 2008: 135–136). For example, the norms of parenting in Norway can create a "referential standard" by which migrant parents are evaluated (Fylkesnes et al. 2018).

Misrecognition and status subordination occur when "institutionalized patterns of cultural value" exclude, make invisible, or constitute some actors as inferior or other (Fraser 2004: 129). Exactly what kinds of misrecognition do findings relating to NCWS indicate migrants may face? Not restricted to migrants, Blomberg et al.'s

[14] See the Human Rights Act (*Lov om styrking av. menneskerettighetenes stilling i norsk rett (Menneskerettsloven)*), §2(4) and §3.

(2011: 42) findings indicate that a "considerable number of referrals" to NCWS tend to be related to "a more general concern about the child's living conditions and a rather low number of referrals concerning abuse and physical neglect." In addition, there is a risk for families with a lower socioeconomic status or income to be pathologized: for example, in NCWS, Kojan (2011) has found that in lower-income families, the family is often problematized, while in higher-income families, it is often the child who is problematized. For families living in poverty, social workers tend to focus on personal flaws and individual failures (Korzeniewska et al. 2019; Ylvisaker et al. 2015). This may relate to child welfare practitioners comparing families to middle-class norms (Vagli 2009). These economic factors are relevant as migrants are overrepresented in families with lower socioeconomic status (Staer and Bjørknes 2015).

Another potential area for misrecognition includes normative understandings. Maboloc (2019) argues structural injustice may occur as a result of prejudices against certain groups which are reinforced through societal norms. Qureshi and Fauske (2010) and Qureshi (2009) found that NCWS's evaluation of the caring situation of children is influenced by dominant values and normative understandings about what is best for the child, relations between parents and children, and children's development. Anna used her own experiences of growing up in another country to reflect on cultural differences between the roles of parents and children and how this could create misunderstandings in Norway:

> *I see – and again because of my background – I see that in terms of expectations, there are lots of expectations on how much a child should do at home as opposed to how much a parent should do, right? Who is responsible for what. Me as a child, I had responsibility for picking up the younger ones at school, the kindergarten all of the time. But here [in Norway], I mean, it is not really a child's duty, it's a parental duty to try and pick up the child... so there is a clear cut – how do I put it – roles here for what a child does as opposed to what a parent should do. So that is different. And for me, working where I do today [at NCWS], I see that and it is very clear really. There are different ways of bringing up a child, depending on where you come from.* (Anna)

There are also cultural understandings relating to children's development which have a potential to create misunderstandings between migrant families and NCWS. When examining child development through a Norwegian normative framework, there are specific things that child welfare practitioners are expecting children to master by specific ages for what is perceived as "normal" development. Anna reflects on different understandings of child development, informed by her background:

> *Here [in Norway] it starts much earlier, you are free to go around, even as a child. Whereas other cultures, you know, they carry the baby until they are like ten months old. They are rarely on the floor. But it is not because they can't crawl, but just because you have to carry them, you still seem them as a baby. But here [in Norway], ten months already they are expected to follow this, and to do that, do that. So it is – for me it is very interesting to see the difference because of my background. I'm like yeah, but those kids that weren't able to crawl at ten months and all that – they are still doing fine. It is just a matter of what culture, or society you belong to really.* (Anna)

The issue of difference in understanding relations between parents and children and children's development is linked to questions of objectivity, relativism, morality, and ethics. Bø's (2015) findings suggest that child welfare practitioners in Norway often lack awareness that their understandings – which inform interpretations in child welfare cases – are largely influenced by "Western" ideology rather than a "universal" understanding; "minority mothers could easily get the impression that their own views about their children's upbringing were disqualified by the Norwegian social workers" (Bø 2015: 570). Bø (2015: 572) argues that "subjects like child welfare, social work and social policy are particularly saturated with ideology" and that "social workers ought to be able to identify the implicit perceptions of 'a good life' and 'good family conditions' that influence the viewpoints, policies and interventions of their own profession." How objectivity, relativism, morality, and ethics further relate to questions of recognition in the case of NCWS is an area which should be continued to be explored.

According to Wodak (2008: 60), institutional exclusion – that is, "*deprivation of access* through means of explicit or symbolic power" – is related to the normative nature of institutions, even when policies, procedures, regulations, objectives, mandates, and legal frames do not necessarily and formally empower them for such an outcome. In this way, institutional exclusion may occur despite policies which appear neutral (Wodak 2008). Occasionally, exclusion will not result from the institution's hard design or defined structure, but from its reiterated or habitual social practice, despite its formal structure and policies. This is compatible with Fraser's understanding of misrecognition being institutionalized (Fraser 2000; Olson 2008). John, a migrant parent and translator, reiterated his concern about Norwegian norms being a guiding factor in NCWS' interactions with migrant families. From his experience as a translator, he perceived little room for cultural nuances in Norwegian legislation and practice:

> *In many ways, the principles in NCWS in theory are good, but in practice they can target different communities. It is very institutionalized.* (John)

Recognition is appropriate in highlighting that institutions may relay misrecognition through practices which impede participatory parity for migrant families. We understand institutions as bodies which reinforce normative understandings, which may result in pathologizing certain groups. This can be further crystalized through the legislation which describes the mandate of NCWS. In the case of migrant interactions with NCWS, recognition may bring attention to the institutional nature of status subordination and highlight potential areas for improvement.

11.6 Equality, Sameness, and Difference in the Context of the Egalitarian Nordic Welfare State

The Nordic countries often characterize themselves as traditionally homogeneous in ethnic and political terms (Keskinen et al. 2019). This can mask significant political and constitutional differences within and between the Nordic nations (Forsberg and Kröger 2011; Langford and Schaffer 2015) as well as disguising past ethnic diversity (Keskinen et al. 2019; Osler and Lybæk 2014). Norwegian anthropologist Marianne Gullestad (2002a) argues that equality is often understood as sameness in Norway, which can result in inequalities for migrants.

"Immigrants" are asked to "become Norwegian," at the same time as it is tacitly assumed that this is something they can never really achieve. "They" are often criticized without much corresponding consideration of "our" knowledge of "their" traditions, or "our" ability and willingness to reflect critically upon "our" own. "We" ("Norwegians") are thus considered more advanced and hierarchically superior to "them" (Gullestad 2002a: 59).

From the outside, this may appear to be the result of a historically homogeneous national context; however, this perceived homogeneity has been the outcome of intense repression and assimilation policies of the Sámi and five national minorities in Norway (Gullestad 2002b; Keskinen et al. 2019). Gullestad (2002b) argues the thinking behind this assimilatory legacy persists with migrants, although not as obvious or extreme as previous policies. John was concerned about pressures for migrants to conform in the case of NCWS:

> *Here in Norway, people tell you that in Norway, we do it this way. It means that it is an assimilative system, even though in the regulation they take care of different cultures; but in practice, if you don't do it the Norwegian way, then you are the loser. It is silent assimilation... In institutions, it is assimilative.* (John)

This presents the question of assimilation as a criterion for access to social services. Policies focused on migration and integration in Norway, pursued in the name of egalitarianism, may have inequal outcomes; "migrants face policies that are supposed to grant equal access to the welfare state, and the same time as they are expected to become 'the same' as the prototypical Norwegian (that is, in accepting... specific forms of parenting...) in order to be recognized as equal" (Bendixsen et al. 2018: 25). Lopez (2007) found that families with minority background interacting with the Family Welfare Services in Norway (Norwegian: *Familievernet*) experienced an indirect pressure to conform to Norwegian majority norms and ideals, where Norwegian understandings of problems and solutions were presented as superior.

In applying this to recognition and the case of NCWS, the context of egalitarianism and equality-as-sameness may be understood as partially informing institutionalized patterns of cultural value within NCWS, as institutions cannot be removed from the society they operate within. Related to Bø's (2015) findings of little awareness of child welfare practitioners regarding embedded ideology behind understandings in child welfare, Fylkesnes et al. (2018) found that from the perspective of migrant parents, a hierarchy of knowledge exists in NCWS, where Norwegian culture and NCWS are presented as the "right way" and homeland practices or minority cultural practices are seen as the "wrong way." Recognition can provide insights here, as the ideal of recognition is "a difference-friendly world, where assimilation to majority or dominant cultural norms is no longer the price of equal respect" (Fraser 2001: 21; Fraser and Honneth 2003: 7). Recognition, thus, may be suitable in highlighting the structural issue of access and assimilation in the case of egalitarian Norway.

Ahmed's (2014) notion of "civilizing" may also be relevant when reflecting on signs of migrants' misrecognition, where migrants are viewed by the receiving society from a position of cultural superiority and are perceived to be in need of instruction on "civilized" values. This process can contribute to the construction of the lower social position as "other," thus evoking ideas of normality and pathology where the higher social position is seen as the norm and expressive of "the ideal," while the other side may be seen as "deficient" (Anthias 2001: 845). Within the context of NCWS, the findings of Johannesen and Appoh (2016) describe how African migrant parents in their study were aware that they were viewed as lacking in Norwegian parenting competence, which may relate to this perception of being in need of instruction on "civilized" values. John expressed how the perception of a homogeneous Norwegian society may be related to notions of "civilizing":

> *Norway is not a multicultural country yet – it is a multicultural country in the making. Here it is a homogenous society in perception. They have the perception of one culture and you have to in some way assimilate. Other cultures are not yet seen as developed.* (John)

Equality, sameness, and difference are also related to the issue of navigating between universalism and cultural relativism – an issue which has gained plenty of attention in NCWS. To what extent should universalist policies, standards, and expectations be enforced on everyone equally and when should a more culturally relativist approach be taken? NCWS has been accused of both not being culturally relative enough *and* of allowing too much space for cultural difference (Bredal 2009; Rugkåsa et al. 2017). While NCWS should be aware of the dangers of cultural relativism in practice – highlighted by Zavirsek (2001) including allowing severe abuses to continue to occur because it is perceived as culturally normative – there is a general understanding that culturally competent social work practice accounts for difference (Pemberton 1999). This was emphasized in the narrative of Anna:

> *The challenge remains that we, as social workers, also have to try to understand that even if it is not Norwegian, or as long as they don't do the same as we do, doesn't mean it is wrong. Not everyone sits at the table when they eat. Some people sit on the ground when they eat; some people prefer to sit on the ground when they eat. Some people, it is not very common to show affection for the child in other ways, but here it is shown in another way – but you know, I think we have to try to accept the difference.* (Anna)

The norm of participatory parity may be a suitable tool by which to try to navigate this tension between universalism and cultural relativism. In this way, practices may be distinguished between those which promote parity and those which do not.

This is an insight which may be valuable to enhance a more culturally competent social work practice.

As mentioned earlier, we understand law to be essential in the quest for recognition and social justice. From the perspective of international human rights law, the issues of equality, sameness, and difference relate to minority rights. The application of minority rights to migrants has been a special challenge within international law. Especially in Europe, migration has often been associated with the idea of a threat to national and local cultures, whereas the value of multicultural societies has been rejected and challenged (Pentassuglia 2009; Morondo Taramundi 2018). Other than individual rights, minority rights often address a set of additional rights to which minorities' members may be entitled. One of the characteristics of cultural rights is that they may circumscribe the individual right to participate in one's culture, which adds a collective dimension to its character of individual rights. In this context, the range and applicability of human rights are limited when it comes to minorities rights and pose a threat to migrants' participation as peers.

11.7 The Dynamic Nature of Culture

Fraser's theory of recognition gives space for culture to be viewed as dynamic. She is critical of identity politics – the affirmation of group identities – in that she expresses concern that identity politics may often simplify and reify group identities while encouraging separatism and ignoring intragroup domination (Fraser 2000; Fraser and Honneth 2003; Olson 2008). Fraser argues that within identity politics, there is little room for cultural dissidence, experimentation, and cultural criticism (Olson 2008). As opposed to identity politics, "what requires recognition is not group-specific identity, but the status of individual group members as full partners in social interaction," providing space for the dynamic nature of culture and not making assumptions that, for example, a child's culture is the same as his or her parents' or that a migrant's culture is unchanged from life in his or her home country (Fraser 2001: 24–25; Olson 2008: 134).

The narrative of Anna, a child welfare practitioner who grew up in a collectivist culture, highlights the dynamic nature of culture; she describes while working with migrant parents who also are used to more collectivist forms of life and points out the importance of integrating more individualistic aspects of decision-making into their parenting as well. Anna highlights it is important for children living in a very individualistic Norway to learn to navigate the system here for their well-being while still having that exposure and seeing the value in more collectivistic approaches. One of the five main issues child welfare practitioners described with multicultural social work in Norway, as described by Bø (2015), is cultural differences in parent-child relations. Regarding child-rearing practices, a tension between individualism and collectivism was articulated as a challenge in all six focus group interviews (Bø 2015: 566). A common theme for Ashley, a migrant parent, was trying to implement parenting aspects from both Norway and her home country:

In some way, I truly believe that I am trying to find this perfect middle between the Norwegian relaxed way and [home country] maybe too stressed. (Ashley)

This narrative highlights that culture is not a static entity for parents and that children may have a different culture than their parents. While a focus on group culture may reify difference, recognition gives space for culture to be dynamic and individuals to be different while still being treated with respect.

As previously mentioned, the dynamic nature of culture implies we cannot take for granted that children have the same culture as their parents. This may present challenges to some understandings of cultural competency or sensitivity, which may generalize characteristics of particular migrant groups. The importance of cultural competency is widely agreed upon in the literature as necessary, but skills needed for cultural competent practice are not clearly defined; therefore, participatory parity may provide a standard by which cultural diversity is recognized and individuals are represented within their culture – giving space for children to have different culture than their parents and for culture to be dynamic – and where the voice of the individual can be heard while taking into account structural factors (Olson 2008).

Considering this dynamic nature of culture, recognition, and the quest for justice in human rights, tensions may arise regarding "the best interest of the child." The UN Convention on the Rights of the Child follows the usual methodology set forth in the Universal Declaration of Human Rights and in the International Covenant on Civil and Political Rights and the International Convenant on Economic, Social and Cultural Rights. These instruments rely on the assumption that everyone is entitled to all the rights and freedoms without distinction of any kind (such as race, color, sex, language, religion, political or other opinion, national or social origin, property, birth, or other status). Such assumption aims at ensuring equal treatment to all children and a no discrimination policy while protecting children's identity, economic, social, and cultural rights. The Norwegian *Child Welfare Act* has the ultimate goal to protect children that are living in conditions which may be harmful to their health and development, ensuring they are raised in a safe environment. The main principle concerning children protection is the "best interest of the child," but among the principles of the child welfare system are also the protection of family ties, continuity in the child's upbringing, and that children should grow up with their parents (UN 1989).

The principle of the "best interest of the child" does not appeal to other specific legal concepts and therefore leaves an open door to a vast array of interpretations. This is particularly relevant for cases in which the child may not entirely share her or his parent's culture, values, behaviors, etc. This has been largely discussed under the label of "minorities within minorities" and the conflict of rights which may arise in such situations (collective vs. individual rights) (Morondo Taramundi 2018; Wahlbeck 2016). A definition of the "best interest of the child," under these circumstances, is a difficult task in a concrete situation. Morondo Taramundi mentions two types of "minorities within minorities": one in which a certain group challenges and revises cultural norms within the minority group and another in which a certain group tries to advance external interests. Only the second hypothesis, according to the author, poses the conflict between the collective right of the minority and the individual right to equality or freedom, given that the move toward external references constitutes an act of will toward a different cultural orientation than the one adopted by the minority group (Morondo Taramundi 2018). Nonetheless, the quest for justice in such situations may possibly become difficult to attain if the interests of the child are hardly made concrete by the child's manifestation of intent. If assessing how oppressive and harmful specific cultures can be toward adults who can elaborate on feelings and perceptions is a very hard task – especially considering the complex nature of the concept of culture – how much harder it will be to do so when children's perceptions, feelings, and best interests are at stake?

While recognition may be suitable for understanding aspects regarding the dynamic nature of culture, it may become murkier in regard to the "best interest of the child" – especially when considering "minorities within minorities." In some cases, participatory parity may serve as a tool by which to help navigate the "best interest." This is an area which should be explored further, with consideration for the child as a subject of justice and participation of the child in interactions with NCWS.

11.8 The Importance of Redistribution and Representation

According to the latest OECD Economic Survey (OECD 2019), Norway has one of the highest standards of living in the world; at a glimpse, this may give an impression that migrants in Norway only suffer from misrecognition. This may entice one to neglect issues of distributive justice; however, a more careful examination into migrants' living conditions in Norway and their interactions with NCWS seems to confirm the relevance of applying Fraser's two-dimensional conception of justice premised on the norm of participatory parity. Fraser describes redistribution as related to class politics as well as socioeconomic transformation regarding racial-ethnic injustice; in this way, (mal)distribution is "rooted in the economic structure of society" (Fraser and Honneth 2003: 12–13). Emma expresses her daily observations of economic differences between many Norwegian and migrant families:

> *Economic factors, you know, the difference in the financial aspect would – usually between the ethnic Norwegian family and an immigrant family is huge. And we know that economic stress impacts a family and... how they are able to meet their children's needs.* (Emma)

Many areas relating to redistribution are vital when considering migrant interactions with NCWS. According to Fauske et al. (2018), evidence supports that there are clear social inequalities between families in NCWS and those who are not. This is substantiated by Staer and Bjørknes (2015: 31) whose findings suggest that an overrepresentation of migrants in NCWS is associated with migrant variations in socioeconomic background. In Norway, children of a migrant background account for 53% of all children in low-income families (UN Committee on the Elimination of Racial Discrimination (UN CERD) 2017). Therefore, when addressing the need for redistribution, we are interested in the socioeconomic status of migrants which is related to many issues including that the unemployment rate for ethnic minorities is more than three times as high as the general population; that invitations to a job interview are 25% lower for persons with foreign names; and that people who may perceived to have a foreign background face discrimination in obtaining employment and receive lower salaries (UN CERD 2019).

Consequently, (mal)distribution plays a fundamental role when analyzing injustice in the interaction between migrant families and NCWS. Therefore, recognition in and of itself is limited in understanding the complicated relations between migrants and NCWS; however, when utilizing Fraser's theory on social justice considering both recognition and redistribution, a more comprehensive understanding may be acquired. Our understanding is that while misrecognition, discrimination, and racism are related to maldistribution, recognition and redistribution are not instances of and cannot be reduced to one another; both institutional misrecognition within NCWS and issues related to maldistribution affecting the socioeconomic status of migrants – related to rates of children living in low-income households, unemployment, additional difficulties in finding work, etc. – are also necessary to be addressed in the quest for social justice. NCWS has the capacity to work in addressing some aspects of redistribution through services which provide economic assistance for families; trust becomes an important issue relating to NCWS and redistribution, as emphasized by John:

> *They [NCWS] have many activities helping single moms or single dads who have economic problems... The problem is that they [the parents] have this negative image and even if they [the family] need help they don't talk to NCWS because it is perceived as something dangerous, because maybe they will create a case and take my kids.* (John)

In more recent works, Fraser also addresses representation as a third dimension of justice (Olson 2008). A third dimension of injustice occurs when individuals are "impeded from full participation by decision rules that deny them equal voice in public deliberations and democratic decision-making," constituting political injustice or misrepresentation (Fraser 2008b: 60). As law plays an essential role in the quest for social justice, we will focus on representation in terms of the political participation of migrants who are over 18 years of age. Child welfare legislation was described in the narratives as a factor which may reinforce or crystallize institutional misrecognition of migrants. We understand voting and political representation to influence the opportunities which migrants have in terms of politically influencing child welfare legislation.

Migrants have the right to vote in local elections in Norway after residing 3 years continuously on a visa which counts toward permanent residence and may vote in the federal elections only after acquiring Norwegian citizenship – a process which normally takes at least 7 years.[15] When it comes to political representation, migrants have low voting turnouts in Norway; in the 2019 local election, 45% of women with

[15] See the Norwegian Nationality Act (*Lov om norsk statsborgerskap (statsborgerloven)*), Chapter 3 §7(e).

a migrant background from Africa, Asia, and Latin America voted, compared to 74% of women without a migrant background (Statistics Norway 2019b). In the last two federal elections, approximately 55% of Norwegian citizens with a migrant background voted, compared to 80% of Norwegian citizens without a migrant background (Statistics Norway 2019a). In this way, migrants are underrepresented voices within the Norwegian legal system, which impacts the legislation by which NCWS operates. Further, it appears migrants have been underrepresented in voicing comments relating to amendments of the *Child Welfare Act*. Considering the limitations of recognition, we argue that a critical theory of recognition becomes stronger when understood within a larger theory of social justice which accommodates issues of (mal)distribution and (mis)representation.

11.9 Conclusion

While the individual narratives shared in this chapter cannot be generalized in and of themselves, concerns expressed by international human rights monitoring mechanisms and adjustments to domestic legislation indicate the current interactions of NCWS with migrant families have room for improvement. Reforms to domestic legislation are promising in terms of increasing participatory parity for migrant families; the effects of these changes should be investigated in the future.

Here, our aim has been to explore the extent to which Fraser's approach to social justice – with a particular focus on recognition – may be suitable for providing insights regarding misrecognition of migrant families in NCWS. Overall, we argue recognition is suitable when considering misrecognition as institutionalized subordination; equality, sameness, and difference; and the dynamic nature of culture. However, we argue the concept of recognition is strengthened when understood as one dimension of social justice, in conjunction with redistribution and representation. In a quest for justice relating to migrant interactions with NCWS, Fraser's conception of justice – accommodating "claims for social equality and defensible claims for the recognition of difference" – seems to be a suitable tool through which to gain insights on understanding the need for change in this case (Fraser 2004: 126). Lifting the focus from individual practitioners, Fraser's approach emphasizes institutional and societal aspects of injustice. The role of societal power relations is an arena which should continue to be investigated in the quest for justice in migrant interactions with NCWS.

Considering human rights monitoring mechanism reports and other research regarding the status of migrants in Norway, the structural nature of injustice against migrants appears to be part of a larger trend that is not solely limited to the case of NCWS. This is connected to problems of misrecognition, maldistribution, and misrepresentation. Fraser's conception of social justice may provide insights that aim toward a more socially just world, including in the case of recognition in migrant interactions with NCWS.

Acknowledgments We would like to extend our deepest gratitude to the editor, the research group Human Rights and Diversities at the University of South-Eastern Norway, and fellow participants at the Workshop on Recognition, Migration, and Critical Theory at the Centre for Ethics and Poverty Research (CEPR), University of Salzburg, for their thoughtful comments on earlier versions of this manuscript. The development of this chapter was enhanced through their critical engagement.

References

Ahmed, Sara. 2014. Not in the Mood. *New Formations: A Journal of Culture/Theory/Politics* 82: 13–28. https://doi.org/10.3898/NEWF.82.01.2014.

Andersen, Barbro. 2020. Ny løsning i barnevernssaker: – Slipper grufulle beskrivelser av familien. NRK Nordland, June 08.

Anthias, Floya. 2001. The concept of 'social division' and theorizing social stratification: Looking at ethnicity and class. *Sociology* 35 (4): 835–854. https://doi.org/10.1177/0038038501035004003.

Archard, David, and Marit Skivenes. 2009a. Balancing a child's best interests and a child's views. *International Journal of Children's Rights* 17: 1–21. https://doi.org/10.1163/157181808X358276.

———. 2009b. Hearing the child. *Child & Family Social Work* 14 (4): 391–399. https://doi.org/10.1111/j.1365-2206.2008.00606.x.

Bendixsen, Synnøve, Mary Bente Bringslid, and Halvard Vike, eds. 2018. *Egalitarianism in Scandinavia: Historical and Contemporary Perspectives*. Cham: Palgrave Macmillan.

Berg, Berit, Veronika Paulsen, Turid Midjo, Gry Mette D. Haugen, Marianne Garvik, and Jan Tøssebro. 2017. *Myter og Realiteter: Innvandreres Møter Med Barnevernet*. NTNU Samfunnsforskning.

Blomberg, Helena, Clary Corander, Christian Kroll, Anna Meeuwisse, Robeto Scaramuzzino, and Hans Swärd. 2011. A Nordic Model in Child Welfare? In *Social Work and Child Welfare Politics Through Nordic Lenses*, ed. Hannele Forsberg and Teppo Kröger, 29–45. Bristol: The Policy Press.

Bø, Bente Puntervold. 2015. Social Work in a Multicultural Society: New Challenges and Needs for Competence. *International Social Work* 58 (4): 562–574. https://doi.org/10.1177/0020872814550114.

Bredal, Anja. 2009. Barnevernet og minoritetsjenters opprør: Mellom det generelle og det spesielle. In *Over profesjonelle barrierer – Et minoritetsperspektiv på sosialt arbeid med barn og unge*, ed. Ketil Eide, Nausahd A. Qureshi, Marianne Rugkåsa, and Halvard Vike. Oslo: Gyldendal Norsk Forlag.

Christiansen, Øivin, and Norman Anderssen. 2010. From Concerned to Convinced: Reaching Decisions About Out-of-Home Care in Norwegian Child Welfare Services. *Child & Family Social Work* 15 (1): 31–40. https://doi.org/10.1111/j.1365-2206.2009.00635.x.

Christiansen, Øivin, Karen J. Skaale Havnen, Anette C. Iversen, Marte Knag Fylkesnes, Camilla Lauritzen, Reidunn Håøy Nygård, Fredrikke Jarlby, and Svein Arild Vis. 2019. *Når barnevernet undersøker. Delrapport 4*. Tromsø: UiT, RKBU Nord.

Dyrhaug, Tone, and Vibeke Sky. 2015. *Barn og unge med innvandrerbakgrunn i barnevernet 2012*. Oslo-Kongsvinger: Statistics Norway.

Ewick, Patricia, and Susan S. Silbey. 1995. Subversive Stories and Hegemonic tales: Toward a Sociology of Narrative. *Law and Society Review* 29 (2): 197–226. https://doi.org/10.2307/3054010.

Fauske, Halvor, Bente Heggem Kojan, and Anita Skårstad Storhaug. 2018. Social Class and Child Welfare: Intertwining Issues of Redistribution and Recognition. *Social Sciences* 7 (9). https://doi.org/10.3390/socsci7090143.

Fitzsimmons, Kevin. 1997. Is There an Inherent Bias in the Tradition of Helping Professions That Can Negatively Impact Multicultural Clients? In *Controversial Issues in Multiculturalism*, ed. Diane De Anda, 135–157. Needham Heights: Allyn and Bacon.

Follesø, Reidun, and Kate Mevik. 2011. In the Best Interest of the Child? Contradictions and Tensions in Social Work. In *Social Work and Child Welfare Politics: Through Nordic Lenses*, ed. Hannele Forsberg and Teppo Kröger, 97–111. Bristol: The Policy Press.

Forsberg, Hannele, and Teppo Kröger. 2011. Introduction. In *Social Work and Child Welfare Politics Through Nordic Lenses*, ed. Hannele Forsberg and Teppo Kröger, 29–45. Bristol: The Policy Press.

Fraser, Nancy. 2000. Rethinking Recognition. *New Left Review* 3 (3): 107–118.

———. 2001. Recognition Without Ethics? *Theory, Culture and Society* 18 (2–3): 21–42. https://doi.org/10.1177/02632760122051760.

———. 2004. Institutionalizing Democratic Justice: Redistribution, Recognition, and Participation. In *Pragmatism, Critique, Judgment: Essays for Richard J. Bernstein*, ed. Seyla Benhabib and Nancy Fraser, 125–148. Cambridge: The MIT Press.

———. 2007. Identity, Exclusion, and Critique: A Response to Four Critics. *European Journal of Political Theory* 6 (3): 305–338. https://doi.org/10.1177/1474885107077319.

———. 2008a. Abnormal Justice. *Critical Inquiry* 34 (3): 393–422. https://doi.org/10.1086/589478.

———. 2008b. *Scales of Justice: Reimaging Political Space in a Globalizing World*. Cambridge: Polity Press.

———. 2010. Who Counts? Dilemmas of Justice in a Postwestphalian World. *Antipode* 41 (1): 281–297. https://doi.org/10.1111/j.1467-8330.2009.00726.x.

Fraser, Nancy, and Axel Honneth. 2003. *Redistribution or Recognition? A Political-Philosophical Exchange*. London: Verso.

Freire, Paulo. 2003. *Pedagogy of the Oppressed*. New York: Continuum.

Fylkesnes, Marte Knag, Anette C. Iversen, Ragnhild Bjørknes, and Lennart Nygren. 2015. Frykten for barnevernet – En undersøkelse av etniske minoritetsforeldres oppfatninger. *Tidsskriftet Norges Barevern* 92: 80–96.

Fylkesnes, Marte Knag, Anette Christine Iversen, and Lennart Nygren. 2018. Negotiating Deficiency: Exploring Ethnic Minority Parents' Narratives About Encountering Child Welfare Services in Norway. *Child & Family Social Work* 23: 196–203. https://doi.org/10.1111/cfs.12400.

Gullestad, Marianne. 2002a. Invisible Fences: Egalitarianism, Nationalism, and Racism. *Journal of the Royal Anthropological Institute* 8 (1): 45–63.

———. 2002b. *Det norske sett med nye øyne. Kritisk analyse av norsk innvandringsdebatt*. Oslo: Universitetsforlaget.

Havnen, Karen J. Skaale, Øivin Christiansen, Eirinn Hesvik Ljones, Camilla Lauritzen, Veronika Paulsen, Frederikke Jarlby, and Svein Arild Vis. 2020. Å medvirke når barnevernet undersøker: En studie av barn og foreldres medvirkning i barnevernets undersøkelsesarbeid. In *Delrapport 5*. Tromsø: UiT, RKBU Nord.

Hennum, Nicole. 2011. Controlling Children's Lives: Covert Messages in Child Protection Service Reports. *Child & Family Social Work* 16 (3): 336–344. https://doi.org/10.1111/j.1365-2206.2010.00744.x.

Ipsos. 2017. *Tillit til barnevernet blant personer med innvandrerbakgrunn*. Oslo: Ipsos.

Johannesen, Berit Overå, and Lily Appoh. 2016. 'My Children Are Norwegian But I Am a Foreigner': Experiences of African Immigrant Parents within Norwegian Welfare Society. *Nordic Journal of Migration Research* 6 (3): 158–165. https://doi.org/10.1515/njmr-2016-0017.

Keskinen, Suvi, Unnur Dís Skaptadóttir, and Mari Toivanen, eds. 2019. *Undoing Homogeneity in the Nordic region: Migration, Difference and the Politics of Solidarity*. Milton Park/Abingdon/Oxon: Routledge.

Kojan, Bente Heggem. 2011. Norwegian Child Welfare Services: A Successful Program for Protecting and Supporting Vulnerable Children and Parents? *Australian Social Work* 64 (4): 443–458. https://doi.org/10.1080/0312407X.2010.538069.

Korzeniewska, Lubomiła, Marta Bivand Erdal, Natasza Kosakowska-Berezecka, and Magdalena Żadkowska. 2019. *Trust Across Boarders: A Review of the Literature on Trust, Migration and Child Welfare Services*. Gdańsk: PRIO.

Kosher, Hanita, and Asher Ben-Arieh. 2020. Children's Participation: A New Role for Children in the Field of Child Maltreatment. *Child Abuse & Neglect*. https://doi.org/10.1016/j.chiabu.2020.104429.

Krommendijk, Jasper. 2015. The Domestic Effectiveness of International Human Rights Monitoring in Established Democracies. The Case of the UN Human Rights Treaty Bodies. *The Review of International Organizations* 10: 489–512. https://doi.org/10.1007/s11558-015-9213-0.

Langford, Malcolm, and Johan Karlsson Schaffer. 2015. *The Nordic Human Rights Paradox: Moving Beyond Exceptionalism*. University of Oslo Faculty of Law Research Paper No. 2013-25.

Lazzeri, Christian. 2009. Recognition and Redistribution: Rethinking N. Fraser's Dualistic Model. *Critical Horizons* 10 (3): 307–340. https://doi.org/10.1558/crit.v10i3.307.

Lopez, Gro Saltnes. 2007. *Minoritetsperspektiver på norsk familievern. Klienters erfaringer fra møtet med familievernkontoret*. Nova: Oslo.

Maboloc, Christopher Ryan. 2019. What Is Structural Justice? *Philosophia* 47: 1185–1196. https://doi.org/10.1007/s11406-018-0025-3.

Mezzanotti, Gabriela. 2018. Entre *non-entrée* e *non-refoulement*: Uma análise crítica do discurso norueguês em sua atual gestão migratória (Between *non-entree* and *non-refoulement*: A critical analysis of the Norwegian discourse in its current migration governance). In Migrantes forçados: conceitos e contextos, eds. Liliana Lyra Jubilut, Fernanda de Magalhães Dias Frinhani, and Rachel de Oliveira Lopes, 507–531. Boa Vista: Editora da UFRR.

Morondo Taramundi, Dolores. 2018. Minorities-Within-Minorities Frameworks, Intersectionality and Human Rights: Overlapping Concerns or Ships Passing in the Night? In *Ethno-Cultural Diversity and Human Rights*, ed. Gaetano Pentassuglia, 256–285. Leiden: Brill. https://doi.org/10.1163/9789004328785_010.

NOAS. 2019. *Norway's Asylum Freeze*. https://www.noas.no/wp-content/uploads/2019/02/Storskog-rapport-februar-2019.pdf. Accessed 24 Apr 2020.

OECD. 2019. OECD Economic Surveys Norway.

Olson, Kevin, ed. 2008. *Adding Insult to Injury: Nancy Fraser Debates Her Critics*. Brooklyn: Verso.

Osler, Audrey, and Lena Lybæk. 2014. Educating 'The New Norwegian We': An Examination of National and Cosmopolitan Education Policy Discourses in the Context of Extremism and Islamophobia. *Oxford Review of Education* 40 (5): 543–566. https://doi.org/10.1080/03054985.2014.946896.

Øverlien, Carolina. 2012. *Vold i hjemmet – barns strategier*. Oslo: Universitetsforlaget.

Pemberton, Deirdre. 1999. Fostering in a Minority Community – Travellers in Ireland. In *Fostering kinship: An International Perspective on Kinship in Foster Care*, ed. Roger Greeff, 167–180. Aldershot: Ashgate.

Pentassuglia, Gaetano. 2009. *Minority Groups and Judicial Discourse in International Law: A Comparative Perspective*, International Studies in Human Rights. Leiden: Brill | Nijohff.

Piña, Darlene L., and Laura Canty-Swapp. 1999. Melting Multiculturalism? Legacies of Assimilation Pressures in Human Service Organizations. *The Journal of Sociology & Social Welfare* 26 (4): 87–113.

Qureshi, Naushad A. 2009. Kultursensitivitet i profesjonell yrkesutøvelse. In *Over Profesjonelle Barrierer – Et minoritetsperspektiv på sosialt arbeid med barn og unge*, ed. Ketil Eide, Nausahd A. Qureshi, Marianne Rugkåsa, and Halvard Vike. Oslo: Gyldendal Norsk Forlag.

Qureshi, Naushad A., and Halvor Fauske. 2010. Barnevernfaglig arbeid med minoritetsetniske familier. In *Integrasjon og mangfold. Utfordringer for sosialarbeideren*, ed. Mehmed S. Kaya, Asle Høgmo, and Halvor Fauske. Oslo: Cappelen Akademisk Forlag.

Rugkåsa, Marianne, Signe Ylvisaker, and Ketil Eide. 2017. *Barnevern i et minoritetsperspektiv – Sosialt arbeid med barn og familier*. Oslo: Gyldendal Akademisk.

Skivenes, Marit. 2014. How the Norwegian Child Welfare System Approaches Migrant Children. In *Child Welfare Systems and Migrant Children: A Cross Country Study of Policies and Practice*, ed. Marit Skivenes, Ravinder Barn, Katrin Kriz, and Tarja Pösö, 39–61. Oxford: Oxford University Press.

Skivenes, Marit, and Astrid Strandbu. 2004. Barn og Familieråd. En analyse av det organisatoriske rammeverket for barns medvirkning i barnevernet. *Tidsskrift for Velferdsforskning* 7 (4): 213–228.

———. 2006. A Child Perspective and Children's Participation. *Child, Youth and Environments* 16 (2): 10–27. https://doi.org/10.7721/chilyoutenvi.16.2.0010.

Staer, Trine, and Ragnhild Bjørknes. 2015. Ethnic Disproportionality in the Child Welfare System: A Norwegian National Cohort Study. *Child and Youth Services Review* 56: 26–32. https://doi.org/10.1016/j.childyouth.2015.06.008.

Statistics Norway. 2019a. *Innvandrere og stortingsvalget* 2017. https://www.ssb.no/valg/artikler-og-publikasjoner/innvandrere-og-stortingsvalget-2017. Accessed 17 Apr 2020.

———. 2019b. Langt flere unge stemte i årets lokalvalg. https://www.ssb.no/valg/artikler-og-publikasjoner/langt-flere-unge-stemte-i-arets-lokalvalg. Accessed 17 Apr 2020.

Steinrem, Ida, Gunnar Toresen, Glorija Proff, and Alexander Proff, eds. 2018. *Barnas barnevern: Trygt, nyttig og samarbeidende for barn*. Oslo: Universitetsforlaget.

Strandbu, Astrid, and Svein Arild Vis. 2008. *Barns deltakelse i barnevernssaker*. Tromsø: Barnevernets Utviklingssenter i Nord-Norge.

Thorud, Espen. 2019. *Immigration and integration 2017–2018: Report for Norway to the OECD*. Oslo: Norwegian Ministries.

———. 2020. *Immigration and Integration 2018–2019: Report for Norway to the OECD*. Oslo: Norwegian Ministries.

UN Committee on the Elimination of Racial Discrimination. 2017. *Combined Twenty-Third and Twenty-Fourth Periodic Reports Submitted by Norway Under Article 9 of the Convention, Due in 2017*. (Report No. CERD/C/NOR/23-24).

———. 2019. Concluding Observations on the Combined Twenty-Third and Twenty-Fourth Periodic Reports of Norway. (Report No. CERD/C/NOR/CO/23-24).

UN Committee on the Rights of the Child. 2010. Consideration of Reports Submitted by States Parties Under Article 44 of the Convention. (Report No. CRC/C/NOR/CO/4).

———. 2018. Concluding Observations on the Combined Fifth and Sixth Periodic Reports of Norway. (Report No. CRC/C/NOR/CO/5-6).

UN Human Rights Council. 2019. *Report of the Working Group on the Universal Periodic Review: Norway*. (Report No. A/HRC/42/3).

United Nations. 1989. *Convention on the Rights of the Child*. New York: UNICEF.

Vagli, Åse. 2009. *Behind Closed Doors: Exploring the Institutional Logic of Child Protection Work*. PhD dissertation, University of Bergen.

van Bijleveld, Ganna G., Christine W.M. Dedding, and Joske F.G. Bunders-Aelen. 2015. Children's and Young People's Participation Within Child Welfare and Child Protection Services: A State-of-the-Art Review. *Child & Family Social Work* 20 (2): 129–138. https://doi.org/10.1111/cfs.12082.

Vis, Svein Arild. 2014. *Factors That Determine Children's Participation in Child Welfare Decision Making: From Consultation to Collaboration*. PhD dissertation, UiT The Arctic University of Norway.

Vis, Svein Arild, and Nigel Thomas. 2009. Beyond Talking – Children's Participation in Norwegian Care and Protection Cases. *European Journal of Social Work* 12 (2): 155–168. https://doi.org/10.1080/13691450802567465.

Vis, Svein Arild, Amy Holtan, and Nigel Thomas. 2012. Obstacles for Child Participation in Care and Protection Cases – Why Norwegian Social Workers Find It Difficult. *Child Abuse Review* 21 (1): 7–23. https://doi.org/10.1002/car.1155.

Wahlbeck, Östen. 2016. True Finns and Non-True Finns: The Minority Rights Discourse of Populist Politics in Finland. *Journal of Intercultural Studies* 37 (6): 574–588. https://doi.org/10.1080/07256868.2016.1235020.

Wodak, Ruth. 2008. 'Us' and 'them': Inclusion and Exclusion – Discrimination Via Discourse. In *Identity, Belonging and Migration*, ed. Gerard Delanty, Ruth Wodak, and Paul Jones, 54–77. Liverpool: Liverpool University Press.

Ylvisaker, Signe, Marianne Rugkåsa, and Ketil Eide. 2015. Silenced Stories of Social Work with Minority Ethnic Families in Norway. *Critical and Radical Social Work* 3 (2): 221–236. https://doi.org/10.1332/204986015X14331614908951.

Zavirsek, Darja. 2001. Lost in Public Care: The Ethnic Rights of Ethnic Minority Children. In *Beyond Racial Divides*, ed. Lena Dominelli, Walter Lorenz, and Halur Soydan, 171–188. Hampshire: Ashgate.

Part III
Recognition and Refugees

Chapter 12
Epistemic Injustice and Recognition Theory: What We Owe to Refugees

Hilkje C. Hänel

Abstract This paper starts from the premise that Western states are connected to some of the harms refugees suffer from. It specifically focuses on the harm of acts of misrecognition and its relation to epistemic injustice that refugees suffer from in refugee camps, in detention centers, and during their desperate attempts to find refuge. The paper discusses the relation between hermeneutical injustice and acts of misrecognition, showing that these two phenomena are interconnected and that acts of misrecognition are particularly damaging when (a) they stretch over different contexts, leaving us without or with very few safe spaces, and (b) they dislocate us, leaving us without a community to turn to. The paper then considers the ways in which refugees experience acts of misrecognition and suffer from hermeneutical injustice, using the case of unaccompanied children at the well-known and overcrowded camp Moria in Greece, the case of unsafe detention centers in Libya, and the case of the denial to assistance on the Mediterranean and the resulting pushbacks from international waters to Libya as well as the preventable drowning of refugees in the Mediterranean to illustrate the arguments. Finally, the paper argues for specific duties toward refugees that result from the prior arguments on misrecognition and hermeneutical injustice.

Keywords Refugees · Forced migration · Recognition theory · Epistemic injustice · Hermeneutical injustice

---

H. C. Hänel (✉)
Department of Political Science, Potsdam University, Potsdam, Germany
e-mail: hilkje.charlotte.haenel@uni-potsdam.de

G. Schweiger (ed.), *Migration, Recognition and Critical Theory*, Studies in Global Justice 21, https://doi.org/10.1007/978-3-030-72732-1_12

12.1 Introduction

Serena Parekh (2020) has recently called for a reframing of the refugee crisis. She argues that framing the discussion in terms of rescue masks the ways in which Western states are connected to the harms that refugees suffer from. Investigating the legal obligations and moral duties that we have toward refugees, political philosophers overwhelmingly focus on whether there are any legal obligations or moral duties such that Western states should rescue refugees from harms that are completely unconnected to them. Parekh argues that we should distinguish between primary and secondary harms. Primary harms are the harms that refugees suffer from in their home country and they are often the reason for refugees to become refugees in the first place. Although, this is not undisputed, many would think that Western states bear no direct connection to these primary harms. Yet, according to Parekh, there are also so-called secondary harms that refugees suffer from. These are harms that "refugees experience in refugee camps, informal urban spaces or as they seek asylum" (Parekh 2020, 22). And, Parekh proceeds to show, these harms are the direct and indirect consequences of the policies that Western states implement—often to keep refugees away as "Fortress Europe" blatantly showed again over the Easter weekend in 2020 (Kulikowski 2020) and as Trump keeps demonstrating with the family separation policy (Aguilera 2019). Hence, when we think about the legal obligations and moral duties that Western states have toward refugees, we fail to capture the whole picture unless we consider the ways in which Western states are causally connected to at least some of the harms that refugees have to endure.

This paper starts from the premise that Western states are connected to some of the harms refugees suffer from.[1] It focuses on the harm of acts of misrecognition and its relation to epistemic injustice and, with the help of three recent examples, shows how refugees suffer from both in their desperate attempts to find refuge. I'll proceed as follows: in Sect. 1, I discuss the relation between hermeneutical injustice and acts of misrecognition drawing on my own arguments in Hänel (2020). I show that acts of misrecognition are particularly damaging when (a) they stretch over different contexts, leaving us without or with very few safe spaces, and (b) they dislocate us, leaving us without a community to turn to. And that instances of hermeneutical injustice are deeply entangled with relations of misrecognition in at least two senses: they (i) are the result of a prior history of misrecognition (especially, hermeneutical marginalization) and (ii) render speakers vulnerable to future

[1] I here concentrate on so-called Western states as the group of relatively wealthy, liberal democratic states, mainly because they play a powerful role in producing and reproducing the harmful policies that refugees suffer from, they are the most discussed destination that some (though less than is often thought, see Betts and Collier 2017) refugees are trying to reach, and they often have the economic, political, and cultural means to fulfill their duties and obligations toward refugees. This is particularly important for my discussion of such duties in Sect. 3, and, hence, when I speak of institutional duties, I refer to Western states as such and when I speak of individual duties I focus on citizens of these states.

acts of misrecognition. Next, in Sect. 2, I consider the ways in which refugees experience acts of misrecognition and suffer from hermeneutical injustice, using three specific examples to illustrate my ideas: the case of unaccompanied children at the well-known and overcrowded camp Moria in Greece, the case of unsafe detention centers in Libya, and the case of the denial to assistance on the Mediterranean and the resulting pushbacks from international waters to Libya as well as the preventable drowning of refugees in the Mediterranean. I argue that all three cases are examples of misrecognition fortified by hermeneutical injustice. Finally, in Sect. 3, I consider our duties toward refugees, focusing on refugees trying to cross the Mediterranean and on refugees located in European and Libyan refugee camps and detention centers, and our specific duties toward them regarding hermeneutical justice and proper recognition.

12.2 The Relation Between Hermeneutical Injustices and Misrecognition

12.2.1 Taking a Look at Misrecognition

Recognition theory is both a normative and a psychological theory; it says something about how we treat people and how we should treat them. For example, if we recognize another human being as a full person, we do not only acknowledge that they have this particular person-status, but we also treat them differently, more respectfully, for having this status. That is, when we recognize someone as a full person, we have certain obligations toward them.[2] So much for the normative side. According to the psychological theory, we are dependent on recognition from others and from society in general to develop our own identities. From very early on, we need the feedback from others to be able to develop fully. Many recognition theorists distinguish between at least three forms of recognition: respect, esteem, and love. These three forms of recognition correspond to the different spheres in which recognition can be given, society in general, interpersonal encounters, family, and friends (cf. Honneth 1992). When we are met with misrecognition from others, when the norms and values of a society depict us in unfavorable ways, or when recognition is absent, we can be seriously hindered in developing our own identities, our values, and our personal projects. Misrecognition or the absence of recognition alienates us from who we are, which is why Taylor describes recognition as a "vital human need" (1994, 26).

What then is misrecognition? Both Taylor and Honneth, currently the most prominent advocates of contemporary recognition theory and both deeply influenced by Hegel, have surprisingly little to say about the complementary concept of *mis*recognition. For Taylor, misrecognition is simply and straightforwardly the

[2] I say more about the relation between recognition and moral obligations in Sect. 3.

opposite of recognition, making misrecognition the absence of recognition (cf. Taylor 1994, 25). According to Honneth, misrecognition is "the withdrawal of social recognition, in the phenomena of humiliation and *disrespect*" (Fraser and Honneth 2003, 134). Yet, this can be interpreted in a number of ways; for example, Laitinen proposes that misrecognition is best understood as failing to adequately respond to another person and their accompanying normatively significant features (cf. Laitinen 2012), Pilapil argues that misrecognition is a failure to appropriately acknowledge another's moral status as a person (cf. Pilapil 2012), Rorty can be read as claiming that misrecognition is "an absence of moral imagination in seeing others as like ourselves" (Martineau et al. 2012, 4), and according to Bernstein, misrecognition is a failure to being "recognized as being of equal worth," a failure of not "being 'accorded' a status" (Bernstein 2005, 311). Not being recognized as of equal worth is a source of moral injury. For the purposes of this paper, we do not need to find an ultimate theory of the precise character of misrecognition; rather, it is enough to understand it as a moral injury due to failures of granting appropriate status or personhood, a failure with lasting consequences for who we (think we) are.[3]

Fanon (1952/2008) has provided a sharp and illuminating account of the relationship between misrecognition and personal identity in regard to racism and colonialism. Fanon employs a psychoanalytic strategy to describe the effects of anti-Black racism on embodiment and selfhood of Black people. He argues for, what he calls, the inferiority complex, his account of how Black people internalize anti-Black racism. It is only when the Black child sees itself "reflected in the mirror of white culture" (Oliver 2001, 33) that its normal psychological development stops and it becomes nothing but skin. Seeing oneself in this mirror brings about experiences of alienation and misrecognition; it takes away Black individuality and forces Black people into an identity that is not their own. A theory that Du Bois cashed out as "double consciousness" (1989). Psychological development and social structures are deeply entangled in so far as the racist social structures create racist psychological structures. Deeply influenced by Hegel, Fanon writes: "As long as he has not been effectively recognized by the other, it is the other who remains the focus of his actions. His human worth and reality depend on this other and on his recognition by the other. It is in this other that the meaning of his life is condensed" (Fanon 1952/2008, 191).

Acts of abuse and rape, denial of rights, and denigration or insults—that is, in short, acts of misrecognition—can lead to the destruction of a person's identity, their capacity to define themselves, make plans, and have personal values.[4] In other

[3] While I think it can sometimes be useful to distinguish between misrecognition, absence of recognition, and, broadly, failures of recognition depending on what needs to be described, for the sake of this paper, I will use them synonymously. The important point to drive home is that all failures of recognition have significant consequences for our self-development and all failures of recognition are fortified by hermeneutical injustice—a point I am making below. I say more about recognition as a moral injury in Sect. 3.

[4] I here assume that different acts of misrecognition can harm a person's identity. For an explanation and a critique of Honneth's view that distinct acts of misrecognition harm distinct parts of one's identity, see Hänel 2020, 7.

words, recognition has a causal relation to self-recognition and therefore for our capabilities to define ourselves and to make and value our own projects (cf. Honneth 1992). Especially if, as is the case with racism and colonialism, such acts of misrecognition are experienced over a long period of time and in different social contexts. Furthermore, the causal relation between recognition and self-recognition has implications for making coherent choices "about what is good or bad, what is worth doing and what has meaning and importance" (Taylor 1989, 28). And this again has consequences for society as a whole; recognition provides the necessary means for social living and moral progress (cf. Honneth 1995).

Not all acts of misrecognition are equally damaging. Instead, whether an act of misrecognition is deeply harmful to one's personal identity and ongoing development of that identity depends on at least two factors: first, the general constitution of the person (including their personal history and mental health), and, second, the source of the misrecognition. Let me briefly provide two examples that can explicate these factors. First, having experienced recognition in terms of love, esteem, and respect from parents, teachers, friends, and the wider society while growing up provides us with self-confidence, self-respect, and self-love. Thus, it helps us build a strong personality that can thwart against future acts of misrecognition. It builds the stamina that is necessary to protect ourselves from others' misrecognition. Furthermore, having had such an upbringing usually implies that we have communities and contexts in which we find attentive listeners, safe spaces, and help to heal, which also implies that the acts of misrecognition we experience are either individual acts or are restricted to some contexts; they do not take over our whole life. For example, the misrecognition that I, as a white woman, might experience in the context of academia is, although hurtful, restricted to one context. And despite the fact that it is not a single, individual instance, but cumulative acts, I count myself lucky to have attentive listeners and safe spaces with my family and friends and within certain subcultures of academic philosophy that help to get through such misrecognition.

Second, the people we respect and the people close to us can often hurt us more than strangers; they shatter our self-recognition and our worldview in more extreme ways. Susan Brison (2002) writes about the losing of trust toward others and the world after a traumatic instance. We lose trust in others and in the view that the world is just—that is, that we will not be violated by others unless we behave morally corrupt—when we have to endure violence from another person. Often this trust is destroyed even more fundamentally when we know the person who is harming us; we do not generally expect those we love to be capable of cruelty.

That said, it might be the case that some cumulative acts of denigration or insult weight heavier on some persons than a single act of assault. Yet, an interplay of acts of assault and acts of denigration and insult over a long period of time can break nearly anyone. This is partly because acts of assault are harder to overcome without the help of others as well as contexts and time of healing. In fact, Brison (2002) argues that ourselves are undone or shattered in trauma by disrupting our memories; past and present are disconnected and, thus, the trauma victim loses the ability to imagine a future (self). Yet, Brison also states that survivors often do master their traumatic memories by finding their own narrative and, with the help of others, becoming a subject of their own speech again. She writes, "[t]he act of bearing witness to the trauma facilitates this shift, not only by transforming traumatic memory into a narrative that can then be worked into the survivor's sense of self and view of the world, but also by reintegrating the survivor into a community, reestablishing connections essential to selfhood" (Brison 2002, 68). To pick up our shattered selves and to be able to regain the strength, the self-respect, and self-love is a process that relies on the help of others' recognition. Hence, we can say that acts of misrecognition are particularly damaging when (a) they stretch over different contexts, leaving us without or with very few safe spaces, and (b) they dislocate us, leaving us without a community to turn to. Examples of the first are well-known and include social structures interwoven with racism, sexism, or ableism (to name only a few). In an ableist society, a disabled person will find only few safe spaces in which they do not suffer from misrecognition. Same for Black people in racist societies and women, trans, or gender non-conforming individuals in sexist or misogynist societies. However, many—though unfortunately not all[5]—have communities in which they find help, recognition, and safe spaces.

12.2.2 Misrecognition and Hermeneutical Injustice: A Causal Relation

The above discussion runs parallel to the discourse on hermeneutical injustice, both in the sense that structural or systematic hermeneutical injustice is particularly damaging in comparison to incidental hermeneutical injustice and in regard to the ways in which different communities can provide forms of knowledge that are missing in the general social structure. Hermeneutical injustice is an epistemic wrong; in Fricker's words, "a wrong done to someone specifically in their capacity as a knower" (Fricker 2007, 1). We are victims of hermeneutical injustice if a significant experience of ours is "obscured from collective understanding owing to a structural identity prejudice in the collective hermeneutical resource" (2007, 155). First, according to Fricker, hermeneutical injustice is systematic if the hermeneutical marginalization of the knowing subject, which yields a discrepancy in whose experiences are represented by the collective hermeneutical resource, "tracks the subject through a range of different social activities besides the hermeneutical" (Fricker 2007, 156). Fricker discusses the example of Joe, the main protagonist of Ian

[5] Black, disabled, trans, or gender non-conforming individuals are statistically in greater danger than others to become homeless or estranged from their communities, and, hence, many indeed suffer from permanent misrecognition *and* dislocation. Yet, Black, disabled, trans, and queer communities are also extremely emancipatory in their abilities to care for each other and provide safe spaces for each other, not because they are better equipped to be caring but because they had a greater need to develop care due to the historically grown systems of racism, sexism, and ableism.

McEwan's novel *Enduring Love*, who is stalked by a young man. While Joe otherwise enjoys social power and privilege, the experience of being stalked alienates him from his intimate relationships as well as the broader society. Similarly with misrecognition, which is particularly damaging for ourselves if it spreads over various contexts and dislocates us from our own communities.

Second, many have pointed out that Fricker's analysis of collective hermeneutical resources misrepresents the fragmented character of such epistemic resources. José Medina (2012) argues for what he calls "polyphonic contextualism"; hermeneutical resources demand a pluralist analysis of different interpretative communities and practices instead of being assumed to uniformly exist for everyone. While Fricker initially assumed that there is one collective pot in which all hermeneutical resources are stored and different social groups are differently positioned in regard to contributing to the collective pot, Medina argues that there are many different pots in different communities to which differently situated people have different ways of contributing and accessing. For example, the queer community has adequate resources to understand and articulate specific forms of sexual violence. However, the wider society might not have access to such resources, and the people from the queer community might not have the social power to contribute to other resources in similar ways. Hence, speaking of a collective pot of hermeneutical resources ignores the ways in which we are all members of different intersecting social groups with different hermeneutical resources.

Kristie Dotson (2014) instead speaks of "dominant epistemic resources." According to her, there are indeed dominant and shared epistemic resources to which most have access to, but these are shaped by the contributions of only some people, bringing it about that a whole range of experiences of others are not represented and cannot be articulated well. Some people count as more credible than others depending on their social positions and social group membership; and those deemed less credible are required to operate with a set of concepts and assumptions which makes communication impossible because it fails to account for the full range of experiences these persons have. A similar thought is articulated by Matt Congdon when he writes that when faced with moral injuries, we often have to make a choice between articulating our "grievance by drawing upon either authoritative yet inadequate conceptual resources or adequate yet non-authoritative conceptual resources" (2016: 820). Framing hermeneutical resources as "dominant" versus "marginal" can illustrate the problem Fricker is interested in without making the problematic assumption that there really only is one collective resource. Furthermore, it highlights the ways in which different communities can provide forms of knowledge that are missing in the general social structure due to dominant but inadequate resources. Similarly with the problem of misrecognition, we do not experience acts of recognition or misrecognition in only one context; rather, we experience them in various contexts. However, some of these contexts are personally more important—communities in which we are "at home"—and some are socially more important, contexts that shape the moral outlook of our dominant frameworks. While misrecognition in the first might be personally more damaging, misrecognition in the second contributes to the systematic or structural character of the misrecognition we receive and, thus, spills over into other contexts.

Instances of hermeneutical injustice are also deeply entangled with relations of misrecognition in at least two senses: they (i) are the result of a prior history of misrecognition (especially, hermeneutical marginalization) and (ii) render speakers vulnerable to future acts of misrecognition (Hänel 2020, 2). Hermeneutical injustice can fortify misrecognition in so far as it renders speakers vulnerable to future misrecognition, and it is often the result of a prior history of acts of misrecognition. Being hermeneutically marginalized (the precondition for suffering from hermeneutical injustice) is partly due to a lack of respect and esteem that we owe persons as knowers. Furthermore, hermeneutical injustice can compromise recognition by others and thus curtail self-recognition. Hermeneutical injustice occurs when one is unable to understand and articulate a (harmful) experience due to a structural identity prejudice that excludes certain groups from properly participating in our knowledge practices. In some cases, when we fail to articulate our experiences to others, we also fail to gain their recognition. We fail to gain recognition in these cases in two ways: (1) we fail to articulate our experience because of a conceptual lack or social hindrance and (2) because the listener does not commit to understanding our experience, a failure to see another as being capable of suffering and, thus, of compromised dignity. In both cases, (1) and (2), it is because of our membership in a specific social group that we lack recognition in a personal encounter (esteem or love) and in the social structure at large (respect).

Furthermore, Fricker argues that hermeneutical exclusion can lead to a further disadvantage. Hermeneutical injustice can result in the social constitution of agents into something they are not, or into something that is not in their interest to be. In other words, hermeneutical injustice can prevent agents "from becoming who they are" (2007, 168). This is especially the case in cumulative and severe acts of hermeneutical injustice, which can harm an agent even in the aftermath of the actual situation of injustice. Accordingly, hermeneutical injustice can "sometimes [be] so damaging that it cramps the very development of self" (2007, 163). I have argued that when hermeneutical injustice prevents agents from becoming who they are, we can adequately express this harm in terms of misrecognition (Hänel 2020). The argument highlights the relation between acts of recognition and what I call "self-recognition"—that is, our self-respect, self-esteem, and self-confidence—and the relation between self-recognition and hermeneutical injustice. Roughly, the harmful experience of hermeneutical injustice not only leads to loss of epistemic confidence but also compromises our self-recognition, who we think we are.

The idea then is that hermeneutical injustice not only compromises recognition—those who suffer from hermeneutical injustice have suffered from misrecognition or the absence of recognition and are vulnerable to future acts of misrecognition or the absence of recognition—but it also compromises self-recognition. In other words, instances of hermeneutical injustice are harmful because they can leave us without self-respect, self-esteem, and self-confidence; they destroy our capability to form an identity. And, because hermeneutical injustice is due to a structural identity prejudice, members of marginalized social groups in particular suffer from past and future misrecognition and are compromised in their personal identity. This is in line with arguments that misrecognition is a form of oppression: it leaves people with a false view of themselves (cf. Taylor 1994; Ferrarese 2009). But this is doubly so insofar as misrecognition targets primarily subjects who are already victims of oppression due to their social group membership. This suggests that there is a closer, probably constitutive, relation between misrecognition and hermeneutical injustice.[6] Yet, this claim would merit closer analysis and has to be postponed for now. The point that I want to make here, concentrating on the claim that acts of hermeneutical injustice render a subject vulnerable to future acts of misrecognition, is that hermeneutical injustice fortifies misrecognition.

I have argued that acts of misrecognition are particularly damaging when (a) they stretch over different contexts, leaving us without or with very few safe spaces, and (b) they dislocate us, leaving us without a community to turn to. And that acts of hermeneutical injustice are deeply entangled with relations of misrecognition. Both misrecognition and hermeneutical injustice are particularly damaging for ourselves when they are structural and, hence, leaving us with few contexts in which to heal and remake ourselves. This, unfortunately, is the case for many refugees coming to Western countries to seek a (better) life. They often experience traumatic assaults such as torture, rape, and slavery, are estranged from their families and communities, and are met with racist misrecognition in the countries they are fleeing to. In the next section, I consider some particular examples and discuss their relation to misrecognition.

12.3 In Detail: Misrecognition and Refugees

12.3.1 From Camp Moria to the Libyan Detention Centers: Three Cases of Misrecognition

In this paper, I am concerned with refugees—or, what other call survival migration or forced migration. The 1951 Convention Relating to the Status of Refugees and the 1967 Protocol Relating to the Status of Refugees define a refugee as *a person who is outside of their country of origin and is unable or unwilling to return to their country of origin due to fear of persecution on the grounds of rape, religion, nationality, and membership of a particular social group or political opinion*. Many scholars have pointed to the fact that this definition is historically outdated and its focus on persecution is arbitrary (cf. Betts and Collier 2017; Miller 2016). In fact, many forced migrants nowadays are fleeing not because of persecution but because they are not safe in their country of origin (or a country they fled into) due to political instability, natural disasters, lack of means to provide for oneself, or violence and

[6] Dübgen (2012) makes this claim: analyzing misrecognition as a form of epistemic injustice.

wars. Furthermore, many forced migrants either do not leave their country but find shelter somewhere else in the same country or others have to flee a country that they once thought they would find shelter in. Very few forced migrants actually have the means to come to Europe seeking safety. Thus, the above definition fails to capture the full range of people that we intuitively consider to be refugees. Many argue instead that a refugee is someone who is forced to flee "to protect their human rights or basic needs, whether the result of direct persecution, indiscriminate warfare, or natural disaster" (Buxton 2020). Similarly, Miller argues that refugees are "people whose human rights would be unavoidably threatened of they remain in the place they inhabit, regardless of whether the threat arises from state persecution, state collapse, or natural disaster" (Miller 2016: 167). The main point here is that there is a threat, not that this threat has a particular source. However, Fine correctly points out that Miller's definition is still artificially narrow and fails to capture the full range of people we want to capture. That is because Miller still refers to the place a person currently inhabits, not the person's country of origin, country of citizenship, or habitual residence. And Miller argues for the threat to human rights as unavoidable. Meaning that a person does not count as a refugee if the threat could be averted in some way without them leaving the country in question. Hence, living in temporary refugee camps, a migrant would cease to be a refugee (cf. Fine 2019: 27). Thus, scholars are far from being in agreement about how to define who counts as a refugee, even though most agree that the Convention's definition is inaccurate (at least in common times). It is beyond the scope of this paper to argue for an adequate definition or to recount the different discussions about this. Fortunately, for the purpose of the paper, it is enough to roughly follow the idea that refugees are persons who are forced to flee to protect their human rights.[7] In doing so, I follow Betts and Collier in their argument that refuge, at its core, "entails the principle that when people face serious harm at home, they should be allowed to flee and receive access to a safe haven, at least until they can go home or be permanently reintegrated elsewhere" (Betts and Collier 2017, 4). Let me now turn to three specific examples of people trying to find refuge.

Children in Moria There are roughly 5200 unaccompanied children in refugee camps in Greece at the moment. Many being sent by their parents and families, hoping that they could follow soon. Unfortunately, many of these children will have to wait years until they are reunited with their parents; some will never see their families again; some will be stranded somewhere along the Balkan route, suffering from violence, abuse, starvation, cold, and emotional misery; and some will die. Many of these children are deeply traumatized due to wars, the conditions of the journey to Greece, and the absence of stability and parents. The conditions in the refugee camps are obstacles to a healing process, often re-traumatizing the children. At the biggest refugee camp on Lesbos, camp Moria, one psychologist is responsible for 4000 traumatized children. In an interview, she reports that she is helping children, sometimes not older than 2 years old, who bite themselves, pull their hair out, or

[7] See also https://blogs.lse.ac.uk/theforum/who-is-a-refugee/.

bang their heads against the wall of the small office. They have so many feelings that they do not know what to do with themselves and their emotions. Even young children are attempting suicide here. Another doctor says that the development of the children here is backward. They regress; they stop talking, stop playing, and stop eating. The lack of stability and safety, of time to be and play, and of people to turn to, leaves these children alienated from others and themselves. (cf. Kögel 2019; Lobenstein 2019).

Moria is a case of misrecognition. European politicians close their eyes and refuse to acknowledge the dehumanizing conditions in the camp, and in doing so, they also refuse to acknowledge that there are human beings living in dehumanizing conditions. Others acknowledge the dehumanizing conditions but simply do not care. Still others believe that it is morally legitimate to close the European borders and detain refugees that make it over the Mediterranean—whatever the costs. The message they send is clear: "You are worth less than we are." It is a message of misrecognition.[8] It is a failure to imagine others as ourselves and to acknowledge their equal standing and worth, their dignity and personhood, and their suffering. And the message targets people who are already suffering from trauma and broken selves and are dislocated and alienated from their families and communities. Furthermore, this message is repeated in every context that the refugees in Moria step into, be it in the camp, during their journeys, or in the context of applying for asylum and shelter. It is the kind of misrecognition that bears the characteristics of structural and systematic misrecognition and, thus, is particularly damaging. Even more so for the unaccompanied children in Moria. They are dislocated from love, from their parents and families, and, thus, from the possibility of recognition that is necessary for them to develop an identity, to become a person in the first place. Recognition from others fosters our ability to gain self-recognition, which in turn allows us to develop ourselves, our values, and future projects. Recognition, thus, is necessary to us as human beings. In comparison to many other cases of misrecognition, the children in Moria are in a particularly harmful situation. They cannot fall back on the identity and the self-recognition they have developed already—before the trauma and misrecognition at the European border. In other words, the misrecognition that the children in Moria experience is not an obstacle to becoming a person; it is the destruction of their entire identity development.

Libyan Detention Centers Many refugees are currently detained in Libya's so-called detention centers. Triq-al-Sikka is one such example. It is located in Tripoli and home to 300 men, several of whom are sick and/or lie motionless on dirty mattresses in the yard, left to die or recover in their own time. There is no medical care and the sanitary conditions are bad; three of the six toilets are blocked with sewage; and there is no soup, and often not even running water. And this is despite the fact that the European Union has since 2016 poured more than £110 m into improving

[8] One might want to object by saying that we should not be concerned with misrecognition in these cases but rather with a denial of (human) rights that these people have. I say more about this objection below. I am thankful to Gottfried Schweiger for raising this point.

conditions for migrants in Libya, yet the conditions are now worse than before. Furthermore, the centers are targets of air strikes in the civil war that is still ongoing in Libya, often killing migrants who are forced to live there. In another center south of Tripoli, Qasr bin Ghashir, 890 men, women, and children were surrounded and fired upon in April 2019, presumably by Haftar's militias for unknown reasons. Several inmates were hurt and killed. The Amnesty International and Doctors Without Borders filed reports about the incident.

And, there are reports of torture, abuse, and rape in detention centers. Oxfam reports that they have collected testimonies of these crimes. In some centers, militias storm in at night, dragging refugees away to be ransomed back to their families or to be sold into slavery (cf. Naib 2018; Gouliamaki 2018; Quackenbush 2017). Spiegel International reports "about mass graves and how [refugees] were chained and forced to work after arriving in Libya." The report continues stating that "they toil as domestic servants, maids and as workers in fields, factories or on building sites. But they seldom receive any wages. [...] Soon after crossing the desert, they are sold at slave markets and so-called 'Connection Houses' in southern Libya. Male victims are mostly forced to work, while the women are forced into prostitution" (Ehlers and Kuntz 2019). Thousands of refugees flee the camps, sleeping rough on the streets. Many women have lost their husbands to the kidnappings of militias, being alone and homeless with their children and babies. Yet, Libyan detention centers are funded by the European Union. According to the Guardian, a "common theme among migrants here is a crushing sense of being unwanted and of no value, seen even by aid agencies as an inconvenience" (Mannocchi 2019). Libyan detention centers are a case of misrecognition: they refuse shelter, food, warmth, and safety—human rights—and dislocate individuals from their communities. In fact, misrecognition could not be more obvious than in cases of abuse, torture, rape, murder, and slavery. But it is a double misrecognition; it is the direct misrecognition in the form of abuse from the hands of guards and smugglers, and the indirect misrecognition from European politicians and institutions that fail to acknowledge the equal worth of refugees and fund abuse against them.

Denial to Assistance on the Mediterranean One in eight refugees who set out from Libya for Italy in the first 4 months of 2019 died (the number is rising; 1 in 31 in 2018 and 1 in 48 in 2017). The Missing Migrants Project reports 99 refugees who drowned in the first 3 months of 2020. Furthermore, the cooperation with the Libyan coastguard means pullbacks even from international waters and, thus, against international law as well as creating dangerous situations for both refugees crossing the Mediterranean and crew members of NGO ships, including shooting at migrants and NGO workers. And there are reported cases of denial to assistance and other dangerous behavior, both from European coast guards and the Libyan coast guard currently being paid by Europe to rescue and keep refugees away from Europe. Furthermore, there are increased reports of denial to assistance of others ships including the US Marine, sometimes leading to drownings. Finally, NGO ships are denied port and assistance, and their crews face legal sanctions from European countries. In summer 2019, Spiegel International reports that "even volunteer maritime rescuers are now being criminalized. Ports are closed to them. Ships are confiscated. Helpers are put on trial. Lifeline captain Claus-Peter Reisch was just sentenced to pay 10,000 euros by a Maltese court because his ship hadn't been properly registered. The Italian Public Prosecutor's Office is accusing 10 volunteers from the vessel Iuventa of illegal immigration. The crew could face up to 20 years in prison." Probably the most famous case of the prosecution of NGO workers is the case of Carola Rakete. After 17 days on board of the Sea-Watch 3 without being assigned a port, Captain Carola Rakete decided to break the port blockade and delivered the 22 crew members and 53 rescued refugees to Lampedusa. She was arrested on the same day. On June 3, 2019, a group of human rights lawyers lodged a complaint against the EU at the International Criminal Court in The Hague. It argues that, through its policies, the European Union is responsible for "the deaths by drowning of thousands of migrants" (cf. Thelen 2018; Lüdke et al. 2019; Hornig 2019). What other explanation could be given for the denial of assistance if not a failure to imagine those drowning to be like ourselves, to have equal worth and standing, to be full persons worthy of respect and dignity? What other explanation if not misrecognition?

All three cases—the case of unaccompanied children in the Moria refugee camp on the island of Lesbos, the case of Libyan detention camps, and the case of denial to assistance of refugees crossing the Mediterranean—are shocking examples of misrecognition. Often these are double cases of misrecognition, where the abuse and denial of assistance is a direct form of misrecognition and the funding of these treatments by European politicians and institutions is an indirect form of misrecognition, both sending the message that the abused and drowned are worth less. These are shocking examples in so far as they are cases of systemic misrecognition of persons already dislocated from their communities and because these cases could be avoided by treating refugees like persons with certain rights. Apart from the drastic and obvious instances of misrecognition that these refugees suffer from, they also suffer from more subtle cases of misrecognition stemming from epistemic injustices. In these examples, cases of epistemic injustice fortify the misrecognition outlined above and render the subjects vulnerable to future acts of misrecognition in other contexts.

12.3.2 An Objection: Misrecognition or Failures of Human Rights?

Before I discuss in detail how these cases relate to epistemic injustices, let me raise a possible objection to the claim that these cases are obvious examples of misrecognition. One might want to say that these are clearly cases in which human rights are denied and, thus, failings of international law, but question either whether these are linked to misrecognition or whether it is fruitful to do so. Let me make a few points for why I think that a normative account of misrecognition is more fruitful in regard to the topic of this paper than an analysis in terms of human rights. As discussed above, most scholars include abuse, rape, and torture to be clear instances of misrespect and, hence, cases of misrecognition. Most refugees experience abuse, rape, or torture on their journeys as becoming particularly obvious in the case of Libya's detention centers. However, I have also argued that other cases—for example, the case of children in Moria and the denial of assistance in the Mediterranean—should equally count as cases of misrecognition and that we can detect both direct and indirect forms of misrecognition in these cases: the first stemming from the acts of those who torture, rape, and abuse and the second from the European politicians and institutions that fund the direct misrecognition and close the European borders. In fact, respect or equal dignity is often thought of as the most basic form of recognition, an idea that is closely related to universal human rights. Scanlon (1998), for example, argues for respect in terms of the general fact that another individual is a person capable of autonomous agency. According to him, respect is the foundation of morality because the "ideal of acting with principles that others (similarly motivated) could not reasonably reject is meant to characterize the relations with others the value and appeal of which underlies our reasons to do what morality requires" (1998: 162); we all are in relations of mutual respect to others. Thus, what it means to recognize another as equal, according to Scanlon, is to recognize other's capacity to respond to reasons, their autonomous agency.

Extreme forms of humiliation and degradation are clearly failures of the normative expectation to treat another with respect, especially when the other is part of a specific social group that is structurally and systematically denied autonomous agency as is, for example, the case in colonialism, racism, sexism, and ableism. Often, in these extreme forms, others are denied autonomous agency by being socially and materially excluded or treated like objects in interpersonal encounters. And, as I have argued before, in these cases, the relation between misrecognition and self-recognition as a necessary part to have agency in the first place becomes shockingly obvious. All of the cases discussed above include social and material exclusion and/or interpersonal disrespect in the form of humiliation and degradation.[9] European policies have as a direct effect the disrespect and misrecognition of refugees. Housing people in camps and detention centers without access to education, health facilities, or work to earn money for one's basic needs means to socially and materially exclude them from a life worth living and the adequate conditions necessary to develop self-recognition. It denies them autonomous agency. Letting people drown deliberately means to deny them autonomous agency and to treat them as mere objects. Moreover, selling people into slavery, torturing, abusing, raping, and killing them means to deny them autonomous agency and to treat them as mere objects. Thus, all three cases above are extreme cases of misrecognition. Yet, it should not come as a surprise that they are also failures of international human

[9] For a different example and a brutal interpretation of the ways in which refugees are robbed of their humanity by being denied autonomous agency, see Boochani 2018.

right laws; in fact, these laws are grounded in the general idea of equal dignity and respect.

While it is important to discuss the denial of human rights, for example, to hold persons accountable for denial of assistance, illegal pushbacks, and abuse, it is also important to have a normative account of the moral wrong of these cases. Such a normative account can, first, ground the importance of human rights in the first place, and second, cast a wider net of who is in fact responsible for cases of misrecognition as discussed above. Furthermore, human rights are grounded in the general idea that all human beings have equal dignity and respect and, thus, deserve equal treatment. But different scholars differ about the relationship between rights and duties. For example, if refugees have a right to shelter, food, and safety, does this mean that there is always someone who has a duty to provide these goods? And how do we decide who that is? Often then the focus lies with states that have agreed to uphold certain rights for their citizens. Yet, discussion of human rights uncritically assumes that states play a role in securing their citizens' rights and overlook the operations of power, first, between different social groups within their border, and second, in the state's determination of membership in the first place.[10] Furthermore, different scholars disagree about what rights are to be understood as rights to. Are human rights importantly rights to be treated in a specific way and, hence, close to recognition or respect? Or rights to achieve well-being? Pursue one's life plan? Have capacities and opportunities with which to make choices about one's life plan? Thus, what is needed in the first place is an analysis of the suffering imposed by acts of misrecognition that can then be used to determine where and when human rights are not upheld. Second, while with human rights often the focus lies particularly on states, a framework of misrecognition shows both how institutions and their representatives fail to acknowledge individuals' personhood as well as individuals in their various acts.

12.3.3 Hermeneutical Injustice and Refugees

In this paper, I am interested in the relation between misrecognition and hermeneutical injustice. However, cases of epistemic injustice are not necessarily marked by the extreme violence and disrespect described above. How can we explain these? Instances of hermeneutical injustice render speakers vulnerable to future acts of misrecognition, and, as I have discussed with regard to the examples above, misrecognition has deeply problematic consequences for our identity development and self-recognition. Hence, hermeneutical injustice can fortify misrecognition. Refugees are harmed both by primary acts of misrecognition and by secondary acts of hermeneutical injustice that render speakers vulnerable to (future) misrecognition. If it is indeed the case, as many have argued (cf. Fricker 2007; Dotson 2014;

[10] For this last claim and an analysis of statelessness and misrecognition, see Staples (2012).

Medina 2012; Mills 2007), that marginalized experiences are underrepresented in the dominant hermeneutical resource or the general narrative framework, then it is not too far-fetched to assume that the experiences of refugees are underrepresented and, hence, that refugees suffer from the inability to adequately articulate their experiences.[11]

Such an inability to adequately articulate an experience can also be due to distorted dominant concepts and narratives. When it comes to refugees and their struggles for survival, the dominant hermeneutical resource that is influenced by racist and nationalistic views blocks alternative narratives such as the narratives and concepts used by refugees. For example, the dominant narrative according to which refugees (or, alternatively, their cultural values) are construed as dangerous masks the existence of other concepts and narratives. Furthermore, this problematic but dominant narrative can also contribute to the misrecognition of refugees; i.e., perceiving refugees as dangerous makes it easier to deny them equal rights. Interestingly, philosophy and political theory are far from immune to such distortions and the failure to listen and equip the dominant hermeneutical resource with alternative narratives as told by refugees themselves (cf. Fine 2019, 2020). This, however, is not to say that upholding such problematic narratives is a conscious or rational process; rather, we often do so unconsciously to legitimate or justify the comfortable lives we have without solidarity for those who suffer.

Arguing for a theory of white ignorance, Mills starts by saying that social epistemology is racialized; people of color "are not seen at all" (Mills 2007: 18). To uphold this view that marginalizes or negates people of color, whites engage in a cognitive dysfunction that hinders them to understand social relations of power, oppression, and domination in which they participate. Ignoring these social relations makes it possible as well as easier to maintain the status quo (Mills 2007). However, upholding white ignorance is not always a willful and conscious process. As with most theories of ideology, white ignorance is a problem of "social privilege and resulting differential experience, a nonrepresentative phenomenological lifeworld (mis)taken for *the* world" (Mills 2005, 172). Whites misinterpret their experiences and interests as universal experiences and interests, impose them on others, and, thus, create dominant paradigms that fit their lives best (Mills 2005). Mills here provides both a theory about our patterns of belief-forming and why whites lack the motivation to correct their ignorance. The idea behind epistemologies of ignorance is this: ignorances are more often than not actively constructed and function to uphold the existing social power structures. In other words, contrary to common

---

[11] The relation between hermeneutical injustice and misrecognition is often further fortified through testimonial injustice; e.g., the inability to adequately articulate an experience is often due to the lack or distortion of the necessary concepts *and* credibility deficit of the speaker due to harmful prejudices about the speaker's social group, and the lack and distortion of the dominant hermeneutical resource is often causally linked to existing prejudices against certain social groups that prevent them from being regarded as credible and capable of contributing to the said resource. In this paper, however, I concentrate on hermeneutical injustice.

assumptions, ignorance is often not a benign gap in knowledge (cf. Alcoff 2007; Sullivan and Tuana 2007; Tuana and Sullivan 2006).

The interplay of forming dominant paradigms according to particular interests and experiences and being situated in powerful social positions and, hence, able to contribute to the dominant hermeneutical resource more and better than others brings it about that the experiences of the powerful are better represented as those of marginalized people and assumed to be the status quo. This leaves out marginalized experiences, either because no adequate concepts are available or because these are not understood by others. Refugees are a deeply marginalized social group, located at the intersection of racist experiences and experiences of diaspora. In fact, refugees often have to make a choice between articulating their "grievance by drawing upon either authoritative yet inadequate conceptual resources or adequate yet non-authoritative conceptual resources" (Congdon 2016: 820). Not being understood or being understood in only inadequate ways due to the distorted dominant hermeneutical resource is a structural injustice.[12] The hermeneutical injustice that refugees suffer from renders speakers vulnerable to future acts of misrecognition in so far as they cannot adequately articulate their moral injuries and, thus, fail to contribute to the dominant hermeneutical resource, and as it diminishes their own self-trust and, thus, has consequences for their moral treatment by others. Nevertheless, even these subtle cases are systematic. They (a) stretch over different contexts, leaving refugees without or with very few safe spaces, and (b) dislocate them, leaving refugees without a community to turn to.

Furthermore, as hinted at above, these cases of hermeneutical injustice have direct and deeply problematic consequences for identity development and the ability to form values and future projects. Hermeneutical injustices as outline above concern distorted resources and the inability to describe significant experiences due to the background of social power and privilege which allows some people but not others to contribute to the dominant hermeneutical resource—that is, the resource that shapes our dominant interpretative and moral frameworks. On top of that, those suffering from hermeneutical injustice also experience mismatches between their own experiences and narratives and the existing dominant narrative. For example, part of the dominant narrative suggests that Western European countries help refugees coming to Europe; the suffering and mistreatment, the conditions in detention and refugee camps, and the denial of assistance are not part of the narrative. Suffering from hermeneutical injustice and experiencing these mismatches between one's own experience and the given dominant narrative—especially when these stretch over different contexts and there is no community that understands one's

[12] One might want to object that misrecognition in the form of abuse, rape, and torture is well understood by nearly everyone. Yet, this is not the case for more subtle forms of misrecognition as present in messages of unequal worth. Furthermore, what is often not understood is the interplay between racism and forms of misrecognition due to white ignorance. Unfortunately, for lack of space, I cannot outline the precise relation between the dominant framework and the denial of assistance, refugee camps, and detention centers. I am grateful to Gottfried Schweiger to have pointed out this objection.

experiences—eat away at our epistemic capacities. We start questioning our own narratives, our capabilities to interpret our experiences and the world; in other words, we start to self-doubt ourselves and loose self-trust. And, following Fricker, hermeneutical injustice can therefore "sometimes [be] so damaging that it cramps the very development of self" (2007, 163). Hence, hermeneutical injustice can fortify misrecognition.

To sum up, hermeneutical marginalization is causally related to misrecognition. According to Fricker, hermeneutical marginalization refers to the problem that not everyone participates equally "in the practices through which social meanings are generated" (Fricker 2007, 6). That is, those that are hermeneutically marginalized participate less (or not at all) in these practices and that in turn bring it about "that our collective forms of understanding are rendered structurally prejudicial in respect of content and/or style" (Fricker 2007, 6). The experiences of those that are hermeneutically marginalized are not part of the dominant resource, cannot enter the dominant resource, and are, thus, ill-understood, often by both the hermeneutically marginalized and others, sometimes only by others (cf. Fricker 2007, 153–61). I have argued that being in a position of hermeneutical marginalization often implies a history of acts of misrecognition and renders a person particularly vulnerable to future acts of misrecognition. We can say that being the victim of hermeneutical injustice and, thus, being positioned as hermeneutically marginalized is causally related to being vulnerable to acts of misrecognition. Further, the relation between hermeneutical marginalization and misrecognition can create obstacles for the privileged to grasp their duties toward those less privileged and/or vulnerable. The fact that some experiences are not part of the dominant resource brings it about that these experiences are often completely ignored or questioned by those whose experiences are adequately understood because they are included in the dominant hermeneutical resource (cf. Mills 1999, Pohlhaus 2012). Ignoring or questioning harmful experiences and, thus, failing to give people their fair share of credibility is an act of misrecognition and, at the same time, only masks further the structural injustices that lead to hermeneutical marginalization.

12.4 What Do We Owe to Refugees with Regard to Recognition?

12.4.1 Recognition and Moral Obligations

One of the advantages of understanding the above-described cases in terms of misrecognition and not merely in terms of failures of human rights is that it allows us to carve out more specifically who is responsible for what and who owes what to whom. I have argued above that recognition theory is a normative theory; it says something about how we treat people and how we should treat them. It is not far-fetched to assume that this implies some sort of duty. And, in fact, Honneth (1997) argues that a theory of recognition comes with moral obligations and duties. While Honneth distinguishes between corresponding duties for the three spheres of recognition—love, esteem, and respect—I here focus on duties resulting from the need for respect. The general idea is that we can assume moral duties for securing the intersubjective conditions necessary for identity formation. According to this idea, morality follows from an objective analysis of the conditions of life needed for human beings and, at the same time, serves to protect the good life following from these conditions.

This can be made apparent when considering the relation between moral injuries and recognition. According to Honneth, an injury becomes a moral injustice "whenever the subject affected has no choice but to view it as an action that intentionally disregards it (the subject) in an essential aspect of its well-being" (2005, 47), be it through abuse, torture, rape, or (messages of) disrespect. It is the fact that another fails to recognize the subject's standing and their self-relation that is the condition for moral injury, a recognition that we all fundamentally dependent on (cf. Habermas 1996; Tugendhat 1986). If, Honneth argues, the core of moral injuries is indeed located in the refusal of recognition, then moral attitudes are related to the exercise of recognition (cf. Honneth 2005, 50); in other words, we are mutually obligated to adopt attitudes of recognition toward others to secure the conditions for personal integrity. And while the moral attitudes we are obligated to adopt differ depending on the particular relation to the other, moral respect "designates a form of recognition that can be expected of all subjects equally" (Honneth 2005, 53), no matter the interpersonal relation.

In other words, we owe recognition to others qua their equal status or worth. When we fail to fully recognize another person, we commit a moral injury. In the last section of this paper, I want to take a closer look at what we owe to refugees. The general idea is that both individuals as well as institutions and their representatives commit moral injuries by misrecognizing refugees. I have argued that both direct abuse as well as indirect messages of misrecognition and funding of the said abuse should count as misrecognition in the cases above; while it is mostly individuals that commit the first sort of moral injury, it is European institutions that are responsible for the second due to their policies that perpetuate detention centers in Libya, closed refugee camps on the borders, and fail to assist in the Mediterranean. There is a causal relation between these policies and acts and the misrecognition of refugees, and, hence, there are strong moral duties to stop both acts of abuse and the policies of "Fortress Europe." Yet, in the remaining section, I want to concentrate on the specific epistemic duties that both individuals and institutions have toward refugees. If hermeneutical injustice fortifies misrecognition, then there are also specific duties to end hermeneutical injustice in these cases.

12.4.2 Epistemic Duties and Obligations

Fraser (2007) argues that the public sphere is a space for the production of public opinion. The public sphere can be inclusive and meaningful or exclusive; that is, it can be guided by "normative legitimacy" and "political efficacy" or not. According to Fraser, normative legitimacy asks whether and to what degree individuals' political voices become public knowledge, and political efficacy asks whether public opinions are politically effective. Thus, deficits in both prevent some social groups from producing meaningful narratives. In other words, we need high degrees of normative legitimacy to be able to contribute well to the dominant hermeneutical resource such that our voices are heard by others. And we need political efficacy to transform public knowledge (aka the dominant hermeneutical resource) into politically fruitful knowledge. Safouane shows how migrants lack normative legitimacy and political efficacy and are thus prevented from contributing to the given narratives or from countering the given narratives with alternative narratives in any meaningful way (Safouane 2019, 71). Obviously, this is not to say that refugees (or migrants in general) lack narratives; rather, their narratives are not heard by others and therefore fail to become public knowledge and/or politically effective (cf. Fine 2019, 2020). Instead, the dominant hermeneutical resource and, thus, the public sphere is dominated by narratives *about* refugees that help to (re-)produce widespread acts of misrecognition, both interpersonally and structurally; e.g., by describing refugees as asylum shoppers, economic migrants, or illegal migrants, the public creates social identities, and by decontextualizing them, they foster misrecognition and dehumanization of these identities.

Acknowledging the ways in which normative legitimacy and political efficacy are needed to have a voice can help to formulate possibilities to do so and uncover our (epistemic) duties, both in interpersonal encounters and structurally. In the following, I want to draw attention to three possibilities of overcoming hermeneutical injustice and, thus, making it harder for future misrecognition to take place. If hermeneutical injustice and misrecognition are deeply entangled, then part of avoiding misrecognition or finding ways of reparation of past misrecognition is specifically epistemic. It is, first, to counteract hermeneutical injustice by giving refugees their fair share of credibility; that is, to believe their reports of violence and discrimination in their home countries, during their journeys, at European borders, and in European camps. This includes distinguishing between dominant hermeneutical resources and alternative hermeneutical resources and to acknowledge that the necessary concepts and narratives are not missing, but, rather, that they are not heard or part of public opinion, or that they are actively suppressed. Here, we can learn from Fraser's idea of "subaltern counterpublics." Second, it includes to take responsibility for the problematic social structures that produce hermeneutical injustices in the first place. Hence, to find ways for those less privileged to contribute to the dominant hermeneutical resource. While the first can be put as an individual duty in the way Fricker argues for specific virtues to counteract hermeneutical injustice, the second is a structural problem and can, thus, not be counteracted with individual virtues. And, third, it means to booster political efficacy of the alternative narratives and subaltern counterpublics, that is, of refugees' voices. Let me tackle each point respectively.

First, part of the problem that refugees cannot contribute to the dominant hermeneutical resource is due to testimonial injustice. Refugees are given less credibility than they deserve by mainstream media, the public, and the authorities; i.e., their narratives, their experiences, their concepts, and their voices are discredited due to their social group membership. Fricker argues that epistemic virtues can counter testimonial injustice (at least to some amount) (2007, Chapter 3). That is, being epistemically virtuous such that we train ourselves to be attentive listeners and question our own biases and stereotypes (to name only a few ways of being epistemically virtuous) can counter testimonial injustice and, thus, pave the way for contribution to the dominant hermeneutical resource. Calling for epistemic virtues focuses on the ways in which listeners can help to cease epistemic injustices. But we should also take into account the many ways in which oppressed social groups have resisted such injustice already. Fraser's subaltern counterpublics are closely related to the idea described in Sect. 1, namely, that there is a distorted but dominant hermeneutical resource that is shaped by contributions of only some (cf. Dotson 2014) and the acknowledgment that there are different hermeneutical resources within subcultural communities (cf. Medina 2012). According to Fraser, subaltern counterpublics are subcultural discursive arenas that develop parallel to the dominant public sphere. Here, in these subaltern counterpublics, "members of subordinated social groups invent and circulate counter discourses to formulate oppositional interpretations of their identities, interests, and needs" (Fraser 1992, 123).[13] Subaltern counterpublics are formed as resistance to the exclusion from the dominant public sphere. That means, that the subaltern counterpublics, according to Fraser, are a step further than the mere existence of hermeneutical resources within subcultural communities. While any kind of community necessarily has their own hermeneutical resources that can be or cannot be in line with the dominant hermeneutical resource, the subaltern counterpublics already presuppose reflection upon one's membership within an oppressed social group and the active pursuing of creating one's own subcultural hermeneutical resource in opposition to the distorted dominant resource. Subaltern counterpublics are part of the resistance, while mere subcultural hermeneutical resources are not, yet mere subcultural hermeneutical resources can turn into subaltern counterpublics. Acknowledging that there are such subaltern counterpublics means to bring to the fore the self-identification of refugees that is necessarily in opposition to the social identities that were given to them by the distorted but dominant public narratives.

Second, hermeneutical injustice is not an interpersonal problem; rather, it is structural. Thus, to counteract misrecognition by focusing on hermeneutical injustice necessarily needs to include structural solutions. In other words, it includes to

[13] It should be noted that, in her theory about the subaltern counterpublics, Fraser borrows the idea of "subaltern" from Spivak (1988) and the notion of "counterpublic" from Felski (1989).

take responsibility for the problematic social structures that produce hermeneutical injustices in the first place. Anderson (2012) argues that some of the problems blocking widespread epistemic justice are such that no individual could tackle them alone. For example, as research has shown cognitive biases such as prejudices and harmful stereotypes are difficult to correct even for the epistemically virtuous individuals (cf. Saul and Brownstein 2016). Thus, to counteract epistemic injustices, we have to "focus on the principles that should govern our systems of testimonial gathering and assessment," and we need to find out "what epistemic justice as a virtue of social systems would require" (Anderson 2012, 163). For our cases, this would mean to set up institutional structures and procedures that ensure testimonial credibility of refugees and provide access to the dominant hermeneutical resources and public discourse. Part of this, as Anderson suggests, is to integrate diverse institutions and persons engaged in the procedures (cf. Anderson 2012). Refugees as well as everyone else deserve credibility not merely in interpersonal situations but also institutionally.

Furthermore, if this is the case and if epistemic injustice and misrecognition are closely related, then we have to investigate what institutional recognition looks like. I have argued that hermeneutical injustices are the result of prior histories of misrecognition. Thus, tackling problematic acts of misrecognition could help prevent acts of hermeneutical injustice. Jay Bernstein (2015) argues convincingly that recognition is (or, should we say) must always also be institutional recognition, i.e., being granted the right to participate in social interactions and social life as full persons. Being able to participate in social interaction and social life implies already being able to do so also epistemically, for example, by being able to contribute to the public sphere and the hermeneutical resource of the public discourse. Yet, as I have discussed above, deficit of credibility in individual encounters and a lack of recognition in the institutional sphere are often the result of structural causes such as problematic stereotypes of our capacities as knowers and our moral standing as full persons. Any duty to give recognition cannot and should not be distorted by problematic underlying assumptions about the social group membership of the recipient of such recognition. However, since this is often the case, we, as individuals and as a society, owe more than mere institutional recognition to make up for the said distortion. We are obligated to prevent or at least diminish harmful biases to enter into our procedures and interpersonal encounters. This is most of all an institutional task and has to include not just national institutions but also international laws and organizations.

Finally, institutional responsibilities are also about setting up structures that make it harder for individuals to lose their self-recognition and therefore be better equipped to fight off acts of future misrecognition or hermeneutical injustice. I have argued that misrecognition is particularly damaging—for its risk to prevent self-recognition—when (a) it stretches over different contexts and leaves individuals without safe spaces and (b) dislocates individuals from their communities. Thus, we can prevent harmful misrecognition at least partly by ensuring that there are safe spaces and that individuals have communities that they can turn to for help. This implies to rethink the ways in which we create "safe havens" for refugees;

concretely, it means to abolish detention centers (especially in unsafe countries such as Libya), refugee camps (especially when overcrowded and without sufficient infrastructure and sanitary areas), detention pending deportation procedures, splitting of families, and other harmful procedures.[14] And it implies to think about those that are unable to seek help and refuge in another country. Ninety percent of the world's refugees remain in the developing world because very few have the money and resources and are physically capable of seeking refuge in a European or Western country. That is, the most vulnerable remain in dangerous and deeply dysfunctional systems (cf. Schweiger 2016). These people hardly get any help at all. While Europe spends roughly $135 on an asylum-seeker in Europe, only $1 is spend on any refugee who stays in the developing world. And only one in ten refugees who fled from Syria to Turkey, Lebanon, and Jordon receives any material support from the United Nations or its cooperating partners. That is, 400.000 refugees receive minimal material support from the overall four million refugees seeking help and refuge (cf. Betts and Collier 2017). Hence, our duties here are twofold: it is about actively counteracting harmful institutions, procedures, and policies as well as taking responsibility for future institutions, procedures, and policies.[15]

Third, we need to booster political efficacy of the alternative narratives and subaltern counterpublics, that is, of refugees' voices. According to Fraser, for individuals to be able to contribute to public opinion in a meaningful way, it is not enough to merely hermeneutically participate. Thus, it is not enough to guarantee structures according to which everyone has the right to participate in social interactions and social life as full persons. We also have to look at which contributions count and which do not. Refugees need to have a voice *and* their voice has to count. So far, I have mostly concentrated on our duties to provide structures within which refugees have a voice. To have their voice count could be achieved by social movements (cf. Anderson 2014), by restriction of hate speech (cf. Langton 1990 and 1993, Maitra and McGowan 2012), or by wide media coverage. Unfortunately, for lack of space, these solutions have to be brief and abstract. Yet, this last step is important in so far as it can counteract not merely hermeneutical injustice and, thus, future acts of misrecognition, but also make it harder for more extreme acts of misrecognition to take place as it counters the objectification and dehumanization of refugees.

[14] See Betts and Collier (2017) for a proposal of how to think about refuge in refugee policies. Rethinking the ways in which we provide refuge also implies to think about our duties toward those that are incapable of finding refuge themselves, for example, children (and their families). For a powerful argument about our duties toward children in conflict areas, see Schweiger 2016.

[15] In this paper, I will not consider the particular responsibility or specific duties that we—as privileged Western citizens—have due to our causal role in specific refugee-producing conflicts, the histories of colonialism, or the illegitimacy of the international order. For a devastating analysis of the deeply problematic involvement to uphold the international order from colonialism to now, see Hickel 2017.

12.5 Conclusion

To sum up, I have argued that we have duties for remedies against hermeneutical injustice, misrecognition, as well as the structural causes that make both phenomena possible right now. Furthermore, I have argued along the lines of future, collective responsibility (cf. Young 2011 and 2003), making sure to change our institutions, procedures, and policies as to prevent or at least make harder future acts of both interpersonal misrecognition and institutional misrecognition. More precisely, it follows from the premise that Western states contribute to the harms refugees suffer from and the entanglement of misrecognition and hermeneutical injustice, that we have the following duties, either individually or institutionally: first, a duty to counteract hermeneutical injustice by giving refugees their fair share of credibility; second, a duty to take responsibility for the problematic social structures that produce hermeneutical injustices; and, third, a duty to booster political efficacy of the alternative narratives and subaltern counterpublics, of refugee's voices.

References

Aguilera, Jasmine. 2019. Here's What to Know About the Status of Family Separation at the U.S. Border, Which Isn't Nearly Over. *Time*. https://time.com/5678313/trump-administration-family-separation-lawsuits/.

Alcoff, Linda M. 2007. Epistemologies of Ignorance: Three Types. In *Race and Epistemologies of Ignorance*, ed. S. Sullivan and N. Tuana, 39–57. Albany: State University of New York Press.

Anderson, Elizabeth. 2012. Epistemic Justice as a Virtue of Social Institutions. *Social Epistemology* 26 (2): 163–173.

———. 2014. Social Movements, Experiments in Living, and Moral progress: Case Studies from Britain's Abolition of Slavery. In *The Kindley Lecture*. University of Kansas. Department of Philosophy, University of Kansas.

Bernstein, Jay M. 2005. Suffering Injustice: Misrecognition as Moral Injury in Critical Theory. *International Journal of Philosophical Studies* 13 (3): 303–324.

———. 2015. *Torture and Dignity: Essays on Moral Injury*. Chicago: University of Chicago Press.

Betts, Alexander, and Paul Collier. 2017. *Refuge: Rethinking Refugee Policy in a Changing World*. Oxford: Oxford University Press.

Boochani, Behrouz. 2018. *No Friend But the Mountains: Writing from Manus Prison*. Sydney: Picador Australia.

Brison, Susan. 2002. *Aftermath: Violence and the Remaking of a Self*. Princeton: Princeton University Press.

Buxton, Rebecca. 2020. *Who Is a Refugee?* APA Blog, Public Philosophy.

Congdon, Matt. 2016. Wronged Beyond Words: On the Publicity and Repression of Moral Injury. *Philosophy and Social Criticism* 42 (8): 815–834.

Dotson, Kristie. 2014. Conceptualizing Epistemic Oppression. *Social Epistemology* 28 (2): 115–138.

Du Bois, W.E.B. 1989. *The Souls of Black Folk*. Bantam Classics.

Dübgen, Franziska. 2012. Africa Humiliated? Misrecognition in Development Aid. *Res Publica* 18 (1): 65–77.

Ehlers, Fiona, and Kathrin Kuntz. 2019. Refugees Suffer as Libya's Civil War Rages On. *Spiegel International*: https://www.spiegel.de/international/europe/libya-refugees-suffer-as-civil-war-rages-on-a-1266032.html.

Fanon, Franz. 1952/2008. *Black skin, White Masks*. New York: Grove Press.

Felski, Rita. 1989. *Beyond Feminist Aesthetics*. Harvard University Press, Cambridge, MA.

Ferrarese, Estelle. 2009. 'Gabba-Gabba, We Accept You, One of Us': Vulnerability and Power in the Relationship of Recognition. *Constellations* 16 (4): 604–614.

Fine, Sarah. 2019. Refugees, safety, and a decent human life. *Proceedings of the Aristotelian Society CXIX* 1: 25–52.

———. 2020. Refugees and the Limits of Political Philosophy. *Ethics & Global Politics* 13 (1): 6–20.

Fraser, Nancy. 1992. Rethinking the Public Sphere: A Contribution to the Critique of Actually Existing Democracy. In *Habermas and the Public Sphere*, ed. C. Calhoun. Cambridge, MA: MIT Press.

———. 2007. Feminist Politics in the Age of Recognition: A Two-Dimensional Approach to Gender Justice. *Studies in Social Justice* 1 (1): 23–35.

Fraser, Nancy, and Axel Honneth. 2003. *Redistribution or Recognition? A Political-Philosophical Exchange*. London: Verso.

Fricker, Miranda. 2007. *Epistemic Injustice: Power & the Ethics of Knowing*. Oxford: Oxford University Press.

Gouliamaki, Louisa. 2018. Migrant slavery in Libya: Nigerians Tell of Being Used as Slaves. *BBC News*. https://www.bbc.com/news/world-africa-42492687.

Habermas, Jürgen. 1996. *Between Facts and Norms: Contributions to a Discourse Theory of Law and Democracy*. Cambridge, MA: MIT Press.

Hänel, Hilkje. 2020. Hermeneutical Injustice, (Self-)Recognition, and Academia. *Hypatia* 35 (2): 1–20.

Hickel, Jason. 2017. *The Divide: A Brief Guide to Global Inequality and its Solutions*. Windmill Books.

Honneth, Axel. 1992. *The Struggle for Recognition: The Moral Grammar of Social Conflicts*. Cambridge, MA: MIT Press.

———. 1995. *The Struggle for Recognition: The Moral Grammar of Social Conflicts*. Cambridge, MA: MIT Press.

———. 1997. Recognition and Moral Obligation. *Social Research* 64 (1): 16–35.

———. 2005. Between Aristotle and Kant—Sketch of a Morality of Recognition. In *Morality in Context*, ed. W. Edelstein and G. Nunner-Winkler. Amsterdam: Elsevier.

Hornig, Frank. 2019. *We Were All in a State of total Despair*. Spiegel International. https://www.spiegel.de/international/europe/interview-with-sea-watch-captain-carola-rackete-a-1276264.html.

Kögel, Annette. 2019. Zweijährige reißen sich ihre Haare aus. *Der Tagesspiegel*. https://www.tagesspiegel.de/berlin/kinderpsychologin-zu-lage-im-camp-moria-zweijaehrige-reissen-sich-ihre-haare-aus/25100582.html.

Kulikowski, Oliver. 2020. Zwölf Tote und eine illegale Rückführung nach Libyen. *Seawatch News*: https://sea-watch.org/zwoelf-tote-und-eine-illegale-rueckfuehrung-nach-libyen/. (English translation: *Alarmphone Press Release*, https://alarmphone.org/en/2020/04/16/twelve-deaths-and-a-secret-push-back-to-libya/).

Laitinen, Arto. 2012. Misrecognition, Misrecognition, and Fallibility. *Res Publica* 18 (1): 25–38.

Langton, Rae. 1990. Whose Right? Ronald Dworkin, Women, and Pornographers. *Philosophy and Public Affairs* 19 (4): 311–359.

———. 1993. Speech Acts and Unspeakable Acts. *Philosophy and Public Affairs* 22 (4): 293–330.

Lobenstein, Caterina. 2019. Und Deutschland schickt Decken. *Zeit Online*. https://www.zeit.de/2019/53/camp-moria-fluechtlingslager-lesbos-eu.

Lüdke, Steffen, Maximilian Popp, Jan Puhl, and Raniah Salloum. 2019. *European Policies Create New Dangers on Mediterranean*. Spiegel International. https://www.spiegel.de/international/europe/european-policy-leads-to-migrant-drownings-in-the-mediterranean-a-1272521.html.
Maitra, Ishani, and Mary K. McGowan. 2012. *Speech and Harm: Controversies Over Free Speech*. Oxford: Oxford University Press.
Mannocchi, Francesca. 2019. *Torture, rape and murder: Inside Tripoli's refugee detention camps*. The Guardian. https://www.theguardian.com/world/2019/nov/03/libya-migrants-tripoli-refugees-detention-camps.
Martineau, Wendy, Nasar Meer, and Simon Thompson. 2012. Theory and Practice in the Politics of Recognition and Misrecognition. *Res Publica* 18 (1): 1–9.
Medina, José. 2012. Hermeneutical Injustice and Polyphonic contextualism: Social Silences and Shared Hermeneutical Responsibilities. *Social Epistemology* 26 (2): 201–220.
Miller, David. 2016. *Strangers in our midst: The political philosophy of immigration*. Cambridge, MA: Harvard University Press.
Mills, Charles. 1999. *The Racial Contract*. Ithaka: Cornell University Press.
———. 2005. "Ideal Theory" as Rdeology. *Hypatia* 20 (3): 165–184.
———. 2007. White Ignorance. In *Race and epistemologies of ignorance*, ed. S. Sullivan and N. Tuana. Albany: State University of New York.
Naib, Fatma. 2018. *Slavery in Libya: Life inside a container*. Al Jazeera. https://www.aljazeera.com/news/2018/01/slavery-libya-life-container-180121084314393.html.
Oliver, Kelly. 2001. *Witnessing: Beyond Recognition*. Minneapolis: University of Minneapolis Press.
Parekh, Serena. 2020. Reframing the Refugee Crisis: From Rescue to Interconnection. *Ethics & Global Politics* 13: 21–32.
Pilapil, Renante. 2012. From Psychologism to Personhood: Honneth, Recognition, and the Making of Persons. *Res Publica* 18 (1): 39–51.
Pohlhaus, Gaile. 2012. Relational Knowing and Epistemic Injustice: Toward a Theory of Willful Hermeneutical Ignorance. *Hypatia* 27 (4): 715–735.
Quackenbush, Casey. 2017. The Libyan Slave Trade Has Shocked the World. Here's What You Should Know. *Time*. https://time.com/5042560/libya-slave-trade/.
Safouane, Hamza. 2019. *Stories of Border Crossers: A Critical Enquiry Into Forced Migrants' Journey Narratives to the European Union*. Wiesbaden: Springer.
Saul, Jennifer M., and Michael Brownstein, eds. 2016. *Implicit bias and Philosophy, Volume 1 & 2*. Oxford: Oxford University Press.
Scanlon, Thomas M. 1998. *What We Owe to Each Other*. Cambridge, MA, Harvard University Press.
Schweiger, Gottfried. 2016. The Duty to Bring Children Living in Conflict Zones to a Safe Haven. *Journal of Global Ethics* 12 (3): 380–397.
Spivak, Gayatri. 1988. Can the Subaltern Speak? In *Marxism and the Interpretation of Culture*, ed. G. Nelson and L. Grossberg. Champaign: University of Illinois Press.
Staples, Kelly. 2012. Statelessness and the Politics of Misrecognition. *Res Publica* 18 (1): 93–106.
Sullivan, Shannon, and Nancy Tuana, eds. 2007. *Race and Epistemologies of Ignorance*. Albany: State University of New York Press.
Taylor, Charles. 1989. *Sources of the Self*. Cambridge: Cambridge University Press.
———. 1994. The Politics of Recognition. In *Multiculturalism: Examining the Politics of Recognition*, ed. A. Gutmann. Princeton: Princeton University Press.
Thelen, Raphael. 2018. *Are Ships in the Med Ignoring Refugees in Distress?* Spiegel International. https://www.spiegel.de/international/europe/ships-in-mediterranean-may-be-ignoring-refugees-in-danger-a-1239495.html.
Tuana, Nancy, and Shannon Sullivan. 2006. Introduction: Feminist Epistemologies of Ignorance. *Hypatia* 21 (3): 1–3.
Tugendhat, Ernst. 1986. *Self-Consciousness and Self-Determination*. Cambridge, MA: MIT Press.
Young, Iris M. 2003. *Political Responsibility and Structural Injustice*. The Lindley Lecture, University of Kansas.
———. 2011. *Responsibility for Justice*. Oxford: Oxford University Press.

Chapter 13
Asylum and Reification

Heiko Berner

Abstract This chapter describes Axel Honneth's concept of reification, which differs significantly from the original version by Georg Lukács. The theoretical approach is adapted and used for sociological consideration of the reifying conditions under which people with a history of flight suffer in Austria. The empirical considerations are divided into two parts. In the first part, the reifying conditions faced by asylum seekers and recognized refugees are presented. These are dominated by the influence of authorities on almost all areas of life. And in the second part, the effects of these reifying conditions on people with experiences of flight are discussed. Finally, an outlook shows how, with the support of social work institutions, they can respond in order to mitigate or even prevent reification.

Keywords Recognition theory · Reification · Asylum · Refugees · Social work · Empowerment

13.1 Introduction

In recent years, it has been repeatedly noted that asylum seekers in Germany and Austria are treated in a desubjectivating manner, be it in the asylum procedures or during their stay in the course of the procedures (Dahlvik 2016; Seukwa 2006; Grönheim 2018). Experiences from the PAGES project – an action research project in Austria with people who have experiences of flight – left a very similar impression on the researchers. In addition to this, there were also various experiences that can be discussed as questions of recognition (Honneth 1994), for example, such as the arbitrary nature of the asylum procedures or limited access to the labor market. In addition, observations were made during the project, which suggested a deeper, further-reaching – in other words desubjectivating – form of denial of recognition, from which many asylum seekers and recognized refugees suffer. The present

H. Berner (✉)
Salzburg University of Applied Sciences, Salzburg, Austria
e-mail: heiko.berner@fh-salzburg.ac.at

G. Schweiger (ed.), *Migration, Recognition and Critical Theory*, Studies in Global Justice 21, https://doi.org/10.1007/978-3-030-72732-1_13

chapter therefore refers to the concept of "reification," which Honneth (2012) reconceptualizes in a recent volume. He integrates reification into the theory of recognition and designates reification as a form of disregarded recognition.

After having previously looked at the work of Georg Lukács (Honneth 1999), Honneth now focuses on reification (Honneth 2012). He reformulates the concept from Lukács' "Consciousness of the Proletariat" (Lukács 2015 [1923]) and embeds it in the theory of recognition. Reification, according to Honneth, takes place where people are not perceived as fellow human beings and where an attempt is made to arrive at an objective judgment about them without prior recognition. Honneth calls this grievance "forgetfulness of recognition" (Honneth 2012, p. 52). From the examples that Honneth chooses in order to illustrate the phenomenon of reification, it is already apparent that he separates this from a pure critique of capitalism. For example, he mentions slavery in a pre-capitalist era. Here, people become commodities in a very direct sense that has nothing to do with the economization of society. Another example concerns "'industrial' mass murder" (ibid., p. 158) during the NS regime. It becomes clear here that, for Honneth, the aspect of the commodity nature of the subject is no longer central; rather, for Honneth, it is the moment of rationalization combined with the total neglect of the fellow human as a human being that constitutes reification.

PAGES (Participation and Health Promotion of People with Experiences of Escape in the State of Salzburg) is an action research project in which a team from the Social Work study program of Salzburg University of Applied Sciences has worked on various participation activities with some 30 people who have experiences of escape between March 2017 and March 2019. The research focus was on the subjective well-being of project participants in relation to participatory activities and on questions of recognition. In this article, the collected data material is evaluated with regard to Honneth's concept of reification. The questions to be considered here include: do those affected (asylum seekers and recognized refugees) suffer from a specific form of recognition deficit, a form of reification, and how do they react to this?

In a first step, this chapter briefly presents the two concepts of Lukács and Honneth and prepares them for an empirical social science survey. It will become clear that there are some outstanding questions. These include the role of capitalism, or rather of the complete economization of society, which for Lukács was still the actual reason for reifying conditions but which plays a subordinate role for Honneth. Another issue concerns the relationship between institutionalized social conditions that foster reification, their effect on the actions of the people responsible in institutions, and the effects on the recipients of these actions (Sect. 13.2).

This is followed by an empirical study in two steps. First, the potentially reifying circumstances are taken into account. These raise the question: does the institutional environment during and after the asylum procedure have a reifying character? "Environment" refers to the executive apparatus that has an ubiquitous effect on those affected: asylum authorities, basic services, the Austrian Integration Fund (ÖIF), the Austrian employment office (AMS), social services, and also the social work measures involved. In this section, it is argued that in the course of neo-liberalization – or more precisely in the shift toward New Public Management – the mode of rationalization has changed in such a way that the actions of authorities and administrations have a reifying effect (Sect. 13.3). The second step is to apply these conditions on the specific case of the asylum system and to examine the perceptions of those affected. Observations from the PAGES project will be used for this. Various specific narratives, remarks, behaviors, decisions, and opinions of people with experiences of flight, which provide an indication of reifying circumstances, are systematized (Sect. 13.4).

Finally, the reactions of those concerned are discussed. Above all, the project participants repeatedly expressed their indignation at their experiences of disregard, which can be interpreted as displeasure at morally wrong actions, according to Ernst Tugendhat (2006). The will to stand up for one's own rights and interests was also derived from this. In conclusion, a few suggestions are made for actions to help counteract reification at various levels (Sect. 13.5).

13.2 Reification

13.2.1 Reification According to Georg Lukács and Axel Honneth

In his 1923 volume "Die Verdinglichung und das Bewußtsein des Proletariats" ("Reification and the Consciousness of the Proletariat," here in reprint, Lukács 2015), Georg Lukács diagnoses a phenomenon driven by increasing capitalism, which he calls reification. According to Lukács, reification takes place on three levels. First, capitalist conditions reduce things to their mere commodity value. The formerly genuine utility value of an object is thus completely transformed by its nature as a commodity and reduced to its exchange value. Second, the same reifying effect applies to human relations in which fellow people are also perceived as objects and no longer perceived as individuals but rather only as an exchangeable extra in capitalist practice. People objectified in this way are then reduced to being economic goods. Third and finally, every person who is exposed to these conditions adopts and incorporates this logic. What this means is a state that is unconsidered and internalized by the individual, which is supposed to run right through the thoroughly economized society.

There are two phenomena that foster reification. On the one hand, it is the thorough economization of all lifeworlds. These new capitalist conditions differ from previous conditions, because they have the same effect on all areas of society, like a "metabolism of a society" (Lukács 2015, p. 14). On the other hand, rationalization, which also extends to the entire society, promotes reifying conditions. "What is new in modern rationalism is that it appears – to an increasing extent in the course of development – with the claim to have discovered the principle of the interrelation of all phenomena that confront the life of man in nature and society" (ibid., p. 51). It takes its scientific-theoretical origin in positivism, which is influential at the beginning of the twentieth century, such that it can be found in the broadest range of scientific disciplines. However, rationalization is also the central mode of operation in administration or bureaucracy as a form of governance, which Lukács states with reference to Max Weber. Overlaps with economics can be found in Taylorism as an economic control principle or in Fordism as a thoroughly rationalized form of industrial production. Connected with this process is the atomization of the single participants in the society. Social relations become less and less important, and an increasingly differentiated society assigns a purely functional place to each individual. The individual here is both interchangeable and at the same time alienated from the result of his or her actions. Lukács thus provides "a theoretical offer of a view of the whole of social existence" (Dannemann 2018, p. 45). Lukács' core statement is that reification is internalized by all participants in capitalist societies and thus becomes a "second nature" to them (Lukács 2015, p. 17).

Axel Honneth (Engl. 2012, German 2015) adapts Lukács' concept of reification against the historical background of the 2000s. He is aware that the social circumstances of the 1920s are outdated; it is no longer the proletariat fighting for its freedom in the face of reifying conditions but rather new categories of inequality and injustice that have emerged in modern society. Honneth, however, does not relate the concept of reification to the changed social conditions but instead makes the effort of transforming Lukács' historical approach into a general concept of reification.

In this respect, he mentions various criticisms of Lukács' text, from the treatment of which he subsequently derives his concept of reification on the basis of the theory of recognition. In Lukács' argumentation, reification emerges only in the course of increasing capitalism, which is why reification in this case denotes a state that must be distinguished from a historically earlier, pre-capitalist state. At the same time, reification, which by its very nature appears as a dehumanizing practice, can be described only in distinction from a correct or true life practice. Lukács remains little definite on this point, which is why Honneth ultimately describes such "true practice" (ibid., p. 26) as a practice "in which humans take up an empathetic and engaged relationship toward themselves and their surroundings" (ibid., p. 27).

Honneth first clarifies that the difference between a true and a reified state cannot be historically described in terms of disappearance of this true state in capitalist society. Rather, a reified state "has merely concealed it from our awareness" (ibid., p. 31). At least latently, the true state is thus still present and can therefore be made analytically visible. Honneth then tackles the problem of the imprecise definition of a true practice of life with reference to Martin Heidegger and John Dewey. From Heidegger, Honneth integrates the concept of "care" (ibid., p. 32). With this, he opts for an approach that describes the relationship to the world not as one in which "humans primarily and constantly strive to cognise and neutrally apprehend reality" (ibid., p. 32) but in which "humans in fact exist in a modus of existential engagement of 'caring' through which they disclose a meaningful world" (ibid.).

The author of the second approach is Dewey, whose:

reflections boil down to the assertion that every rational understanding of the world is always already bound up with a holistic form of experience, in which all elements of a given situation are qualitatively disclosed from a perspective of engaged involvement. (ibid., p. 36)

More than involved, this compassion should also be supportive and positive. In the following, Honneth calls this "primordial form of relating to the world [...] 'recognition' in its most elementary form" (ibid., p. 37). This special form of recognition precedes the three dimensions of recognition – love, rights, and social esteem (Honneth 1994). It is an "elementary recognition" (Honneth 2012, p. 152) of others as "fellow human beings" (ibid.):

The implication for the structure of my own theory of recognition is that I must insert a stage of recognition before the previously discussed forms, one that represents a kind of transcendental condition. The spontaneous nonrational recognition of others as fellow human beings thus forms a necessary condition for appropriating moral values in the light of which we recognize the other in a certain normative manner. (ibid., p. 152 et seq.)

The essential aim of the change in recognition theory is therefore to justify rejection of the separation into subject and object and – by contrast – to give priority to a perception of the world that always presupposes this elementary recognition over mere cognition. To substantiate this thesis, Honneth refers the individual's general relationship to the world to intersubjective relationships between individuals. He does this by using findings from empirical studies on the development of autistic children. Their limited perception of the world does not result from a lack of cognitive ability but rather from a lack of emotional involvement with other people – and here especially and fundamentally with their primary caregivers. The conclusion from these observations is then: Without "emotional identification with others" (ibid., p. 43), no knowledge about them can be gained.[1]

From this observation, Honneth deduces that "recognition and empathetic engagement necessarily enjoy simultaneously genetic and categorical priority over cognition and a detached understanding of social facts" (ibid., p. 52). This means we cannot regard "the antecedent act of recognition [...] as the contrary of objectified thought but as its condition of possibility" (ibid., p. 54). Objectification in this sense has a reifying effect only if the step of recognition before objectification is missing.

The problem that arises from Lukács' equivalence of reification and objectification is that every society would have to be interpreted as comprehensively reified, because in every form of institutionalization, "we neutralize our original act of recognition" (ibid., p. 54). Since this position seems to be too undifferentiated, Honneth suggests overcoming it and distinguishing such insights preceded by recognition from those that try to manage without this preceding affective relationship. "This kind of forgetfulness of recognition can now be termed 'reification'" (ibid., p. 56).

[1] At this point, a change in the theoretical understanding of compassion comes into play: it need not be affirmative or positive, as described by Dewey. Any form of emotional reference is sufficient (Honneth 2012, S. 51).

This theoretical derivation or reformulation of the reification approach primarily refers to intersubjective relations. Reification occurs where people in communicative situations do not recognize others in a certain way; they do not recognize them as fellow human beings or rather they treat them as objects without having first recognized them as fellow human beings. Honneth also includes the other two reference dimensions in a theoretical framework of recognition: the relationship with the world will not be covered in detail here;[2] what is more relevant here is the relationship of a person to himself. With recourse to various forms of self-perception, Honneth establishes that reification of the self can occur when a person perceives his own desires and inner states as ontically independent units, detached from the self. This can be done in three ways: first, when a person follows a cognitivist self-image, his own wishes appear to be objects that are outside the self and can be perceived by observation; second, Honneth attributes this fact to a constructivist self-understanding. The latter seems less intuitive but can be explained in such a way that wishes, once they have been created by their articulation, also appear to the originator as independent objects. The third way is so-called expressionism, which borrows from the two aforementioned approaches but rearranges them. "We neither merely perceive our mental states as objects nor construct them by manifesting them to others. Instead, we articulate them in the light of feelings that are familiar to us" (ibid., p. 71). Only when mental states appear worthy of being perceived do they appear to the self and can they be expressed. Honneth calls this aspect "self-recognition" (ibid., p. 72). With regard to reification, he then concludes that it is precisely the aforementioned forms of self-observation or articulation that can be described as pathological. Both cognitivism and constructivism treat self-parts as objects: in one case as given and observable and in the other case as self-produced. Yet both have a reifying character that denies "access to one's own state of mind" (ibid., p. 90). The concrete relationship between reifying circumstances and self-reification is discussed below.

13.2.2 Survey of Reification: Adaptation of the Models

13.2.2.1 On the Relationship Between Intersubjective Reification and Self-Reification

If reification is based on the forgetfulness of recognition, "its social causes must be sought in the practices or mechanisms that enable and sustain this kind of forgetting" (ibid., p. 79). Honneth distinguishes between two forms of forgetfulness of recognition – intersubjective relationships and self-relationships – since each has its

[2] This step will not be pursued here, as it is of little relevance for the study to be prepared here. To mention only briefly: participation in the world is achieved through processes described by the term "decentration." If a person is able to put himself into the world view of his counterpart, he attains a differentiated picture of the world, which goes beyond his own ego.

own causes. To make the first dimension analytically comprehensible, he mentions two issues: (1) the purpose of an interaction becomes an end in itself and sight of the relationship to the interaction partner(s) is lost and (2) an interaction partner represents a dehumanizing ideology (see ibid.), for example, a racist ideology.

It is also essential with regard to institutionalized forms of reification that the forgetfulness of recognition may also become apparent when people "continuously contribute to a highly one-sided form of praxis, that necessitates abstraction from the 'qualitative' characteristics of human beings" (ibid., p. 155). Permanence ultimately leads to habitualization of practice (ibid., p. 157).

Honneth describes the connection between reifying social conditions only cautiously, as they can be explained by intersubjective relations and their potentially reifying effects on the self of those affected (Honneth 2015, p. 94).

In this regard, for the empirical study, a distinction can be drawn between three stages. First, there are the institutionalized reifying conditions, such as can occur, for example, in rationalized administrative procedures or in the penetration of an economic logic into various social areas. Second, there are the manifest reifying practices on an interactive level by those who act under these conditions. Third, it is the individuals who in some way have to deal with these conditions and everyday interactions and who are in danger of being reified in their own selves.

At least two forms of such self-reification, which can arise from external pressure, should be distinguished here. (1) Deficient bonding experiences that took place in the course of child development, such that the child was unable or had only limited ability to enter into an affective relationship with him/herself and with others. This form could be explained by attachment theories (see Bowlby 2006). (2) Reification of the (also adult) individual is a consequence of reifying practices by individuals, institutions, social procedures, structures, or discourses to which the individual is exposed. In the research interest pursued here, the second form is more relevant.

In terms of the link between institutional reifying practices and the reification of individuals, Honneth remarks that it is possible "to search for the causes of reifying behaviour in social practices that are connected with the self-presentation of subjects in the broadest sense" (ibid., p. 82). This issue plays an important role during the asylum procedure, as shown below. However, other practices that Honneth mentions, such as racism or treatments that reduce the purpose of an interaction to "an end in itself," also occur.

13.2.2.2 Embedding in Historical References

The present study looks at experiences of institutionalized forms of reification that occur in relations between individuals and institutions or establishments, which are referred to here as social-pathological forms of reification (Honneth 2012, p. 84). Institutionalization here refers to authorities or (e.g., social, educational, care) institutions with which refugees are confronted to a huge extent in their daily lives. This restriction makes it possible to relate Lukács' broad concept to a specific context of reification. At the same time, Honneth's concept, which is very general and therefore sometimes vague, becomes historically embedded.

Since Honneth refrains from systematic historical embedding of reification, he finds it difficult to identify examples of it within social normality. He cites examples that appear only on the fringes of society and writes that he has "realized that it is only in rare and exceptional cases, only at the zero point of sociality that we find a true denial of antecedent recognition" (ibid., p. 157). He then cites "modern forms of slavery, such as are found in the sex trade" (ibid., p. 158), but it is industrialized mass murder in particular that can be regarded as the most extreme form of reification (see ibid.). These extreme examples can certainly be regarded as evidence of reifying practice, but where is the strength of Lukács' model that facilitated making reifying conditions visible in everyday life as a usual practice within social normality? If this selective view of extremes were taken as a guideline, Honneth's concept of reification would not be suitable for the analysis sought here, since – even though they are not an everyday occurrence in state enforcement – asylum procedures are regulated internationally and nationally.

In the following remarks, the historical reference is reinforced, and Lukács' basic assumption that reification is a consequence of spreading capitalism is regarded as having more significant importance again. However, Honneth's criticism of Lukács, that there is a need for precise justification of how the economic part of the economy, in interaction with rationalization, extends to other areas of society, must be taken seriously. A summary of the development of administrative actions is therefore provided below. This shows that capitalist influences are a reality in twenty-first-century Austrian public administrations and that they represent or create reifying relations.

13.2.2.3 Objectification and Reification

Honneth's concept of reification has the precondition that persons within a social practice are primarily not recognized as fellow human beings or at least treated in a reifying manner over a long period of time. By contrast, he does not accept as reification the simple fact of objectification of people in the course of a "'depersonalized' relationship of commodity exchange" (ibid., p. 76), because people who participate in barter transactions must first be considered "as accountable exchange partners" (ibid.). However, according to Honneth, this means that the condition of the forgetfulness of recognition and consequently of reification is not given. Contractual partners cannot objectify one another since, in order to be able to enter into a contract at all, they must recognize one another as legal entities. David Ellerman (2013) argues that employees, who enter into a contractual relationship through a contract of employment, can be reduced to their value as providers of their labor. According to Ellerman, an employment relationship is contractually similar to slavery, except that we no longer belong entirely to the slave owner but that we or our labor is hired. The contractual constellation thereby deprives us of fundamental rights. With Ellerman's argument, it is impossible to uphold Honneth's idea that contractual partners generally cannot act in a reifying manner. Honneth himself weakens his argument when he provides examples in which "the legal substance of the labor contracts" is undermined and thus "institutionalized barriers that have prevented a denial of our recognitional primary experiences are threatening to collapse" (Honneth 2012, p. 80). Under this premise, however, it is no longer possible to distinguish categorically between objectification and reification. There is rather a gradual difference.

This does not mean that the priority of recognition over cognition and the criterion of forgetfulness of recognition should be abandoned. What seem to be more essential are the external circumstances that accompany practices such as the conclusion of contracts, which presuppose the recognition of the other person as a (legal) subject. If these external circumstances promote reifying content in such practices, no categorical distinction between objectification and reification should be made here. On the basis of the empirical material, it is certainly possible to show how such circumstances can arise.

The following empirical study comprises two parts:

1. General evaluation of rationalization and economization of administrational practices. Starting from Max Weber, the modern administrative logic introduced since the 1990s as New Public Management is traced.[3] Here, the influence on (social, educational, advisory) institutions that are financed by the public sector also plays a role. This overview refers to the institutional background, which develops a reifying potential for asylum seekers and recognized refugees.
2. Consideration of specific circumstances concerning the issue of asylum. This part is divided into two steps. The first step shows that, on the level of justice, asylum seekers and recognized refugees do not have the status of recognized legal subjects. The second step is an empirical analysis of the data material from the PAGES project. This is examined with regard to reception and the way in which refugees deal with reifying circumstances. This part of the analysis focuses on the level of the individual and is intended to provide indications of how institutionalized reifying practices affect the people.

The order of Honneth's concept of forgetfulness of recognition is thus reversed in this empirical study. First, it is shown that administrations act in an objectifying way. Then the extent is shown to which prior recognition as a (legal) subject – which is a necessary condition for reification – is denied.

[3] An explanation is required of why the focus here is exclusively on state administration and why the judiciary is not considered relevant. This is because, in Austria – in addition to the basic provision of asylum seekers, which is regulated by the social departments of the federal states – asylum procedures themselves also lie in the hands of the state executive and do not fall within the jurisdiction of the judiciary. In the first instance, the Federal Office for Asylum decides on applications; in the second instance, it is the Administrative Court. This is part of the administration and the administrative judges are administrative officials rather than judges in the strict sense.

13.3 Rationalization and Administration

State administration is the essential social sphere from which reification emanates, in relation to the subject under discussion here. It is in contact with the authorities that the roots of the reifying experiences of asylum seekers and recognized refugees lie.

The Austrian executive, which comprises the subject under discussion here, consists of:

- Asylum authority (Federal Office for Asylum, which is located at federal level and operates in the federal states; this decides on asylum applications in the first instance)
- Federal Administrative Court (which decides on asylum applications in the second instance)
- Basic services (which are represented by the Austrian federal states and govern accommodation, basic financial services, and counseling)
- ÖIF (a unit belonging to the Federal Ministry of the Interior, which is primarily responsible for recognized refugees: initial counseling on jobs and housing, offering so-called value courses)
- Labor office (in Austria: AMS, which the people concerned encounter at the regional level in the regional offices of AMS)
- Social welfare office (at the level of the municipalities or cities)
- Social work measures involved (usually financed by the federal states in the course of the application process, as part of basic services: mainly counseling and care for asylum seekers; for recognized refugees, various measures are used depending on specific problem situations)

Below, the procedural logic of modern state administration is briefly traced[4] in order to show how it is interwoven into the economic and social realm and how this influence – in the sense meant by Lukács – has a reifying effect.

13.3.1 Rationalization According to Max Weber

For Weber, rationalization and the development of bureaucracy are closely linked: "Rationally socialised communal action of a governance structure finds its specific expression in 'bureaucracy'" (Weber 1980 [1922], p. 551). However, Weber not only refers to bureaucracy in terms of state administration; he understands it as a form of organization or rule that is applied in the most diverse social areas, whether in politics, the economy, or state administration. Max Weber observed that bureaucracy is indispensable in large state structures: "The fact that the longer, the more technically dependent the modern large state is on a bureaucratic basis par

[4] A more detailed history of bureaucracy can be found in Becker 2016.

excellence, the larger it is, and above all the more it is or becomes a great power state, the more unconditional it is, is tangible" (ibid., p. 561). At the same time, bureaucracy is increasingly rationalized in the modern state system. Rationalization makes it possible to guarantee equal treatment for all citizens. "For the specific character of modern loyalty to an office, it is decisive that in the pure type it does not establish a relationship to a person – e.g. such as the loyalty of the vassal or disciple under a feudal or patrimonial governance relationship – but is rather devoted to an impersonal and *functional* purpose" (ibid., p. 554). It is precisely this strict orientation of official action toward a factual purpose that is connected with the reduction of the counterpart to one case or to the nature of the same as an applicant – relieved of personal characteristics – that ensures equal treatment of all citizens in the state: "In this case, 'factual' settlement primarily means settlement 'without respect for the person', according to calculable rules" (ibid., p. 562), Weber writes. This applies in particular to democratic constitutional states that claim to treat all citizens fairly, regardless of who they are: "'Legal equality' and the demand for legal guarantees against despotism require *formal* rational 'objectivity' of the administration, by contrast with the personal free will and favour of old patrimonial governance" (ibid., p. 565). This is especially true in the light of the fact that these same citizens can indirectly employ the civil servants in a democracy, such that they are ultimately their representatives: "A civil servant appointed by election of the subjects is essentially independent of the officials superior to him in the successive stages, since his position is derived not 'from above' but rather 'from below'" (Weber, p. 555).

The critical theory of the postwar period takes a stand against the rationalized administration that questions the neutrality of rationalized administrative action. After the experiences of National Socialism, there was a greater focus on the goal of administrative action. According to Theodor Adorno, rationalized administration is not harmful, as long as it serves the people: "When we criticise the administration, we are not criticising rationality. We are not criticising the fact that human conditions are planned as such in order to reduce the suffering that is certainly present through the blind play of social forces" (Adorno et al. 1989, p. 127). On this point, the authors agree with Weber. Administration becomes harmful only when it serves individual interest groups, which moreover pursue specific, quite irrational interests. In recent years, however, the goals of administrative action have increasingly been forgotten, as shown in the following, on the reorientation of administration in the form of New Public Management.

13.3.2 New Public Management

In the 1990s and early 2000s, radical reorganization of administrations took place, which began in the Anglo-American area and later continued in Central Europe. Administrations were reorganized in the form of New Public Management. This reorientation is usually justified by the social change that formed the stimulus for reorientation of the public administration:

> Since the beginning of the century, when Weber conducted his study of bureaucracy, society has changed from an industrialised society marked by years of war and crisis to a modern information and consumer society. (Schedler and Proeller 2011, p. 27)

Added to this was the lack of money available to the state, combined with the impression that administrations were acting in a costly manner and sometimes with little ability to solve problems (ibid., p. 29). While Weber argued that only the administration made it possible to maintain large state structures in the long term, the administration now seemed to have become more of an obstacle to state development. It therefore seemed essential to carry out administrative reform. As liberating as this service-oriented approach may have been in the 1990s, it reads critically, because "the service-orientation of the administration, despite all efforts, has its limits, which are not to be found in the attitude of the civil servants but in the logic of the procedures" (Becker 2016, n. pag.). Particularly with regard to the object considered here, which is in the area of welfare state services, this internal administrative logic must be criticized, as shown below.

The main difference in the new form lies in the reversal from input to output orientation. This is compensated for by predefined results, which the respective administrative unit is required to achieve. However, this is also associated with the need for controllability of performance and the provision of services. One condition of payment for the service is that the result is precisely defined, so that it can be made measurable. In the field of social services, this process means the advance of management logic into the field of social professionalism. A gap arises from the requirement of management to define controllable (i.e., pre-formulated) goals, which contradicts the requirement of professionalism to develop goals only in cooperation with the people affected (Pantuček-Eisenbacher 2006). Ideally, results should now be evidence-based (Albus et al. 2018). Evidence in this context means what is scientifically verifiable. As a rule, the concept of "evidence-based practice" is founded on a positivist understanding of science that is based on causal assumptions (Kleve and Früchtel 2011), so that the characteristic of what Lukács called rationalization reappears here by a roundabout route. However, this positivist understanding is inconsistent with a reflexive understanding of profession in the social sphere, which envisages the formulation or development of goals in dialogue with those affected (Otto et al. 2010). This gives rise to an internal contradiction of this form of control, at least in the social sphere; one of the demands on the new form of control is to leave a large part of the responsibility for the success of the result to the citizen, entirely in the tradition of increasing individualization. The key concept here is an "activating" welfare state. At the same time, however, the citizen is not involved in formulation of the desired results (output); rather, following the idea of evidence-based research, these should be derived from scientific research, albeit under the premise of a specific form of science. New Public Management has characteristics that are very similar to Lukács' description of a thoroughly economized society (calculability of action, positivist understanding of science as a motive for action). Also in the Weberian tradition of bureaucracy, the consequence is desubjectivation of the individual. Moreover, where the formulation of goals, i.e., determination of the output content, is concerned, nothing whatsoever is said about who sets these goals. Who connects which interests with the achievement of specific goals? In the case of scientific studies, who commissioned the study that is used as a basis for evidence? Who finances it? (see Webb 2010).

In New Public Management, for the first time in the history of administrations, the direct interconnection of economy and administration becomes an empirical reality; "'colonization' through principles of the capitalist market" (Honneth 2015, p. 94) takes place in real terms. The advantage of equal treatment of all citizens "regardless of their person" no longer exists, because effect-oriented governance also means giving preference to those citizens who are able to navigate a rationalized world with self-responsibility. The disadvantage is for those who – whether because of their personal history or because of structural disadvantage or discrimination – cannot subsist in this system. The achievement of objectives is not oriented toward their needs and interests but rather toward evidence-based settings and to the economic efficiency of the procedures.

According to Adorno et al. (1989), one justification for rationalized administration lies in its objectives, as long as it serves the citizens. However, the political discussion about goals has been forgotten as a result of commoditization of the administration, namely, through New Public Management and the associated evidence-based setting of seemingly objective necessities. Political decision-making also means formulating goals in accordance with the needs and interests of the people affected. However, because of the advanced scientific nature of goal-setting, this claim is no longer the focus of attention. Input has been replaced by output. As a result, political decisions are now dominated by supposed constraints and efficiency.

13.4 Reification and Asylum

It could be shown that, in addition to the rationalization inherent in administrative action, explicit commoditization has also been introduced. Through these new economized conditions, the addressees of administrative action are increasingly desubjectivated. Through the focus on output and efficiency, the addressees become objects of measurement. This circumstance can easily be described as objectification. However, Honneth's claim concerning reification is formulated in a more sophisticated way: to be able to speak of reification, one must forget about recognition, i.e., prior recognition as a fellow human being or at least as a legal subject; only if this is the case, then reification can occur subsequently. The further argumentation aims to show that this preliminary recognition does not exist for people with a history of flight.

13.4.1 Inequality of Rights

Once they have arrived in Austria, refugees file an asylum application or, more precisely, an application for recognition as a refugee according to the Geneva Convention. This is the condition for the asylum procedure to be initiated. What seems essential here is that the application is not for recognition of the reasons for flight but for refugee status. This formulation is tantamount to invoking status "as a refugee" in the sense of Judith Butler (Grönheim 2018).[5] Through this process, people who all belonged to very different groups before flight, whether in social rank, ethnic/national origin, or age group, are homogenized and lumped together in one group.

This attribution was also a topic in the PAGES project, which provided the data material for the study. For example, one project participant said that after about a year of cooperation, he no longer wanted to participate in the project, because he was participating "as a refugee." Although he was quite positive about his experiences in the project and leaving the project meant that he would lose new friends, he preferred this consistent step, because he saw no other possibility of escaping the label "refugee." Hannah von Grönheim arrives at a similar conclusion and remarks:

> If refugees are made refugees because they are classified as such as individuals when they enter the German asylum system, they are thereby exposed to a disempowerment and objectification of the subject. (Grönheim 2018, p. 252)

In addition to this discursive or symbolic level, the allocation of status also has a very tangible effect, according to Henri Louis Seukwa, because through the asylum procedure, "they [the asylum seekers, note HB] become objects of a legal construction [...] which denies them unrestricted status as legal subjects" (Seukwa 2006, p. 256). The author justifies this by the fact that there is a "barely comprehensible system of legal regulations (including social legislation, right of residence, asylum law)" (ibid., p. 257), which makes it impossible for those affected to find their way around and actively develop a life perspective.

In everyday life, the system is reflected in comprehensive official influence over almost all areas of life, whether in housing, work, education, childcare, or basic financial provision. On the one hand, the ubiquitous presence of the administration increases the reifying effect of the administrative action; on the other hand, it produces this effect against the backdrop of the legal inequality of asylum seekers and recognized refugees compared with citizens of the state. One of Max Weber's remarks was that it makes a difference that the civil servant is appointed "by election of the subjects" (Weber 1980, p. 555; see above). However, this is precisely not the case for people with a history of flight. The next step is to show that in principle,

[5] Hannah Arendt already describes this phenomenon very impressively in her essay "We Refugees," which she begins with the words: "In the first place, we don't like to be called refugees" (Arendt 2016 [1943], ibid., p. 9). Online: https://www.documenta14.de/de/south/35_we_refugees [06.05.2020].

in the case of asylum seekers and recognized refugees, there is no legal equality vis-à-vis citizens.

Even if asylum seekers are integrated into an existing legal system by virtue of their status and thereby experience a fundamental form of just treatment – the asylum decision is made according to the guidelines of the Geneva Convention and they have access to basic services and security (Forst 2012, p. 165) – they do not participate in the decisions on distribution of these resources. In Rainer Forst's approach, the political framework within which justice takes place is the state, although he includes in his perspective the relations between states that are subject to the question of just relations and their consequences (Forst 2012). Forst does not systematically pursue the question of the rights of people who have been forced to migrate from one state to another as a result of war, disaster, or persecution. However, he remains open in his considerations and, with regard to his central justice criterion of "reciprocal and general justification" (Forst 2005, p. 28), speaks of a "noninstitutional perspective of moral rights and duties which apply to every member of the human moral community regardless of political settings" (Forst 2012, p. 172). In doing so, he follows an understanding of justice that is in line with human rights and the basic justice of distribution (ibid., p. 169). Central to this are the concerns of the "worst off in a society" (ibid., p. 29). To a lesser extent, the statements apply not only to asylum seekers but also to recognized refugees, since they can also be counted as part of this group:

> It is essential that the members of these groups [the worst-off in society, HB] are not regarded as objects, but as subjects of justice; the first task of justice is to enable them to participate and participate genuinely in the institutions of society. (ibid., p. 29)

It is now open to debate, whether asylum seekers should be able to participate in political decision-making processes. For this group, however, participation does not even take place at the preliminary stages of participation, which begin with informing people (Wright 2010). Participation, according to Wright, takes place only when people are able to have a say in their own living conditions. However, neither sufficient information nor codetermination with regard to concrete everyday decisions take place for asylum seekers. All this contributes to jeopardizing their status as politically recognized subjects. In the sense of moral justice, asylum seekers and recognized refugees do not have the status of legal subjects, which they could acquire only if political decisions that concern their lives were justified to them. Participation or at least a certain level of participation would provide them with basic status as people with a minimum set of citizens' rights, but they are denied this.

Both of Honneth's conditions of reifying practices are given: (1) there is no basic recognition as a (here: legal) subject and (2) as shown above, administrations act – through rationalization and economization – in an objectifying way.

13.4.2 Manifestations of the Reification of People with Experiences of Flight

This chapter shows the specifics of reification, to which people with a history of flight are exposed in Austria during and after their asylum procedure. The basis for this is the data material collected during the action research project PAGES. This comprises the experiences of 30 people, collected and recorded in various ways including guideline-based interviews, observation notes, research diaries, WhatsApp or email communication, written reports from participants, photos, or experiences recorded as drawings. From this complex material, four examples that exemplify the specific facets of reification are recounted and arranged systematically.[6]

13.4.2.1 Reduction to a "Case"

Yakub,[7] a young asylum seeker living in a refugee camp in a village in 2016 and 2017, tried to find a job. As he had no possibility of finding a legal, paid job as an asylum seeker, he asked around the neighborhood whether he could provide help – also unpaid – for example, with gardening. However, none of the neighbors wanted to accept this offer, Yakub said. The PAGES employee who accompanied Yakub had the idea of asking the nearby institution for mentally handicapped young people whether voluntary work or an internship in one of the workshops was possible. Yakub was interested because he had worked as a welder before and the facility included a locksmith's shop. The two of them – project staff member and Yakub – went to the facility in May 2017 and were quickly given a friendly welcome by the person responsible in the Human Resources department. After this initial meeting, however, a weeklong process of gathering information from the relevant offices in the state of Salzburg began. The employees in the Human Resources department turned out to be extremely tenacious and tried to find a form in which an employment relationship was possible. The whole process took until the end of July, when Yakub was finally able to start as an intern in the locksmith's shop at the institution. During the process, however, he was completely at the mercy of the procedure and depended solely on the actions of the Human Resources department, which supported him. Owing to the complex legal framework and the barely comprehensible responsibilities of the various contact people involved on the side of the state, Yakub was not in a position to act on his own but was dependent on the HR department. The Human Resources department was very sympathetic to him but – because of the circumstances – treated him not so much as an individual but rather as a refugee case. Fortunately, this changed with the start of the internship, when Yakub had the opportunity to appear as an individual.

[6] Since the author worked with a smaller group of four participants, the majority of the examples come from this group. These are supplemented by examples from other participants.

[7] All proper names have been changed.

This example shows that the person concerned had no opportunity to get involved for himself or even to influence the general conditions. This began with the choice of accommodation to which he was assigned without being asked and which – owing to the village environment – provided him with no alternative employment. He had exhausted his only means of addressing neighbors directly and was then dependent on help from others; after that, he was **perceived as a "refugee case".** This corresponds to the criterion for objectification, as set by Honneth, who writes that we are dealing here "with a process in which cognitive goals have become completely detached from their original context, with the result that our cognitive stance has become rigid or overemphasized" (Honneth 2012, p. 60). In the field of social work, such a professional practice is considered normal in which not the clients are regarded as cases but their problems or the situations in which they find themselves are described as cases (Müller 2009). However, this is precisely what cannot be seen here. The reduced perception of being an object of the actions of others and the resulting helplessness outraged Yakub. He articulated his displeasure several times during the long wait for a response to his application.

This example could be interpreted as objectification (not yet as reification), which is the result of an NPM-oriented administrative practice. Yakub's experience becomes a case of reification only against the backdrop of the second condition described above: as an asylum seeker, he is not a fully-fledged legal subject and he has no opportunity to participate in the process. In this sense, he is not an agent of the process but becomes reduced to a purpose as "an end in itself" (Honneth 2012, p. 79).

With his positive decision on asylum, the experience was repeated. Yakub, who now had access to the labor market through his new status, spoke to the regional labor market service. There, he was again advised and referred to the labor market without having any background knowledge or information, without being able to participate actively in events for himself. In an interview he said later:

> She [the AMS employee, note HB] is looking for an apprenticeship or if she finds one, she will let me know when we have found one. I don't know until now, what does the AMS do. What is their work and what do they do? But today, we made an appointment and should I go there. (Interview with Yakub)

13.4.2.2 Two Dimensions of Recognition: Justice and Emotional Affection

Another example, this time in the context of the Yakub's social care, shows how social institutions also can contribute to the desubjectivation of asylum seekers. Yakub had meanwhile moved to a small town, because the opportunity had arisen to take part in a compulsory school-leaving course. His German teacher had helped him to attend this course, i.e., again he would not have had the opportunity by himself to obtain information about this offer, to register for it, or to apply for a move to an institution near the course. Since the course was at least three-quarters of an hour's walk from his home and the autumn had brought continuous rain, Yakub asked his social worker whether her organization could finance a bus ticket for him.

He had heard earlier that this was possible. Her immediate reaction was incomprehensible to Yakub, but he remembered the way she talked to him. He says in an interview:

> Not friendly, no, not friendly, and the one time when I say, 'I need help,' she says, 'Yeah, okay, fine.' And just like that. And she talks so much. She says when I want to say something, she says, 'Stay. I want to talk. You just listen.' And that was bad. Yeah, that was not friendly. (Interview with Yakub)

What exactly the social worker wanted to convey to him is no longer clear, but it becomes apparent that Yakub felt like he was not treated with sufficient sensitivity. Whether this was due to the personality of the social worker or to the difficult working conditions to which she herself was subject is hard to judge. Certainly, the external conditions and especially the high level of care, owing to the efficiency of the financing authority, contributed to the fact that clients could not be offered a sufficiently personal response.

Another characteristic can be observed in this example: Yakub received a bus ticket after a month. For him, however, this manifest aspect was less important in the description of the experience – we could generalize it by referring to access to resources – than the infantilizing treatment by the social worker. The case therefore refers to two dimensions of recognition: on the one hand is the legal recognition associated with the right to the bus ticket and on the other hand is the reifying treatment that takes place at the level of emotional recognition. Yakub was particularly indignant about the second level in his observation of the process. The fact that he had to wait a month for the ticket and had to walk to the course in the rain during this time was no reason for him to be annoyed. This **phenomenon of entanglement of two dimensions of recognition** occurred again and again in the course of the project. The participants did not always take offence at the disregard of recognition within only one of these dimensions. However, they were often able to separate the different levels, as the following example shows. Through the interconnection of refusal of recognition on different levels within an interaction, an intersectionality of disregarded recognition could be detected.

13.4.2.3 Self-Presentation and Self-Reification

A third example describes another aspect of reification. Amin fled from his homeland because of his conversion to Christianity. His application for asylum had been rejected by the asylum authorities in the first instance, and he was summoned to appear before the administrative court of second instance following his appeal. He went with his lawyer and a friend who was able to attend the trial as a personal escort. Amin and his friend later independently described very similar impressions of the trial. During the trial, which lasted several hours, the administrative judge did not address Amin personally. She spoke only to the interpreter. When Amin quickly answered a question about the Christian faith, the judge turned to the interpreter with a sarcastic comment, in the sense of "look, he has learnt it well by heart." If, however, he hesitated with an answer, she purported to detect uncertainty, which made his reasoning appear implausible. Amin had the impression that no matter what he said, it would be interpreted against his intentions. Immediately after the hearing, the administrative official said that the application would not be granted. The written rejection decision followed weeks later.

Such desubjectivation by the asylum authorities is by no means an isolated case. This is a result of Julia Dahlvik's study on asylum procedures in Austria:

> However, a turning away from the individual (de-personalisation) can also be observed in the interrogation situation, which in itself can be a very intimate face-to-face communication, as asylum seekers are also questioned about their private lives. For example, it has been observed that a speaker speaks to the interpreter about instead of directly to the asylum seeker. (Dahlvik 2016, p. 197 et seq.)

The indignation expressed by Amin and his companion was partly due to the lack of recognition on the legal level, in this case with regard to the asylum application. However, the treatment or rather nontreatment or disregard for the applicant by the official was also a reason for their indignation. Again, as in the last case study, refusal of recognition can be observed on two levels: on the legal level and on the personal level.

Another factor in this example was also striking. This is the element of arbitrariness in the evaluation of Amin's statements. No matter how he expressed his knowledge of the Catholic faith – hesitantly or fluently – he could not convince the official. A similar phenomenon occurred in the asylum procedure with regard to Amin's efforts to integrate. In the first instance, the negative asylum decision stated:

> It has been established that you left Iran because of the economic situation, better educational opportunities in Europe or for other reasons. [...] with regard to your situation in the case of your return: [...] you would be willing and able to accept any kind of work in Austria. It cannot be said that you could not do the same in Iran. (Amin's asylum decision)

With his asylum application, Amin had enclosed various certificates and confirmations from internships and from institutions where he volunteered. Amin's willingness to make an effort to integrate was used as an argument for rejection. Amin reported after the hearing in the second instance:

> And our judge said that we have no integration, although we had so much integration in Austria. But perhaps this integration, the contact with people, congregation or Church, that is not the opinion of this court. Maybe she meant that we have to earn and take care of ourselves, and that's why I always try to work. (Interview with Amin)

Now he was accused of having made too little effort to integrate. The unpredictability of the procedure is also evident in this case. On the one hand, there seem to be criteria for evaluation in the asylum procedure, but on the other hand, it is not clear which characteristics within these criteria are claimed to be positive and negative.

The interrogation by the asylum authority and in the second instance by the administrative court requires the people concerned to **tell their life story in a preformed way**. This narrative must meet the assessment criteria of the deciding authorities and therefore follows the basis of reifying forms of self-representation. In Honneth's words:

> Institutions that latently compel individuals merely to pretend to have certain feelings or to give them a self-contained and clearly contoured character will promote the development of self-reifying attitudes. (Honneth 2012, p. 83)

One aggravating factor is that it is **impossible for applicants to predict the outcome** of the evaluation because, as the examples show, evaluations can be arbitrary.

In the three case studies reported so far, the main focus is on cases of recognition in the field of law or social esteem. Yet they are always accompanied by rejection of the people concerned as fellow human beings. This can be seen in disregard or in inability to act in a situation, because the necessary information is lacking and the person concerned is overwhelmed and cannot foresee the consequences of his or her own actions. This is also where the fact comes into play that the individual feels treated as a case.

13.4.2.4 (Everyday) Racism

As an important reifying practice, Honneth mentions treatment of people that follows a dehumanizing ideology. This includes racist ideologies. The fourth example deals with a specific form of racism, to which participants in the project were repeatedly exposed. What is meant here is not only overt racism but also a more implicit, sometimes unintentional, form of everyday racism (Terkessidis 2004). Mark Terkessidis defines various characteristics of everyday racism, which – in addition to devaluation of those affected – initially include desubjectivating attitudes or actions. Terkessidis calls the central desubjectivating practice "Entantwortung" (ibid., p. 186 et seq.), which means "de-responsibilization." This happens when one person looks at another exclusively on the basis of stereotypical characteristics and thus removes him or her from individuality. Such treatment is difficult to classify for those affected, because the practice is often diffuse and not clearly expressed.

At a project meeting, Amin reported for the first time explicitly about the racist experiences he had experienced at the AMS (the Austrian job center). He recounted a conversation with the AMS adviser. When asked by the PAGES project researcher, he emphasized that the adviser had not only been reluctant to provide information but also deliberately refused to pass on information. This situation concerned seasonal work, to which asylum seekers have access. The example of the negative asylum decisions already showed that Amin was very keen to participate in the society. Taking up seasonal work would therefore have offered him the opportunity to earn money himself, but the aspect of integration was also important to him. He was even more frustrated by the open rejection of the adviser.

The PAGES participants repeatedly reported racist experiences during the course of the project. Sarah, for example, took part in a joint project activity in which the participants produced handicraft items and sold them at a weekly market in Salzburg.

There, she was insulted by a passerby because she wore a headscarf. Another example was given by Ben, whose friend had been called a monkey by a clerk at the asylum office.

Another of Yakub's experiences shows that an experience of racism can be diffuse and difficult to classify. When asked by a project staff member about an application course, he replied that he was not sure whether he should continue to attend this course. He wrote on WhatsApp:

> I was at the AMS and I already told you about it, but my adviser didn't do anything and she said I had to do this course. The teacher is from Salzburg and he always speaks dialect and I don't understand him and I said three times 'Please, talk clearly', and he said 'You have to learn dialect'. But I need time, and last time he explained about a lesson and I didn't understand and he was angry. (Yakub, WhatsApp message)

Yakub was not sure whether he had been racially excluded here but he was outraged at the teacher's behavior. He had the impression, he said later in conversation, that the teacher deliberately excluded him without directly devaluing him. Yakub's self-assessment is characterized by self-confidence in his own achievements. The teacher's generalized assessment that he did not make enough effort at language acquisition therefore hit him particularly hard.

These few examples are representative of many such experiences gathered in the course of the project. All these experiences of racism took place in different areas of life and in different forms; sometimes they were subtle in the sense of everyday racism, and sometimes they were explicit. However, the reifying effect was particularly pronounced in the context of authorities or social institutions, where everyone expected to be treated fairly.

13.5 Conclusions: Summary and Reactions

The present chapter deals with the question of whether reifying conditions can be assumed in the context of asylum in Austria and if this is confirmed, how such conditions affect people with experiences of flight.

To begin, Axel Honneth's concept of reification, which he developed in contrast to Georg Lukács' approach, was described. In Honneth's understanding, rationalization remains an essential component of reification. However, the reduction of people to their commodity value – the central issue for Lukács – has been classified as a historical circumstance. Honneth defines "forgetfulness of recognition" as a prerequisite for reification. According to this, there must be prior rejection of the counterpart as a subject in order for reification to occur subsequently in the course of objectifying treatment.

Later, the theoretical approach was adapted to the empirical survey. It was assumed that reification is not only a marginal phenomenon of (here: Austrian) society but in the course of asylum procedures that it also appears as part of social normality. By contrast with Honneth, it was then argued that a reifying hierarchy can also arise between contracting parties of equal legal status, if one contracting party is not free in the decision to conclude a contract, owing to existing power relations.

In the empirical study, the general practice of the authorities under conditions of New Public Management was first described as thoroughly rationalized and thoroughly economized. Thus, one condition for reification is present. For the group of people affected here, it is also the case that they cannot be described as having equivalent legal status to citizens. This circumstance represents the second condition for reifying relations.

In summary, various manifest forms of practice that appear under these reifying conditions were described and structured:

- *Reduction to the "case"*: This means individuals are not perceived in their individuality but rather treated as a case and therefore represent only the purpose of an institutionalized practice.
- *Interaction of several dimensions of recognition*: Here, it becomes clear that desubjectivation (interpreted here as a refusal of recognition) on an emotional level and refusal of recognition on another level (of law or social esteem) often occur together. It is suggested that this coincidence be named reification. However, these occurrences are evaluated independently of each other by the people affected. Sometimes, the negative effects reinforce one another; in which case, we may call the phenomenon intersectionality of recognition relationships.
- *Self-reification as an incremental component of the asylum procedure*: In the asylum procedure, asylum seekers are forced to recount their lives according to predetermined criteria. This promotes self-reification. A further complicating factor is that, owing to arbitrary, incomprehensible decision-making practices, they cannot predict whether their self-representation will produce a positive result.
- *(Everyday) racist experiences in contact with authorities*: Subtle or explicit racist experiences are suffered by many people in Austria who have a history of flight. They find it particularly outrageous when these experiences occur in contact with authorities or in the course of social interventions, because they expect to be treated objectively and fairly here.

In conclusion, some considerations are presented with regard to how such reifying circumstances can be handled.

So far, the indignation of project participants has been expressed as a response to this objectifying treatment. Ernst Tugendhat interprets indignation as a response to an action that is considered to be morally wrong. Tugendhat says "that if one member violates the commonly accepted norms, the others react with a negative affect and this affect can be called indignation" (Tugendhat 2006, p. 19). Yakub's reaction to being treated "as a case" and Amin's indignation at being ignored in court and at the randomness of the decisions concerning his application should therefore be understood not merely as anger but rather as an expression of moral outrage at the official practice that made one person an impotent recipient of assistance and the other an object of proceedings. At the same time, however, their outrage indicates that the two see themselves as members of the social community; otherwise, they might not have felt entitled to outrage. Alone, they are barely heard. The connection between reification and indignation seems to be clear. Several project participants repeatedly verbalized this immediately after such or similar experiences of disregard.

During the project, several participants also became apathetic, a state in which they were completely unable to act. However, it is difficult to prove a direct connection between this and reifying practices. There can also be other reasons for such feelings of powerlessness: frustration over negative decisions, the absence of family members, or a lack of job opportunities. These can be discussed in terms of recognition theory but are not necessarily related to reification.

One observation, however, could be made in connection with reifying experiences: in those affected, they sometimes awakened awareness of structural disadvantages and consequently a spirit of resistance. This begins with the feeling and expression of indignation, which can be a first step toward resistant practice. A final quote from project participant Anika shows that consideration of collective experiences can have the effect of raising awareness. She said:

> 'You are in the asylum process, you are nothing, you need nothing, you have no value.' It's different with you [with the PAGES researcher, note HB] and the project: as if I belong, I am a part of here, of the society in Austria, of everything that is done here. [...] That is only with you; we have only experienced and felt this with you, with the project and with the activities. (Interview with Anika)

Desirable responses on the part of the executive and social services, which can be inferred from these results, can be divided into three levels. (1) First, they concern the practice of social work. Wherever possible, this should try to create a balance against the reification of the official treatment. This can be achieved by social workers meeting their clients as individuals and not as cases (Müller 2009). However, resources are necessary for this: time for the individual person and for relationship work (Ansen 2009) and the opportunity to reflect on experiences in supervision or racism-sensitive professional training, to name but a few. In addition, an attempt should be made to empower those affected and encourage them to participate politically within the bounds of possibility, whether through exchange, reflection, or support with articulating their interests. Access to information about legal options and legal representation are also possible means. (2) In order to reduce reprehensible practices from the outset, the state executive should be oriented toward similar principles and consider just and social goals, for the achievement of which appropriate input is necessary. Output is certainly important but the central question should be which socially desirable, just goals must to be pursued (as output)? (3) Finally, the political participation of asylum seekers and recognized refugees is an essential step that helps to prevent their reification from the outset, since this is the only way that forgetfulness of recognition can be prevented.

References

Adorno, Theodor W., Max Horkheimer, and Eugen Kogon. 1989. Die verwaltete Welt oder: Die Krisis des Individuums. In *Gesammelte Schriften. Volume 13. Nachgelassene Schriften 1949–1972*, ed. Max Horkheimer, 121–142. Frankfurt a.M.: Fischer.

Albus, Stefanie, Heinz-Günter Micheel, and Andreas Polutta. 2018. Wirksamkeit. In *Handbuch Soziale Arbeit. Grundlagen der Sozialarbeit und Sozialpädagogik*, ed. Hans-Uwe Otto, Hans Thiersch, Rainer Treptow, and Holger Ziegler, 1825–1832. Munich: Ernst Reinhardt Verlag.

Ansen, Harald. 2009. Beziehung als Methode in der Sozialen Arbeit. Ein Widerspruch in sich? *Soziale Arbeit* 10 (2009): 381–389.

Arendt, Hannah. 2016 [1937]. *Wir Flüchtlinge*. Stuttgart: Reclam Verlag.

Becker, Peter. 2016. *Bürokratie. Docupedia Zeitgeschichte*. https://docupedia.de/zg/Becker_buerokratie_v1_de_2016. Accessed 9 July 2020.

Bowlby, John. 2006 [Orig. 1969]. *Bindung*. Munich/Basel: Ernst Reinhardt Verlag.

Dahlvik, Julia. 2016. Asylanträge verwalten und entscheiden. Der soziologische Blick auf Verborgenes. Eine Forschungsnotiz. *Österreichische Zeitschrift für Soziologie* (2016) (Suppl 2) 41: 191–205.

Dannemann, Rüdiger. 2018. Georg Lukács' Verdinglichungstheorie und die Idee des Sozialismus. In *Der aufrechte Gang im windschiefen Kapitalismus*, ed. Rüdiger Dannemann, Henry W. Pickford, and Hans-Ernst Schiller, 37–66. Wiesbaden: Springer Verlag.

Ellerman, David. 2013. On the Renting of Persons. *Social Science Research Network (SSRN)*. https://ssrn.com/abstract=2344920 or https://doi.org/10.2139/ssrn.2344920. Accessed 9 July 2020.

Forst, Rainer. 2005. Die erste Frage der Gerechtigkeit. *Aus Politik und Zeitgeschichte 37/2005*: 24–31.

———. 2012. Towards a Critical Theory of Transnational Justice. *Metaphilosophy* 32 (1/2, January 2001): 160–179.

Honneth, Axel. 1994. *Kampf um Anerkennung. Zur moralischen Grammatik sozialer Konflikte*. Frankfurt a.M.: Suhrkamp.

———. 1999. *Die zerrissene Welt des Sozialen. Sozialphilosophische Aufsätze*, Extended New Edition. Frankfurt a.M.: Suhrkamp.

———. 2012. *Reification. A New Look at an Old Idea*. New York: Oxford University Press.

———. 2015. *Verdinglichung. Eine anerkennungstheoretische Studie*. Frankfurt a.M.: Suhrkamp.

Kleve, Heiko, and Frank Früchtel. 2011. Die Wirkung der Sozialen Arbeit. Ein Dialog zwischen Heiko Kleve und Frank Früchtel. *Sozialmagazin. Die Zeitschrift für Soziale Arbeit, 36th Year, Issue 3,* March 2011: 32–37.

Lukács, Georg. 2015. *Die Verdinglichung und das Bewußtsein des Proletariats*. Bielefeld: Aisthesis Verlag.

Müller, Burkhard. 2009 [1993]. *Sozialpädagogisches Können: Ein Lehrbuch zur multiperspektivischen Fallarbeit*, 6th ed. Freiburg i.Br.: Lambertus Verlag.

Otto, Hans-Uwe, Andreas Polutta, and Holger Ziegler, eds. 2010. *What Works. Welches Wissen braucht die Soziale Arbeit. Zum Konzept evidenzbasierter Praxis*. Opladen/Farmington Hills: Verlag Barbara Budrich.

Pantuček-Eisenbacher, Peter. 2006. *Soziale Diagnostik. Verfahren für die Praxis Sozialer Arbeit*. Wien: Böhlau Verlag.

Schedler, Kuno, and Isabella Proeller. 2011. *New Public Management*, 5th ed. Bern/Vienna: Haupt Verlag.

Seukwa, Henri Louis. 2006. *Der Habitus der Überlebenskunst. Zum Verhältnis von Kompetenz und Migration im Spiegel von Flüchtlingsbiographien*. Münster/New York/Munich/Berlin: Waxmann Verlag.

Terkessidis, Mark. 2004. *Die Banalitität des Rassismus. Migranten zweiter Generation entwickeln eine neue Perspektive*. Bielefeld: Transcript.

Tugendhat, Ernst. 2006. Das Problem einer autonomen Moral. In *Ernst Tugendhats Ethik. Einwände und Erwiderungen*, ed. Nico Scarano and Mauricio Suárez, 13–30. Munich: C.H. Beck.

von Grönheim, Hannah. 2018. *Solidarität bei geschlossenen Türen. Das Subjekt der Flucht zwischen diskursiven Konstruktionen und Gegenentwürfen*. Wiesbaden: Springer.

Webb, Stephen A. 2010. Zur Validität von evidenzbasierter Praxis in der Sozialen Arbeit. Einige Überlegungen. In *What Works – Welches Wissen braucht die Soziale Arbeit. Zum Konzept evidenzbasierter Praxis*, ed. Hans-Uwe Otto, Andreas Polutta, and Holger Ziegler, 188–201. Opladen/Farmington Hills: Verlag Barbara Budrich.

Weber, Max. 1980. *Wirtschaft und Gesellschaft. Grundriß der verstehenden Soziologie*. Obtained from Johannes Winckelmann. Student Edition, Tübingen, 5th ed. http://www.zeno.org/nid/20011439785. Accessed on 9 July 2020.

Wright, Michael T., ed. 2010. *Partizipative Qualitätsentwicklung in der Gesundheitsförderung und Prävention*. Bern: Huber.

Chapter 14
The Structural Misrecognition of Migrants as a Critical Cosmopolitan Moment

Zuzana Uhde

Abstract Transnational migrants and their struggles have become central for rethinking cosmopolitanism from below. This chapter builds on the theoretical and empirical arguments of the critical cosmopolitan perspective and proposals for methodological cosmopolitanism, which shifts the angle from which the social sciences look at social reality. Who is regarded as a relevant social actor to put forth cosmopolitan claims is crucial. Nevertheless, the author suggests that equally important is what struggles are taken into consideration. She suggests that cosmopolitan critical social theory can be usefully oriented by the concept of recognition toward the experiences of harms and wrongs as pre-political motivations for social struggles and the related articulation of claims. Migrants' lived critique is an expression of their struggles against structural misrecognition that is mediated by the geopolitics of borders and the structures of global capitalism, and the claims they voice that arise from these struggles need to be taken into consideration in the process of articulating cosmopolitan norms. In the first part of the chapter, the author offers a critical explanation of the geopolitics of borders within capitalist globalization in order to outline the social relations and practices that bring about the structural misrecognition of forced transnational migrants. In the second part, she examines the lived critique of forced transnational migrants through the concept of recognition. She argues that while forced transnational migrants do not necessarily share a cosmopolitan consciousness, they can be defined as cosmopolitan actors if conceptualized as a structural group. In the concluding part, she compares the viewpoint of migrants' lived critique with that of organized migrant protests that have obtained political visibility but may provide only partial foundations for cosmopolitan critical social theory. She suggests that the claims arising from migrants' lived critique expand the normative horizons of cosmopolitan imaginaries to include a more radical critique of global capitalism. In this sense, it engages in struggles also for the benefit of those who do not migrate.

Z. Uhde (✉)
The Czech Academy of Sciences, Institute of Sociology, Prague, Czech Republic
e-mail: zuzana.uhde@soc.cas.cz

G. Schweiger (ed.), *Migration, Recognition and Critical Theory*, Studies in Global Justice 21, https://doi.org/10.1007/978-3-030-72732-1_14

Keywords Cosmopolitan critical theory · Migrants' lived critique · Forced transnational migration · Global capitalism recognition

14.1 Introduction

The idea of cosmopolitanism has in recent years gained renewed attention, especially within the Anglo-American tradition of political thought since the 1990s. This discussion of cosmopolitanism has, however, largely taken place from the dominant perspective of liberal-normative political theory. Its version of moral and political cosmopolitanism offers "cosmopolitanism from above," which is based on individualist and universalist foundations and is uncritically defined from a position of power (Ingram 2013; Sager 2019). The liberal-cosmopolitan focus on individuals as citizens of the world reflects the lifestyle of a mobile transnational elite, but it does not sufficiently challenge the geopolitical hierarchies embedded in global capitalism. However, the new millennium also ushered in an array of critical perspectives on cosmopolitanism (Delanty 2009; Beck 2006; Pieterse 2006; Kurasawa 2004, etc.). Transnational migrants and their struggles have become central in a rethinking of cosmopolitanism from below (Sager 2018; Caraus and Paris 2019; Eze 2017). Who gets to be regarded as a relevant social actor who can put forth cosmopolitan claims is indeed key. Nevertheless, I suggest that equally important is what struggles are taken into consideration. To pay exclusive attention to traditional forms of political protest (demonstrations, marches, campaigns, etc.) omits from the research focus a substantial part of protest in society. The pre-political everyday struggles for recognition expressed in the lived critique of migrants need to be taken into consideration in the process of articulating cosmopolitan claims.

Many organized migrant protests demand the right to entry and to unconditional recognition as equal moral beings entitled to equal rights. In the current political climate, these seem like radical or even utopian demands. Moreover, under the social conditions of severe existential suffering and social distress, organized migrant protests tend to focus more on goals that are achievable in the here and now. Nevertheless, ultimately, they may reaffirm the cognitive bias of methodological nationalism and an institutional and legal framework defined around the nation-state from which migrants were excluded in the first place.[1] The scope of their claims is often confined to seeking the universal validity of demands for open borders and equal rights. However, given the current global power hierarchies and the structurally unequal inclusion of world macro-regions into global capitalism, open borders

[1] The book on migrants' protests edited by Tamara Caraus and Elena Paris (2019) presents several examples of migrant activism that from their very foundations do not challenge the institution of the nation-state as the authority defining the dividing line between inclusion and exclusion, such as Sans-Papier, the Dreamers, A Day Without Us marches, etc.

reinforce and reproduce existing vulnerabilities and the subordinate inclusion of marginalized migrants as disposable cheap labor. The open borders within the EU Schengen zone shed a clear light on this dynamic, which gained public visibility when interior EU borders were abruptly closed to stop the spread of COVID-19. As a result of the anti-pandemic measures, wealthier EU countries have faced a shortage of care workers and seasonal agricultural workers, most of whom come from central and eastern European countries, and this has exposed the mechanism by which freedom of movement acts as an essential tool for exploiting the mobility of circular migrants, pushing down wages, and extracting more profit. The illusion of a borderless Europe is built on formalized paths for a subtle combination of inclusion of EU migrant workers through access to labor market and their exclusion from some social and labor rights protection. The borders are open for people to cross as though there were no borders, but the structural positioning of different European regions and member states in the EU macro-regional arrangement and the ways in which they are integrated into the global economy serve to maintain the everyday power that borders have to categorize people and reproduce the existing geopolitical hegemony. Open borders for human mobility do not mean a borderless world.

Achille Mbembe argues that we are experiencing an intensification of the fundamental dialectics of opening and closure, that is to say, of globalization and de-globalization tendencies, which are aggravated by the global character of the capitalist form of social relations (Mbembe 2018). Amidst the global interconnection and mobility of capital, goods, and privileged groups, border controls are being increasingly outsourced and externalized in order to avoid the political responsibility for racialized border violence and in order to shrink the category of migrants allowed to enter to the smallest group possible. At the same time, migration management has become a highly profitable enterprise. The global security market is one of the fastest-growing industrial sectors, boosting capitalist globalization despite current nationalist tendencies. Transnational migration is an inherent part of the global economy, and the current forced migration (in a broader sense which I will explain later) is in some respects a direct and foreseeable and in other respects an unintended consequence of late-modern capitalist modernity. Local conflicts and political, economic, social, or ecological hardships are co-produced by transnational practices, and globally produced risks have localized impacts. From the perspective of methodological nationalism, migration is regarded as a problematic deviation from the norm. In contrast, adopting the perspective of methodological cosmopolitanism shifts the angle from which we look at social reality and foregrounds the struggles of transnational migrants.

In this chapter, I build on the theoretical and empirical arguments of critical cosmopolitanism, which situates transnational migrants as the quintessential cosmopolitan subjects. In the first part of the chapter, I offer a critical explanation of the geopolitics of borders within capitalist globalization. The goal is to outline the social relations and practices that lead to the structural misrecognition of forced transnational migrants. In the second part, I focus on the contribution of the critical theory of recognition and examine the lived critique of transnational migrants. I present the argument that, while marginalized migrants do not necessarily share a cosmopolitan consciousness, they can be defined as cosmopolitan actors if conceptualized as a structural group. In the concluding part, I compare the viewpoint of migrants' lived critique with that of organized migrant protests. Organized migrant protests have attained some level of visibility but may provide only partial or distorted foundations for cosmopolitan critical social theory, as they are impeded by a receiving state bias and a need to translate their claims into the geopolitically biased language of the migration agenda.[2] I then outline the contours of the cosmopolitan critical theory of recognition. I suggest that embedding cosmopolitan critical social theory in the migrants' everyday struggles for recognition serves to expand the normative horizons of cosmopolitan imaginaries beyond freedom of movement and respect for migrants' human rights.

14.2 The Geopolitics of Borders and Global Capitalism

The system of global capitalism requires opening of borders for the flow of capital, goods, and the transnational capitalist class. However, it also requires the enforcement of borders, which are used to ensure that the impacts of proliferating crises and globally produced risks remain localized and to extract profit through legal offshoring practices or illicit financial flows.[3] Global capitalism requires borders that are porous and fading and at the same time reified and exclusionary. These dual bordering processes also involve the differentiated categorization of people and their transnational mobility regime. Transnational professionals and the transnational capitalist class embody the ideal of freedom of movement. The categorization of transnational migrants into political migrants (toward whom states have some responsibility to provide protection) and economic migrants (who are left subject to the benevolence of individual states) is then necessary in order to reconcile the inherent contradictions of capitalist globalization and its accompanying liberal narrative: to uphold the ideal of freedom of movement and human rights within escalated global inequalities mobility has to be controlled, and some categories of people must be classified as a

[2] Migration is one of the examples that reveal the co-optation and neoliberal reframing of human rights. Aleksandra Ålund and Carl-Ulrik Schierup (2018) analyzed the evolution of the Global Forum for Migration and Development and pointed to the strategies of the selective inclusion of human rights arguments as signs of their pacification and co-optation. They argue that it is manifested in a shift from migrant labor rights toward moral migrant rights and by the marginalization of labor unions. However, the mobilization of moral arguments shifts attention to protection and partial improvement of migrant conditions at the expense of migrants' claims for global social justice and of systemic changes of the structures producing forced migration (see also Likić-Brborić 2018).

[3] Offshoring practices and illicit financial flows are regarded as a fundamental accompanying effect of global capitalism. However, its premise is not a borderless world but a selectively bordered world. Borders make it possible to establish different legal jurisdictions, which allows mobile capital to escape public control. While the state is no longer the main organizing principle, it is still the executive power and enabler of global capitalism (Robinson 2014).

threat. Even the agenda for refugees' protection is defensive. It maintains the problematic distinction between political refugees and economic migrants in order to preserve the limited protection of refugees, which is now recognized in the international legal order, but this opens the door to a deterioration of the human rights regime as it pertains to transnational migrants in general as a result of the externalization of so-called border management in an effort to stop "economic migrants." Borders become a space of the violation of rights, the responsibility for which states try to avoid by externalizing and outsourcing border controls through bilateral agreements and the involvement of private companies. Julia Schulze Wessel argues that in these "externalized border zones," migrants become the rightless persons described by Arendt, and are so despite the development of international protections for refugees since the Second World War (Wessel 2016: 54). The need to absorb some migrants – as border controls can never be fully executed and in some cases entitlement to asylum protection is undeniable – meets a demand for cheap labor in some sectors of the labor market in wealthier countries (mainly in the fields of care and agriculture), in what Branka Likić-Brborić identifies as a "neoliberal reframing" of migrant protection and the "'developmental approach' to the problems of refugees." She shows that this weakens the protection of refugees as it introduces a business-run model of finding niches in the transnational labor market for refugees and repositioning them as "weakly protected economic migrants" (Likić-Brborić 2018: 771).

In Europe, economic migration is symbolically represented by the figure of the African migrant, who is presented as a threat to the European way of life and the socioeconomic standards of the middle and poorer classes.[4] Economic migration is depoliticized as an isolated problem that needs to be controlled and managed. The manufacturing of a culture of fear of migrants and moral panic mobilizes the racial category of blackness, which is a product of transatlantic modernity shaped through colonial expansions and the advent of capitalism and reshaped through global capitalism (Mbembe 2017; Quijano 2000; De Genova 2018). Based on a historical analysis of the formation of global power structures and economic dominance through colonial expansion, in which Europe has emerged as a dominant geopolitical identity as opposed to racially classified "non-Europeans" – Indians, blacks, mestizos, and yellow – Aníbal Quijano identifies the "coloniality of power" as a pervasive system for classifying the world population, determining the division of world resources, the social relations of capital, and production of knowledge that continues to shape the current geopolitical landscape in the era after colonial empires (Quijano 2000). Sabelo J. Ndlovu-Gatsheni (2013) has further developed the notion of the "coloniality of power" to analyze the position of the African macro-region within the current global capitalist system by showing different stages of Euro-American dominance over the African macro-region, from colonialism and the strategies of colonial rule, the Cold War split of Africa soaked in violence, to the

[4]As only 2% of all migrants living in Europe in 2017 came from Africa (UN 2017), this media representation sheds a clear light on the racial formation of "European" whiteness built in opposition to the blackness as a product of colonial expansions and as an inherent part of modernity.

neoliberal turn which uses coloniality to advance structural adjustment programs, and the geopolitical War on Terror. These histories have determined the ways in which the African macro-region has been incorporated into the capitalist global economy. It is important to take this perspective into account to understand the structural sources of contemporary migration from and within Africa. It also, however, sheds light on the representation of African migrants and echoes Chinua Achebe's 1975 analysis of the racism in Joseph Conrad's *Heart of Darkness* (Achebe 2016). Achebe showed that the principal problem for Conrad arose when black Africans left their designated place as savages and claimed recognition as equal moral beings. Achebe argues that it is this claim for the recognition of Africans outside Africa that generates difficulties, fears, and resistance. We can see the same sentiments behind the political slogan "we must help migrants in their country of origin," which reveals the underlying racist resentment against recognizing Africans as equal human beings the moment they move from the role and place that has been defined for them. Thus, what is called migration management is in reality a manifestation of the coloniality of power and its structural violence, exploitation, and dispossession. In this context, Achille Mbembe has highlighted that the language of security and border management is a translation of the racialized violence of the current geopolitics of mobility (Mbembe 2018).

Migration management is increasingly tied to the enforcement of borders through securitization, militarization, and outsourcing and is defined by the interests of the wealthier "receiving" states. However, migration management is still part of the dynamics of global capitalism, even though it is premised on the existence of borders and nation-states' claims to territorial sovereignty. It focuses on controlling human mobility in order to keep the global structures of power intact and to preserve the transnational practices that benefit some macro-regions but impoverish others. The statist rhetoric is also instrumental to the unleashing of immense resources from public budgets and from migrants as well. While originally the term "migration industry" was used to describe a network of different actors assisting migrants and often taking advantage of their vulnerable position, Ninna Nyberg Sørensen and Thomas Gammeltoft-Hansen have expanded the meaning of migration industry to include both the facilitation of migration and its control. They "define the migration industry as encompassing not only the service providers facilitating migration, but equally 'control providers' such as private contractors performing immigration checks, operating detention centers and/or carrying out forced returns" (Gammeltoft-Hansen and Nyberg Sørensen 2013: 6). They identify different actors within the migration industry: transnational corporations, which are mainly involved in the externalization of border controls and the criminalization of migration; the agencies that facilitate legal migration; the smaller businesses that assist migrants; the illegal networks involved in smuggling or human trafficking; and the NGOs and humanitarian organizations that may be involved for nonprofit reasons but are still part of the industry, some of which advocate on behalf of migrants, while others work to legitimize governments' restrictive and securitization lenses (Gammeltoft-Hansen and Nyberg Sørensen 2013: 10–12). We can add to this list also some governments in countries that host large numbers of refugees and use this fact as a source of foreign financial capital through humanitarian assistance. Hosting refugees is clearly a heavy financial burden for poorer countries, but it can also facilitate the influx of investment negotiated by governments, to boost the local formal and informal economy.[5] These networks serve to embed transnational migration in the system of capitalist globalization, while the growing commodification of migration ties it up in complex ways with global capital. First, global capital benefits from the economic order that produces the structural sources of transnational migration, and second, it benefits from the political responses to migration through the privatization and outsourcing of a migration management in that it pours huge investments into the securitization and militarization of border controls.[6] Moreover, new surveillance technologies applied in humanitarian assistance for refugees reinforce the power of borders to categorize people and serve as a way of testing and developing new biopolitics based on big data and digital surveillance (Lemberg-Pedersen and Haioty 2020). The dynamics of this can best be grasped with the concept of a "global border industrial complex." While the migration industry comprises a wide variety of actors, including small-scale enterprises and the shadow economy, I understand the global border industrial complex to refer to the network of powerful individuals and transnational corporations that are linked to governments' migration and security policies and funds but that also have ties to the global financial capital.[7] The global financial capital makes use of borders to shuffle off both responsibilities for social reproduction of the local labor force and responsibilities arising from its role in a production of global risks. Some forms of borders are profitable for transnational economic practices and global capital as they facilitate various illicit financial flows, such as tax evasion, trade mis-invoicing, capital flight, and also the ever-rising profits generated by border management. The complex refers to the connection between different industries and global financial capital that forms networks and lobby groups tied to local, national, and supranational political institutions.

[5] Globally, the vast majority of refugee, according to the conventional definition of the term, are hosted by developing countries. In 2018, under the common understanding of the term, there were 25.9 million international refugees, 3.5 million asylum-seekers, and 41.3 million internally displaced people, 84% of whom lived in developing countries, while 33% were being hosted in the least developed countries worldwide (UNHCR 2019).

[6] While the global migration agenda operates in a nation-state-centered institutional framework, which takes the perspective of the state and its claim to territorial sovereignty as a starting point, non-state actors are increasingly influencing the discursive framing of human mobility and migration policies (Likić-Brborić 2018; Betts 2013). Moreover, the enforcement of migration management requires an elaborate system of surveillance of human mobility, militarized border controls, and deportation channels. On the surface, it seems that the security and military industries respond to the states' demand. However, Martin Lemberg-Pedersen shows how in subtle ways private security companies under the guise of expert consultancy became the key actors in formulating the EU's border control policies (Lemberg-Pedersen 2013; Lemberg-Pedersen et al. 2020).

[7] The report More Than a Wall by Todd Miller analyses the emergence and functioning of the border industrial complex in the USA (Miller 2019). However, despite focusing on border enforcement, the complex is not strictly tied to an institution of the nation-state as the involved transnational corporations and transnational financial actors operate globally.

A conventional understanding of migration distinguishes between forced migration as a reaction to political repression, violence, and other disasters and economic (or labor) migration motivated by the search for a better life or for a temporary solution by which to access economic resources. The migration of transnational professionals is usually regarded as a separate issue. However, the distinction between forced and economic migration falsely assumes that while the first group has no other option but to migrate, the other group freely decides to leave its country of origin. Even though migration is a decision migrants actively make as part of their coping strategies, it is not a decision that is made free from structural constraints, and life in migration is a continuation of their struggles against injustice and misrecognition. Such a depoliticized approach to economic migration ignores the embeddedness of economic and social factors in the global economy: local problems are connected to transnational practices through direct and unintended consequences. Raúl Delgado Wise has expanded the concept of forced migration beyond the conventional understanding that refers to refugees, asylum-seekers, and internally displaced people and argues that the dynamics of global capitalism produce structural conditions in which "migration has essentially become a forced population displacement." In his view, forced migration specifically includes migration due to violence, conflicts, and catastrophes; human trafficking; dispossession, exclusion, and unemployment; and deportations (Delgado Wise 2018: 750–751). So-called economic migration is a form of forced migration. But not all migrating people are forced migrants. When I speak about transnational migrants, I refer specifically to groups of migrants who are to various degrees forced to leave for different reasons largely linked to global structural injustice. The distinction then lies rather between people who cross borders in response to different kinds of conflict and hardship and people who are in an advantaged position because of their skills and economic, social, or political status and who have access to a significantly more flexible transnational mobility regime.

The global border industrial complex fuels the nation-state-centered securitization and enforcement of borders, which is simultaneously driving dual bordering processes. It comprises a variety of actors, who make use of borders, not only the geographical but also the symbolic ones between "us" and "them," but also in different ways circumvent them. We can analyze the internal structure of the global border industrial complex through the prism of the four fractions of the transnational capitalist class proposed by Leslie Sklair (2003) which is key to understanding the relation between the state and other actors in today's form of capitalist globalization.[8] Through the migration industry and the global border industrial

[8] Leslie Sklair (2003) redefined the classic definition of the capitalist class based on ownership of the means of production to include other forms of capital besides economic, i.e., political, knowledge, and cultural capital. According to Sklair, the transnational capitalist class today includes not only the owners of the major corporations and the managers who run them (the corporate fraction) but also globalizing politicians and bureaucrats at the international, national, and local levels who align with global capital (the state fraction), professionals in the global labor market (the technical fraction), and actors in control of the media (the consumerist fraction) (Sklair 2003: 17–23).

complex, the contemporary politics of transnational migration is inherently connected to transnational networks of capital accumulation and the institutional workings of global capitalism, and this is the case even though global capitalism is usually seen as defying borders and the industry is based on the reproduction of the concept of borders – both geographical (although not necessarily nation-state borders) and symbolic. By enforcing an ostensible solution to migration driven by the crises of global social polarization and ecological unsustainability that global capitalism has generated (Robinson 2014; Sklair 2002), global capital uses borders to delay the crisis of overaccumulation by opening new markets of outsourced and privatized border control. The bordering processes selectively target marginalized migrants as racialized subjects. They do not target the mobility of capital or globally privileged classes.

Jan Nederveen Pieterse (2006) calls this capitalist cosmopolitanism,[9] which rests on the coloniality of power (Quijano 2000; Ndlovu-Gatsheni 2013). I will refer to the current globalizing social reality as capitalist globalization or global capitalism and reserve the term "cosmopolitanism" for critical and normative thinking. Against the backdrop of a critical analysis of capitalist globalization, it is only possible to start examining different cosmopolitan practices that break free from the ideological bias of Euro-American-centrism and seek instead to challenge global socioeconomic and political structures with a practical interest in emancipation. Critical cosmopolitanism (Sager 2018; Beck 2006; Fine 2007; Delanty 2009) has been variously termed "emancipatory cosmopolitanism" (Pieterse 2006), "cosmopolitanism from below" (Kurasawa 2004), "local cosmopolitanism" (Chan 2018), "transversal cosmopolitanism" (Hosseini et al. 2016), or "abject cosmopolitanism" (Nyers 2003). The common denominator is that it is regarded as a process of continuous redefinition and a search for a balance between a universal conception of humanity and solidarity, on the one hand, and local bonds, on the other hand, and its aim is to direct specific attention to marginalized actors and their ongoing struggles, which constitute what Ulrich Beck (2006) calls the really existing processes of cosmopolitanization. Robert Fine argues that cosmopolitanism is not a ready-made idea; rather, it is a research agenda that develops a perspective that seeks to address an existing social reality that he calls the age of cosmopolitanism, in which the chances for a cosmopolitan future are open though not yet fully realized (Fine 2007).

[9] Beck (2006) calls it the false cosmopolitanism of a transnational capitalist class and global elites, who merely instrumentalize cosmopolitan arguments to reproduce and consolidate the current geopolitical and economic arrangements on a global scale.

14.3 Forced Transnational Migrants as Cosmopolitan Actors

To overcome the shortcomings of the mainstream social sciences, Beck proposes applying the analytical perspective of methodological cosmopolitanism as opposed to methodological nationalism[10] (Beck 2006; Beck and Sznaider 2006). In the social sciences, methodological cosmopolitanism focuses on the global interactions that bring about new forms of sociability and transform the role of nation-states and forms of transnational economic, political, and cultural practices, and on their unintended consequences and associated global risks. Sandro Mezzadra's and Brett Neilson's (2013) methodological proposal to understand borders not only in their geographic sense but also as an epistemic perspective and apply it to study the "proliferation of borders" and the changing dialectic between inclusion and exclusion, or between opening and closure, that structure the relation between political power and global capital is an angle of analysis conducive to methodological cosmopolitanism. In line with the methodology of critical theory, Beck argues that the social sciences must focus on the really existing processes of cosmopolitanization – on the emerging cosmopolitan tendencies in transnational forms of life, practices, norms, and institutions – and search for actors' cosmopolitan critiques as the necessary first step toward any normative proposal for cosmopolitan political arrangement. However, from the perspective of methodological cosmopolitanism, it is also possible to examine the local and national dynamics in their social complexity and in a dialectic relationship with the evolving cosmopolitan social reality. Methodological cosmopolitanism does not preordain the subject of research interest but rather the analytical starting points from which a particular issue is explored. Transnational migration cannot be fully understood within the prevailing paradigm, which approaches global relations through the world's political division into nation-states. We can even postulate that the cognitive bias of methodological nationalism is a form of epistemic misrecognition of migrants. While transnational migration practices constitute really existing processes of cosmopolitanization in that they challenge the legitimacy of borders as a means of categorizing people, to argue that transnational migrants – before they are organized into collective struggles and movements – are indeed cosmopolitan actors is a proposition that requires further substantiation. Gerard Delanty (2009) notes that a cosmopolitan imagination involves more than the transnationalization and pluralization of the forms of life. He argues that it is a reflexive and internalized openness to the world, the result of a hermeneutic and cognitive process of learning that transforms one's identity and

[10] Although methodological nationalism is presented as a neutral approach, it is based on concealed ideological assumptions that have to do with the territorial sovereignty of nation-states and the conceptualization of society as a social unit that overlaps with the territory of the modern nation-state. As a result, it operates with a naturalized idea of nation-state and borders. The cognitive bias of methodological nationalism dominates in real politics and also in the social sciences, migration studies, and political theory. It distorts the view of the social reality of migration with a receiving state bias and a predominant focus on immigration, which frames migration as a problematic deviation from the norm (Wimmer and Glick Schiller 2002; Sager 2018; Castles 2010).

one's outlook on others. A cosmopolitan imagination refers to a reflexive critique and the normative horizon of ideas about alternative society and actors' claims for justice. Alex Sager argues that "it is not that individual migrants conceive themselves as or necessarily are cosmopolitan actors. Migrants are diverse with many wishing to join national communities. Nonetheless, the act of migrating, whatever the intention, is a *de facto* cosmopolitan act, causing a ripple or a rupture in the national fabric" (Sager 2019: 176). The act of migration is not usually a conscious act of resistance and may not necessarily lead to the emergence of a cosmopolitan consciousness. However, the experience of arbitrariness and the wrongs of borders that exclude some based on their place of birth and include others based on their political, economic, or social status transforms migrants' subjectivity and perspective. The experience of misrecognition opens the way for critical intuition.

Theories of recognition generally draw on the argument that individuals owe their personal integrity to intersubjective relations with others, which are, however, often mediated by institutionalized forms of interactions. The aim of critical theory is to expose, using a critical imagination, the emancipatory and normative potential of certain elements of reality that are grounded in existentially experienced conflicts between ideals and practice as forms of injustice. These existentially experienced harms can be approached through the concept of recognition. Actors' subjective experiences of misrecognition entail an emancipatory potential as pre-political motivations for social struggles and claim-making. In a contextualized and critical explanation of these social struggles, critical theory then articulates actors' claims for expanding normative horizons so as to close the gap between ideals and practice (Young 1990, 2011; Fraser and Honneth 2003; Hrubec 2012). According to Honneth, recognition order arises out of the historical process of struggles against misrecognition and the related articulation of claims for reformulating the normative principles that govern society or a specific sphere of society – intimate, legal, and economic spheres as three historically institutionalized but not discrete spheres (Honneth in Fraser and Honneth 2003). However, focusing only on the psychological aspects of individuals' needs for recognition in intersubjective relations with others does not provide explanations on which basis it would be possible to conceptualize collective protest and actors' claims without presupposing their shared identity and collective consciousness. Consequently, without presupposing a collective subject, it is not possible to conceptualize misrecognition as a systemic and institutionalized process. This is also Nancy Fraser's argument against Honneth's theory of recognition. She argues that in order to identify institutionalized subordination without presupposing a shared collective identity, the concept of recognition needs to be reformulated as a matter of social status and not a psychological relation to others and the self (Fraser in Fraser and Honneth 2003). However, Honneth argues that her status model of recognition not only limits recognition to the cultural sphere, but by bracketing out subjective experiences, it also limits recognition struggles to existing social movements and organized political protests. He makes a valid point against Fraser's curtailed understanding of recognition; however, Fraser's critique of undue emphasis on psychological identity cannot be entirely overlooked. Moreover, Honneth's approach is premised on methodological nationalism, as

Fraser and others have critically pointed out. David Ingram (2018) expanded Honneth-inspired critical theory of recognition in order to engage with global injustice. However, he still approaches the issue of migration with an overly static understanding of people's need for recognition and a sense of belonging tied to a person's place of origin and a person's presential attachment. I will argue later that under more favorable social and economic conditions, an identity of belonging can be reconstructed in a transnational or cosmopolitan way. Marek Hrubec (2013) proposes a more apposite Honneth-inspired approach that develops the perspective of recognition at the global level, taking into account the everyday struggles of the global poor as a critical response to contemporary global interactions. In my view, Marek Hrubec and Iris M. Young present approaches by which critical theory can conceptualize a collective subject of social change without essentializing the group's identity while still including not yet politically articulated struggles against misrecognition and their claims for global justice. I will build on Young's theory in order to articulate forced transnational migrants' lived critique of global structural injustice.

Iris M. Young argues that critical theory's "normative reflection arises from hearing a cry of suffering or distress, or feeling distress oneself" (Young 1990: 5). Even though her empirical reference points are acting and experiencing subjects, she works with an analytical framework that focuses on group differences not for the sake of identity politics but to scrutinize repressive social structures, their modes of reproduction, and their material, political, or cultural effects. Young differentiates between cultural and structural groups (Young 2000). While cultural groups are brought together by language, everyday practices, forms of sociability, and aesthetic or religious conventions, which offer their members certain means of shared expression and communication and create an environment of mutual affinity, structural groups are related to material or psychological aspects and social status. According to Young, "a structural social group is a collection of persons who are similarly positioned in interactive and institutional relations that condition their opportunities and life prospects" (Young 2000: 97). Examples of the structural differences described by Young include relations constituted on the basis of gender, "race," class, sexuality, and disability. These categories refer to "a particular form of social positioning of lived bodies," a specific structural link between institutional conditions, individual life possibilities, and their realization (Young 2005: 22). The ways of coming to terms with these structures are therefore changeable. These categories do not refer to an individual identity which is always unique, but specify the conditions in which individual identity forms.

Although migration is often discussed in terms of intercultural interactions, differences between migrants' and the majority's cultural norms and practices, or the conflicting policy goals of assimilation, inclusion, and exclusion, these perspectives appear less important if the prime focus is on structural injustice. According to Young, cultural differences become a political issue if they are, as they often are, linked to structural inequalities and structurally embedded misrecognition. She stresses that many situations presented as cultural conflicts are actually sociopolitical or economic conflicts because they are based on contests over territory, resources, decision-making power, or positions in the division of labor. The bulk of the research in migration studies focuses on identity formation and their material social forms (e.g., remittances, transnational networks, transnational forms of families and other social relations, transnational diasporas), emphasizing differences between groups of migrants diversely positioned with respect to their cultural or ethnic background and country of origin. There are differences between diverse groups of migrants; however, their structural position tends to produce similar outcomes. I suggest that it is important not to lose sight of the fact that forced transnational migrants share a specific position in social structures and institutionalized relations that makes them vulnerable to marginalization, exploitation, violence, material and social suffering, and psychosocial harm, and which constrain their ability to fulfill and develop their capacities, express their opinions or experiences, and participate in defining the conditions of their lives (Uhde 2019; Pinzani 2019). If we understand their structural position as a defining characteristic, it exposes the inadequacy of the distinction between political and economic migrants. They form a structural group of people who do not share a collective identity but are all exposed to structural misrecognition mediated by the geopolitics of borders and the structures of global capitalism. While forced transnational migrants face the injustice of repressive border policies, unjust or missing international laws, border violence, and the harmful actions of diverse actors in the migration industry and the global border industrial complex, Young's concept of structural injustice[11] makes it possible to trace also injustice in the structural sources of migration, the historically constituted geopolitical order that has culminated in the current global capitalism that rests on the coloniality of power.

While forced transnational migrants' everyday struggles for recognition are seemingly only individual, if we see them as a structural group, their experiences of misrecognition become structurally generalizable. A cosmopolitan critical social theory can then articulate the collective features of these struggles and their generalizable claims for justice. It overcomes the individualist bias of liberal cosmopolitan theories. To articulate the normative reflection rooted in these struggles, it is important to look at migration as a process that starts with the migrants' decision to leave but continues throughout their lives as they continue being labeled migrants, with a range of economic, political, and cultural consequences. The circumstances that force migrants to leave are transnational in their scope as they are part of a globally entangled world, yet there is a lack of transnational and global institutions through which people from disadvantaged regions can claim their rights. Migration is often a forced choice in response to land grabbing and dispossession, the conflicts and violence that are linked in one way or another to geopolitics and the interests of global capital, exploitation, a lack of development, ecological disasters, etc. Yet it is usually falsely interpreted as the exclusive result of internal state dynamics

[11] Young argues that "structural injustice is a kind of moral wrong distinct from the wrongful action of an individual agent or the repressive policies of a state. Structural injustice occurs as a consequence of many individuals and institutions acting in pursuit of their particular goals and interests, for the most part within the limits of accepted rules and norms" (Young 2011: 52).

(corruption, failed state, etc.) or civil and ethnic conflicts. However, this problematic understanding of local conflicts is only possible through the prism of methodological nationalism.[12] Migration is a survival strategy or an active take on life; it is, in other words, a struggle for recognition. Migrants are compelled to manipulate available resources and legal options or use illegal means in order to navigate in the system and cross borders. When they reach a destination, their struggle against structural misrecognition continues, as they are treated as second-class people and denied access to rights and their social contribution is systemically diminished. Transnational migrants develop a myriad of everyday strategies to circumvent legal limitations, reclaim respect, and rebuild their lives, and some of these strategies are classified as illegal. All these little everyday struggles are an expression of transnational migrants' "lived critique."

This perspective on migrants' everyday struggles echoes James Scott's concept of "weapons of the weak," through which he made visible everyday acts of resistance and non-confrontational protest as a form of class struggle (Scott 1985). For cosmopolitan critical social theory, the psychosocial needs for recognition are relevant as a motivational driver of migrants' lived critique and not as an end in itself. Through lived critique, forced transnational migrants express their claims, which question the legitimacy of frames of reference for migration laws that are centered on the perspective of nation-states, in which they are caught up, and the existing global economic system. A normative reflection of their lived critique points out to their demands to extend the scope of recognition beyond the nation-state – not only in terms of including now excluded groups (open borders) but also in terms of redefining the normative principles of recognition in a way that would transform global economic and geopolitical structures (the right to development). In other words, their lived critique is an expression not only of their struggle for legal recognition but also, and more importantly, of their more fundamental struggle for social recognition, which demands a radical restructuring of the global capitalist mode of production and its logic of accumulation. José A. Zamora convincingly argues that ignoring this aspect is a common denominator of liberal theories of global justice, usually proposing some form of global redistribution. However, it does not address the core of the problem as "sustaining [capitalist] accumulation today requires forms of expropriation and looting that cause population flight and massive displacement" (Zamora 2019: 88). The mobility of forced transnational migrants is part of the really existing processes of cosmopolitanization. Even though migrants do not necessarily internalize a collective cosmopolitan consciousness in a normative sense, their lived critique is not merely an individual expression of psychosocial harm, nor does it represent conformity with the normative horizon of *inter*national

[12] Alison J. Ayers has criticized the concept of civil war, which, according to her, is not only ideologically convenient but also "rests upon the highly problematic conception of the state as a reified entity, with interests and capabilities analytically separate from the totality of global social relations within which states inhere" (Ayers 2010: 155). However, the causes of such economic or political conflicts are historically embedded in global geopolitics – a colonial past, the geopolitics of the Cold War, or the subsequent War on Terror – and are exacerbated by today's global capitalism.

social reality. A structural group of forced transnational migrants constitutes a collective subject of cosmopolitanism. However, when pre-political claims are being translated into an organized protest and activism, they may be adjusted to fit the mainstream political referential framework in order to be heard, or they may be tailored to conform to short-term political goals. Traditional forms of political protest (demonstrations, marches, strikes, blockades, campaigns, etc.) and confrontational collective actions often take place at borders, in detention centers, or in destination countries, and this hampers abilities to maintain a holistic perspective on migration as a process. Moreover, the visibility of these protests is generally obscured by the receiving-state bias of, in most cases, the Euro-American part of the world, even though about half of all migrants globally stay outside of these macro-regions.[13]

14.4 From Lived Critique to Critical Cosmopolitanism

Although methodological nationalism is still a prevailing perspective in the social sciences, many critical scholars now argue that to access the critical and transformative potential of transnational migration, it is necessary to abandon the vantage point of 'receiving' states and set out instead from the standpoint of migrants as mobile subjects. However, I argue that in order to do so, the critical cosmopolitan perspective cannot conceptualize migrants' struggles solely in terms of organized movements and political collectives. The majority of protests that gain visibility are already located in destination states or at their borders, and their claims to be granted access or to achieve the legalization of their status often outweigh the more radical claims that arise from migrants' lived critique and that are sometimes present at the beginning when they come together to form an organized political collective. Thus, equating migrant protests with organized political protests is a distortion that results from receiving state bias. Tamara Caraus and Elena Paris note that: "Remaining within methodological nationalism, scholars cannot but observe the so-called paradox of migrant protests: migrants formulate radical claims such as 'No One is Illegal' and end up asking for legalization in a certain nation-state, thus reconfirming the very institutions that they contest." And they go on to argue "that the failure is not only of migrant activism. Migrant activism experiences this paradox also because there is no alternative, non-statist, institutionalized way of addressing their claims" (Caraus and Paris 2019: 10). They are right in pointing out that the goal of these movements to improve the situation of its participants foreordains them to forfeit certain more radical claims that under current conditions cannot be realized in the short term. They suggest that theories need to provide an outline and

[13] Out of the total number of 258 million "international" migrants worldwide in 2017, about 30% (78 million) were living in Europe and 21% in North America. Moreover, intracontinental migration prevails over intercontinental migration mobility. For example, in 2017, 67% of all European migrants remained in Europe, while 53% of all African migrants were in Africa (UN 2017).

conceptual tools for thinking about and institutionalizing a world beyond borders. However, I propose that first and foremost critical cosmopolitan theories have to broaden the understanding of political protest to include migrants' lived critique and their everyday critical intuition.

Social theories often fail to see forced transnational migrants as political actors unless they collectively organize. This is the case of the otherwise inspiring book edited by Tamara Caraus and Elena Paris on migrant protests and their radical cosmopolitics (Caraus and Paris 2019) or Ariadna Estévez's account of migrant protests as a response to the denial of recognition, where it is only in violent action and conflicts that she identifies migrants' struggles for recognition (Estévez 2012). But this is also the case of accounts that focus on nontraditional migrant protests, such as lip-sewing (Bargu 2017). Among these protests, probably the most radical in terms of transgressing the nation-state framework and articulating cosmopolitan claims beyond borders are the *No Borders*[14] movement and *No One Is Illegal (NOII).*[15] Frédéric Mégret (2019: 32) distinguishes three kinds of cosmopolitan claims present in migrants' protests: "a cosmopolitanism of law and human rights" (equal treatment for migrants), "a cosmopolitanism of inclusion and hospitality" (the right to be included), and "a cosmopolitanism of freedom of borders" (the right to migrate). He argues that *No Borders* and *No One Is Illegal* put forth the third claim for open borders. Both movements are against border controls and against categorizing migrants according to a different legal status, which is seen as a means to divide, control, and exclude some categories. The first of the two movements is anchored more in the anarchist tradition and the second in the socialist tradition. The manifestos of both movements criticize global capitalism as the source of the structural relations that cause most human mobility. NOII also criticizes the conditions under which migrants are included in the social welfare, labor market, etc., thus including the argument not only of open borders but also under what conditions. However, the primary focus of both movements is on abolishing border controls and questioning the legitimacy of borders. In the end, even these movements run the risk of adopting the mainstream referential framework that treats migration as a legal issue. As radical as their claim for open borders is, it softens the critique of global capitalism as an inseparable aspect of the critique of the situation of forced transnational migrants. In other words, it approaches borders only in their territorial sense and overlooks how global capitalism, which rests on the coloniality of power, creates symbolic and economic derivations of borders within one location and globally. Open territorial borders do not necessarily overcome the problem of the subordinate inclusion of individuals who have little choice other than to give in to being exploited in a wealthier country, when that is still an improvement in their situation compared to the lack of opportunities in poorer countries. Development is then reduced to an individual strategy (also through remittances, etc.).

---

[14] http://noborders.org.uk/news/no-borders-manifesto

[15] http://www.tacticalmediafiles.net/articles/3238/No-One-is-Illegal_-Manifesto;jsessionid=7BF90D89B801E0BE84C525FEA7FAEF7F

I argue that the migrants' lived critique advances cosmopolitan claims, and it is an important task of cosmopolitan critical social theory to articulate these claims. Firstly, it targets global capitalism through a critical reflection of the structural conditions people face before migration, the critique of which is an inherent part of their lived critique because structural exploitation, dispossession, violence, and oppression by global capitalist forces are at the roots of the existentially experienced harm and misrecognition that precedes their decision to migrate. After experiencing the activities of transnational corporations or the effects of structural adjustment policies firsthand, these people have critical intuition about the wrongs of global capitalist forces. They may blame their governments for some wrongs and for not taking action to address problems, but on their migration route, they also piece together a broader picture of the system that condemns them to a position as outcasts, and they develop an everyday global awareness. In the organized migrant protests that take place in destination countries, this aspect of migrants' lived critique is sidelined. Migration is a form of struggle against misrecognition which manifests as the lack of the right not to migrate (right to development). Secondly, their lived critique also targets the worldview of bordered communities, which contradicts their lived experience of social struggles against circumstances that are transnational in scope and force them to migrate, and it targets the lack of transnational institutions through which they could effectively claim remedies and their rights. This second aspect of migrants' lived critique that targets and challenges the nation-state-centered legal and institutional framework in which they are required to mold their lives is much more pronounced. Moreover, while organized protests challenge the moral legitimacy of the claim for nation-state sovereignty over borders, the vantage point of migrants' lived critique sheds light on transnational capitalist practices, which in most states (except global powers) effectively vanquish their chances of exercising their sovereignty, and this exposes the ideal of sovereignty as a weakening concept. I argue that the claims that arise from migrants' lived critique expand the normative horizons of cosmopolitan imaginaries to include a more radical critique of global capitalism. In this sense, their lived critique works also to benefit others who do not migrate.

This is, of course, not to say that we should disregard migrants' organized political protests. Rather, it means that these protests cannot be our only empirical referential points. Migrants' organized protests are practically and epistemically courageous projects. In order to dispel the risk of migrant protests being assimilated within an inclusion claim, Sager argues that, unlike other marginalized groups' social protests, it is not possible to fully co-opt migrants' protests into a nation-state narrative because "migrant exclusion from the nation-state is necessary, not contingent." He goes on to state that "the migrant is defined as an outsider or stranger; gaining full membership depends on effacing one's identity as a migrant" (Sager 2019: 179). However, in my view, he overstates the subversive power of the identity of the migrant. Although migrants challenge the national narrative, the figure of the migrant can be fully included based on meritocratic arguments, and the reference to one's identity is not sufficient as the state can include within its borders a multitude identities of belonging. I recall Young's argument that identity is always the specific and exclusive characteristic of an individual, and as such, it is not a basis on which it is possible to articulate collective political claims (Young 2000). The subversive power lies rather in the existentially experienced harms that are caused by the structural misrecognition of forced transnational migrants, which is what motivates critical reflection and resistance. A migrant who is a member of the elite group of transnational mobile professionals does not necessarily challenge the power of borders to exclude and categorize people. In this respect, while the figure of the African migrant is associated with the label of economic migrant, Afropolitans as a symbolic representation of African modernity experience some racially motivated misrecognition; they also enjoy some privileges as a mobile African elite. The concept of Afropolitanism emerged under the influence of the 1990's cosmopolitan discourse. Mbembe talks about Afropolitanism as "a way of being in the world," characterized by cultural hybridizations and cross-border circulations, and Afropolitans are seen as new African migrants.[16] On the one hand, this image is clearly connected to an elite transnational mobility regime, and in this sense, it is an element of neoliberal political economy and global consumerist culture, with a false aspiration to cosmopolitanism (Kasanda 2018). On the other hand, it shows how one's belonging and the psychosocial need for recognition is not necessarily tied to one's original location and can be reconfigured under favorable economic and sociocultural conditions into cosmopolitan belonging while retaining a partial attachment to a specific political community or culture. These illustrations echo Beck's view that the really existing processes of cosmopolitanization do not necessarily lead to a positive cosmopolitan arrangement. They bring about critical reflexivity, global awareness, and a sense of belonging, which can, nevertheless, still be in the tow of capitalist globalization and its sharpening inequalities.

At the same time, not all claims and expectations put forth by forced transnational migrants as cosmopolitan actors are justified. To make such a distinction, we need a theoretical articulation of the normative criteria for critique, articulated from the position of methodological cosmopolitanism. Transnational migrants' claims are usually disregarded on the basis of categorizations and boundaries that seem legitimate only from the perspective of methodological nationalism. But there is a need for a cosmopolitan justification of limitations of the rights of other groups.[17] In this sense, it is not legitimate to deny equal rights and social recognition of some groups based on arbitrariness of their place of birth, but it is also not legitimate to claim the inclusion of some groups of migrants deliberately on the condition that

[16] Achille Mbembe's essay on Afropolitanism was published in 2005 and is available at http://africultures.com/afropolitanisme-4248/

[17] I leave aside the discussion on overall norms of global justice. In this chapter, I focus only on one aspect of global justice which requires radically altering global economic and geopolitical structures. Today's configuration of borders brings about structural misrecognition of forced transnational migrants not only in terms of limiting their mobility but also, and perhaps more pressingly as it concerns majority of people who do not migrate, in terms of limiting their self-development in places they are forced to leave as the borders are functional for global capital to escape taxation, regulation, and public control.

other groups of migrants are excluded.[18] This implies that the construction of distinct categories of deserving migrants (refugees) and undeserving migrants (economic migrants) is not defensible from the perspective of the theory of global justice, which is grounded in the concept of recognition, even though such categorization is legally codified today. While some groups of migrants are more vulnerable and may need special protection, this should not be contingent on the general exclusion of others who are able-bodied adults but in a perhaps less life-threatening situation. However, critical cosmopolitanism does not require that all values are universally shared but rather an openness to others and reflexive merging of universalism with particular solidarities and local bonds (cf. Turégano 2019). Claims arising from the lived critique of forced transnational migrants are not an absolute criterion of global justice, but they present a vector that the theorizing of global justice needs to follow. To take seriously the migrants' lived critique means to acknowledge that the critique of global capitalism precedes the critique of mobility restrictions.[19] Forced transnational migrants make visible the destructive impacts of global capitalism and bring them to the doorstep in wealthier countries. But not everyone affected by these impacts migrates. Open borders will not remedy these structural injustices without the global regulation of capital and structural changes.

Critical cosmopolitanism is empirically embedded in these really existing processes of cosmopolitanization and the lived critique of forced transnational migrants as cosmopolitan actors who foreground the need for a structural critique of global capitalism. This approach thus crucially differs from liberal cosmopolitanism, which is based on an abstraction that not only hides the particularistic foundations of a universalism that is defined from a position of geopolitical power but also presupposes a global capitalist order as a taken-for-granted institutional framework for political cosmopolitan proposals (usually involving some form of redistribution). In contrast, cosmopolitan critical social theory formulates normative reflections of global justice from the vantage point of collective social subjects defined by actors' positions within global social structures. While it questions the overly individualist foundations of liberal cosmopolitan theories, it places the emphasis on individuals' psychosocial needs for belonging and recognition in intersubjective relations in order to understand pre-political social struggles and interpret the normative claims they give rise to. Forced transnational migrants are not the only marginalized cosmopolitan actors, but as a structural group, they make acutely visible the inherent contradictions and failures of global capitalism and reveal how borders operate to maintain global structural injustice.

---

[18] Arguments for inclusion can be made gradually, but the exclusion of others should not form the foundational logic of the argument.

[19] However, it is important to point out that there may be very different motives and foundations behind the political argument "to help migrants in their home country." Even the concept of the "root causes" of migration gets distorted in political debates, usually patching up manifestations (such as poverty or unemployment) of deeper structural problems.

14.5 Conclusion

Critical cosmopolitanism is not a normative prescriptive theory. It is a response to an unjust global geopolitical order that is trying to find cosmopolitan (global) remedies to global problems. The normative dimension of critical cosmopolitanism can be empirically embedded in and oriented by the concept of recognition, which foregrounds the experiences of harms and wrongs that motivate everyday struggles against misrecognition. The concept of recognition directs attention at the pre-political claims expressed through migrants' lived critique. This critique includes not only a critique of repressive migration law but also of the structural conditions that define migrants' lives before the decision to migrate. A normative reflection of migrants' lived critique calls attention to their demands to extend and redefine the normative principles of recognition in a way that would transform structures of global capitalism that rests on the coloniality of power. Since most people's interpersonal relations in late-modern capitalist societies are mediated through institutionalized interactions, most misrecognition arises from interactions between people and institutions or people representing these institutions. It is important to mention that forced transnational migrants face misrecognition from action by many actors within the migration industry and the global border industrial complex and from the direct consequences of the law, but they also face misrecognition as a result of the global structural injustice that arises from an institutionalized global order that systematically benefits global capital, the transnational capitalist class, and elite groups scattered around the world as well as selected global macro-regions and their inhabitants (although to a different degree). In many respects, the selective inclusion of migrants as second-class people and a precarious labor force not only works to maintain global capitalism, but it also serves as a way of giving moral legitimacy to the principle of meritocracy the system presumably rests on. However, amending migration law and opening borders would not eliminate all forms of structural misrecognition of forced transnational migrants, even though it would substantially improve today's brutal and alarming situation.

Organized migrant protests by their very nature act to confront the immediately harmful and brutal effects of immigration laws, detention centers, border controls, and the consequences of their externalization and outsourcing. But, as I argued, these protests do not necessarily represent the full scope of the lived critique of forced transnational migrants as a structural group, i.e., a group of people who are similarly positioned within the social structure and are forced to migrate in response to the direct or indirect consequences of the logic of capitalist accumulation and the transnational relations of production. Cosmopolitan critical social theory needs to go a step further if it is to keep up with the premise that the emancipatory potential of the really existing processes of cosmopolitanization should be taken as its empirical referential point.[20] I suggest that focusing on migrants' lived critique as an

[20] In the next step, cosmopolitan critical theory needs to elaborate an institutional proposal for putting these normative claims in practice. In my view, Iris M. Young's (2011) model of differentiated global political responsibility is a fruitful starting point.

expression of their struggles against structural misrecognition, which starts before the decision to migrate is made, moves cosmopolitan critical social theory in this direction. Although not all migrants' claims are legitimate, the goal is to analytically distill progressive cosmopolitan normative claims arising from these everyday struggles and grasp their emancipatory cosmopolitan moment.

References

Achebe, Chinua. 2016. An Image of Africa: Racism in Conrad's Heart of Darkness. *The Massachusetts Review* 57 (1): 14–27.

Ålund, Aleksandra, and Carl-Ulrik Schierup. 2018. Making or Unmaking a Movement? Challenges for Civic Activism in the Global Governance of Migration. *Globalizations* 15 (6): 809–823.

Ayers, Alison J. 2010. Sudan's Uncivil War: The Global-Historical Constitution of Political Violence. *Review of African Political Economy* 37 (124): 153–171.

Bargu, Banu. 2017. The Silent Exception: Hunger Striking and Lip-Sewing. *Law, Culture and the Humanities*. https://doi.org/10.1177/1743872117709684.

Beck, Ulrich. 2006. *Cosmopolitan Vision*. Cambridge: Polity.

Beck, Ulrich, and Natan Sznaider. 2006. Unpacking Cosmopolitanism for the Social Sciences: A Research Agenda. *The British Journal of Sociology* 57 (1): 1–23.

Betts, Alexander. 2013. The Migration Industry in Global Migration Governance. In *The Migration Industry and the Commercialization of International Migration*, ed. Thomas Gammeltoft-Hansen and Ninna Nyberg Sørensen, 45–63. New York: Routledge.

Caraus, Tamara, and Elena Paris, eds. 2019. *Migration, Protest Movements and the Politics of Resistance. A Radical Political Philosophy of Cosmopolitanism*. Routledge.

Castles, Stephen. 2010. Understanding Global Migration: A Social Transformation Perspective. *Journal of Ethnic and Migration Studies* 36 (10): 1565–1586.

De Genova, Nicholas. 2018. The 'Migrant Crisis' as Racial Crisis: Do Black Lives Matter in Europe? *Ethnic and Racial Studies* 41 (10): 1765–1782.

Delanty, Gerard. 2009. *The Cosmopolitan Imagination. The Renewal of Critical Social Theory*. Cambridge: Cambridge University Press.

Delgado Wise, Raúl. 2018. Is There a Space for Counterhegemonic Participation? Civil Society in the Global Governance of Migration. *Globalization* 15 (6): 746–761.

Chan, Stephen. 2018. The problematic Non-Western Cosmopolitanism in Africa Today: Grappling with a Modernity Outside History. *Human Affairs* 28: 351–366. https://doi.org/10.1515/humaff-2018-0029.

Estévez, Ariadna. 2012. *Human Rights, Migration, and Social Conflict. Toward a Decolonized Global Justice*. New York: Palgrave Macmillan.

Eze, Michael Onyebuchi. 2017. I am Because You Are: Cosmopolitanism in the Age of Xenophobia. *Philosophical Papers* 46 (1): 85–109. https://doi.org/10.1080/05568641.2017.1295617.

Fine, Robert. 2007. *Cosmopolitanism*. Oxon: Routledge.

Fraser, Nancy, and Axe I. Honneth. 2003. *Redistribution or Recognition? A Political-Philosophical Exchange*. London and New York: Verso.

Gammeltoft-Hansen, Thomas, and Nyberg Sørensen, Ninna, ed. 2013. *The Migration Industry and the Commercialization of International Migration*. New York: Routledge.

Hosseini, S.A. Hamed, Berry K. Gills, and James Goodman. 2016. Toward Transversal Cosmopolitanism: Understanding Alternative Praxes in the Global Field of Transformative Movements. *Globalizations* 14 (5): 667–684. https://doi.org/10.1080/14747731.2016.1217619.

Hrubec, Marek. 2012. Authoritarian Versus Critical Theory. *International Critical Thought* 2 (4): 431–444.

———. 2013. An Articulation of Extra-Territorial Recognition. In *Global Justice and the Politics of Recognition*, ed. Tony Burns, 271–295. New York: Palgrave.

Ingram, David. 2018. *World Crisis and Underdevelopment. A Critical Theory of Poverty, Agency, and Coercion*. Cambridge University Press.
Ingram, James D. 2013. *Radical Cosmopolitics. The Ethics and Politics of Democratic Universalism*. New York: Columbia University Press.
Kasanda, Albert. 2018. *Contemporary African Social and Political Philosophy. Trends, Debates and Challenges*. London and New York: Routledge.
Kurasawa, Fuyuki. 2004. A Cosmopolitanism from Below: Alternative Globalization and the Creation of a Solidarity Without Bounds. *European Journal of Sociology* 45 (2): 233–255. https://doi.org/10.1017/S0003975604001444.
Lemberg-Pedersen, Martin. 2013. Private Security Companies and the European Borderscapes. In *The Migration Industry and the Commercialization of International Migration*, ed. Thomas Gammeltoft-Hansen and Ninna Nyberg Sørensen, 152–172. New York: Routledge.
Lemberg-Pedersen, M., and E. Haioty. 2020. Re-Assembling the Surveillable Refugee Body in the Era of Data-Craving. *Citizenship Studies* 24 (5): 607–624. https://doi.org/10.1080/13621025.2020.1784641.
Lemberg-Pedersen, M., J. Rübner Hansen, and O. Joel Halpern. 2020. *The Political Economy of Entry Governance*, AdMiGov Paper D1.3. Copenhagen: Aalborg University. http://admigov.eu/upload/Deliverable_D13_Lemberg-Pedersen_The_Political_Economy_of_Entry_Governance.pdf. Accessed 23 Feb 2020.
Likić-Brborić, Branka. 2018. Global Migration Governance, Civil Society and the Paradoxes of Sustainability. *Globalizations* 15 (6): 762–778.
Mbembe, Achille. 2017. *Critique of Black Reason*. Durham and London: Duke University Press.
———. 2018. *The Idea of Borderless World*. Lecture at the University of Augsburg, Germany, May 9, 2018. Available at: https://www.youtube.com/watch?v=cUAcfDkLAx4.
Mégret, Frédéric. 2019. Migrant Protests as a Form of Civil Disobedience: Which Cosmopolitanism? In *Migration, Protest Movements and the Politics of Resistance. A Radical Political Philosophy of Cosmopolitanism*, ed. Tamara Caraus and Elena Paris, 29–50. Routledge.
Mezzadra, Sandro, and Brett Neilson. 2013. *Border as Method, or, the Multiplication of Labor*. Durham and London: Duke University Press.
Miller, Todd. 2019. *More Than a Wall. Corporate Profiteering and the Militarization of US Borders*. Transnational Institute. Available at: https://www.tni.org/files/publication-downloads/more-than-a-wall-report.pdf
Ndlovu-Gatsheni, Sabelo J. 2013. The Entrapment of Africa Within the Global Colonial Matrices of Power. Eurocentrism, Coloniality, and Deimperialization in the Twenty-First Century. *Journal of Developing Societies* 29 (4): 331–353. https://doi.org/10.1177/0169796X13503195.
Nyers, Peter. 2003. Abject Cosmopolitanism: The Politics of Protection in the Anti-Deportation Movement. *Third World Quarterly* 24 (6): 1069–1093. https://doi.org/10.1080/01436590310001630071.
Pinzani, Alessandro. 2019. Migration and Social Suffering. In *Challenging the Borders of Justice in the Age of Migrations*, ed. Juan Carlos Velasco and Maria Caterina La Barbera, 139–156. Springer.
Robinson, William I. 2014. *Global Capitalism and the Crisis of Humanity*. New York: Cambridge University Press.
Sager, Alexander. 2018. *Towards a Cosmopolitan Ethics of Mobility. The Migrant's-Eye View of the World*. Cham: Palgrave Macmillan.
———. 2019. Reclaiming Cosmopolitanism Through Migrant Protest. In *Migration, Protest Movements and the Politics of Resistance. A Radical Political Philosophy of Cosmopolitanism*, ed. Tamara Caraus and Elena Paris, 171–185. Routledge.
Scott, James C. 1985. *Weapons of the Weak. Everyday Forms of Peasant Resistance*. New Haven and London: Yale University Press.
Sklair, Leslie. 2003. *The Transnational Capitalist Class*. Oxford: Blackwell Publishing.

Turégano, Isabel. 2019. Ethical Dimensions of Migrant Policies: A Critical Cosmopolitan Perspective. In *Challenging the Borders of Justice in the Age of Migration*, ed. Juan Carlos Velasco and Maria Caterina La Barbera, 95–116. Springer.
Pieterse, Jan Nederveen. 2006. Emancipatory Cosmopolitanism: Towards an Agenda. *Development and Change* 37 (6): 1247–1257.
Quijano, Aníbal. 2000. Coloniality of Power, Eurocentrism, and Latin America. *Nepantla* 1 (3): 533–580.
Uhde, Z. 2019. Claims for Global Justice: Migration as Lived Critique of Injustice. In *Challenging the Borders of Justice in the Age of Migrations*, ed. Juan Carlos Velasco and MariaCaterina La Barbera, 183–204. Springer.
UN. 2017. *International Migration Report* [highlights]. New York: OSN. http://www.un.org/en/development/desa/population/migration/publications/migrationreport/docs/MigrationReport2017_Highlights.pdf.
UNHCR. 2019. *Global Trends. Forced Displacement in 2018*. UNHCR. https://www.unhcr.org/statistics/unhcrstats/5d08d7ee7/unhcr-global-trends-2018.html
Wessel, Schulze Julia. 2016. On Border Subjects: Rethinking the Figure of the Refugee and the Undocumented Migrant. *Constellations* 23 (1): 46–57.
Wimmer, Andreas, and Nina Glick Schiller. 2002. Methodological nationalism and beyond nation-state building, migration and the social sciences. *Global Networks* 2 (4): 301–334.
Young, Iris M. 1990. *Justice and the Politics of Difference*. Princeton: Princeton University Press.
———. 2000. *Inclusion and Democracy*. Oxford: Oxford University Press.
———. 2005. Lived Bodies vs. Gender: Reflections on Social Structure and Subjectivity. In *On Female Body Experience: Throwing Like a Girl and Other Essays*, ed. Iris M. Young, 12–26. Oxford: Oxford University Press.
———. 2011. *Responsibility for Justice*. New York: Oxford University Press.
Zamora, José A. 2019. Human Mobility and Borders: The Limits of Global Justice. In *Challenging the Borders of Justice in the Age of Migrations*, ed. Juan Carlos Velasco and MariaCaterina La Barbera, 73–92. Springer.

Lightning Source UK Ltd.
Milton Keynes UK
UKHW020627110722
405680UK00005B/431